Strategies for River Basin Management

Environmental Integration of Land and Water
in a River Basin

The GeoJournal Library

Series Editor: Wolf TIETZE

Editorial Board:

Strategies for River Basin Management

Environmental Integration of Land and Water in a River Basin

Edited by

Jan Lundqvist
Ulrik Lohm
Malin Falkenmark

Porfessors, University of Linköping, Department of Water in Environment and Society

D. Reidel Publishing Company

A MEMBER OF THE KLUWER ACADEMIC PUBLISHERS GROUP

Dordrecht / Boston / Lancaster

Library of Congress Cataloging in Publication Data
Main entry under title:

Strategies for river basin management.

(The GeoJournal library)
Includes index.
1. Watershed management. 2. Water resources development — Environmental aspects. 3. Land use —
Environmental aspects. I. Lundqvist, Jan, 1942—   II. Lohm, Ulrik, 1943—   III. Falkenmark, Malin, 1925—
TC409.S776 1985   333.91'15   85—18293

ISBN 90-277-2111-4

CIP

Published by D. Reidel Publishing Company
P.O. Box 17, 3300 AA Dordrecht, Holland

Sold and distributed in the U.S.A. and Canada
by Kluwer Academic Publishers,
190 Old Derby Street, Hingham, MA 02043, U.S.A.

In all other countries, sold and distributed
by Kluwer Academic Publishers Group,
P.O. Box 322, 3300 AH Dordrecht, Holland

Printed in FR Germany

TABLE OF CONTENTS

Editorial

Land and water constitute two of the basic resources in the lifesupport system on which man depend. With increasing standard of living, growing populations and geographical concentration of human activities, the pressure on these resources is being intensified at an alarming rate in large parts of the world. There is a rapidly growing concern about the environmental hazards associated with this development if it is allowed to continue unchecked. The concern is related both to a depleting resource base as well as a deprivation of the long-term carrying capacity of our environment.

The most noticeable response has been in the increasing attention paid to management issues. Management seems to be perceived as a principle which assures not only an efficient recource utilization in the contemporary setting, but also one which encompasses the necessary conservation of the productivity of these resources in a long-term perspective. In particular, water management has become a matter of primary concern. This applies to policy formulations at the international level right down to local level. Management of land resources is obviously a more composite issue. Apart from land-use aspects, it includes management of inputs, concervation precautions etc. It is also to be noted that due to the complementarity of land and water, management of land resources should be a decisive consideration in water conservation and management.

Principles for the management and conservation of land and water resources are consequently being reformulated and, hopefully, implemented. In this volume of Geo-Library a number of perspectives and principles in this process are discussed in the River Basin perspective.

In chapter 1, it is argued that with increasing socio-economic development it is necessary to pay due attention to environmentally sound planning and evaluation procedures. By applying an eco-system view on resource utilization, the potentials and limitations can be rationally assessed with regard to a sustainable utilization. A sound conceptual framework is relevant not only from a strict environmental point of view. Such a perspective is also called for in the efforts of designing new projects and adjusting existing ones so that human efforts are not drained and financial investments become cost-effective.

In chapter 2 the close interrelationships between water and land, or rather land-use, are analysed. The articles illustrate the paramount need for improvements in management and conservation in large parts of our world. In many developing countries, human pressure on land and water resources in combination with adverse climatological conditions, i.e. unreliable and sometimes very heavy rainfall, in sensitive surroundings have seriously threatened the environment. In the first part of chapter 2, the need for substanstive conservation measures are cogently discussed. For India, China, Java and Kenya severe erosion problems are presented, which invariably seem to be the outcome of intensification in land-use, without proper attention being paid to conservation measures. In the following articles, special attention is given to the complicated nature of upstream and downstream interactions and of water and soil conservation and utilization in some of the world's largest river basins. The strong political element in the development of international river systems is illustrated.

The last three articles of chapter 2, deal with water quality aspects in industrialized countries. In these countries, extensive industrialization and urban pressure in general, as well as high-input modern agriculture, may lead to considerable changes in water quality characteristics in urban areas as well as in rural areas. The needs and principles for monitoring such changes are discussed as well as cases where noticeable impact from waste water treatment measures have been achieved.

Various conflicts over water and their management are discussed in chapter 3. Besides formal procedures, a number of informal tools in terms of economic sanctions and the use of non-formal institutions may be of considerable importance both for avoiding and for the handling of conflicts. Conflict management is, however, complicated by a number of circumstances. Illustrations of practical experiences from developed as well as developing countries show that political interference and well established routines in the administrative set-up may, indeed, underline conflicts or hamper a rational management. Likewise it is argued that unpredictable conflicts may complicate a smooth conflict management. It is also shown that large scale water regulation projects, e.g. dams, may create conflicts between proponents for a utilization of the regulation capacity for hydro-power generation policies etc. and those favouring the use of the regulating capacity for agricultural purposes. Natural climatological fluctuations are not always taken into due consideration purposes. Natural climatological fluctuations are not always taken into due considerations in this context. As a rule, conflicts which concern the environment are difficult to handle through the strong emotional and ideological perceptions involved.

In chapter 4, solutions to the administration of resources in a river basin context are discussed. It is noted that the idea of river basin authorities attracts a lot of attention. In practice there are, however, a lot of circumstances which impede a proper realization of such authorities. Interbasin transfer of water, or long distance transfer of energy from hydro-power plants will e.g. transcend the division of entities based on the water divide. Likewise it is noted that the Western perspective today of river basin is largely in terms of river systems engineering. Administration of the geographical regions of basins are in the contemporary socio-economic-political setting in industrialized countries influenced by urban oriented policies and analyses. Regional development thinking in developing countries, is obviously to a large degree based on a realization of the close interrelations between land and water resources in a river basin perspective. The option to establish River Basin Authorities is consequently seriously perceived.

In the last two articles, models for resource analyses are presented. It is argued that policy design should be based on a systems approach, where the identification of the system and its parts, its behaviour and management are important steps. A systems approach seems to be particularly challenging in an analysis of urban pressure in a river basin context. It is to be noted that although rural development is a predominant exponent for water and land utilization in terms of its sectoral demand and areal extension, the intensification of land use, or changes thereof, is quite often a result of urban pressure and overall population increase.

This edition of GeoLibrary is an outcome of an international seminar: River Basin Strategy; the relevance of river basin approach for coordinated land and water conservation and management. Based on contributions to the seminar and discussions during the seminar, a number of conclusions were arrived at. These conclusions as well as a synthesis of the seminar, are presented at the beginning of this volume.

The seminar was held at the University of Linköping, Sweden, 4—8 June, 1984. About 50 participants from more than 20 countries and several international organizations took part in the sessions. The contributions to the seminar were made possible through generous grants from UNEP, SIDA, Swedish Ministry of Agriculture and Swedish Council for Planning and Coordination of Research. We are grateful for the financial and enthusiastic support which made the seminar and this publication possible. Particular thanks goes to Mrs. Berit Starkman and Mrs. Ulla Häggström for their patient work in preparing all the manuscripts for publication including their talents for organizing the work within a tight time schedule.

Jan Lundqvist, Ulrik Lohm and Malin Falkenmark

Lundqvist, J.; Lohm, U. and Falkenmark, M. (eds.):
Strategies for River Basin Management, pp.005–017
© *1985 D. Reidel Publishing Company and UNEP*

1 RIVER BASIN DEVELOPMENT AND ECOSYSTEM APPROACHES

River Basin Strategy for Coordinated Land and Water Conservation: Synthesis and Conclusions

Lundqvist, J.; Lohm, U.; Falkenmark, M., Dept. of Water in Environment and Society, University of Linköping, S-58183 Linköping, Sweden

Background

Human activities are closely tied to land and water: they call for interventions to increase the socio-economic benefits from these resources, but are at the same time impeded by the factual results produced by such interventions.

The water divide constitutes a natural physical boundary for examining the interdependence between land and water. This gives the river basin special relevance from the perspective of integrated land and water conservation and management. River basin being an originally hydrological concept, river basin development has been understood as the development of water resources in order to better adapt water availability to water needs in time and space.

The tie between land and water is twofold: on one hand land use is to a large extent water dependent, on the other water use may be disturbed by water-impacting land use. It would be expected that the dual land/water interdependence and interaction were clearly reflected in environmental management. However, the general tendency seems to be to treat land and water as separate issues: water planning being approached as a question of siting and sizing of structures for conveyance and storage of water; land use planning as a question of allocating land for different uses.

The Linköping Seminar was convened to discuss river basin strategies in the field of land/water integration in environmental management. Based on a number of case studies, four selected issues were addressed:

— Obstactles to coordinated land and water conservation and management
— Legal and administrative tools as incentives/disincentives
— Problems due to growing urban systems as seen in a river basin perspective
— River basin as an ecosystem

The Seminar was asked to propose criteria for environmentally sound planning and management, allowing for land/water integration necessary for the maintenance of the productive capacity of the complementary life support systems of land and water.

Three main Global Problems

The world suffers at present from a number of severe global problems related to the natural environment which cause a large-scale deterioration of the natural resource base for present and future generations. The problems are caused by the absence of a clear policy for decision-making at different levels in the hierarchy, based on a recognition that natural systems obey laws of their own. Development is often understood as a socio-economic-political phenomenon only, neglecting the fact that it takes place in a certain natural environment/resource context. Sounder decisions with sustained productivity of natural life-support systems in mind would call for a balance between political objectives and the function of such natural systems.

The Seminar discussions took place against the background of three main global problems, all related to the interdependence of land and water: in developing countries, a massive land and water degradation over large areas in combination with a dramatic urban growth, impoverishing life quality directly and indirectly; and in industrialized countries a serious water quality deterioration.

Land degradation in developing countries

Out of the total food-productive land of about 1500 million ha the present annual losses have been assessed at roughly 15 million hectares. The major manifestation of land degradation in most developing countries is soil erosion caused by extensive clear-felling, over-grazing and the improvident use of marginal lands for agriculture. The erosion of the top soil not only results in progressive losses of productivity of the lands affected, but also takes a heavy toll of the economy in other ways. Thus, it causes the premature siltation of irreplaceable water reservoirs constructed to support irrigation, hydropower production, low flow increase, and flood reduction. Under conditions of better land management, the fresh water lost to the sea as floods could have been retained as ground water and put to beneficial use in areas remote from rivers and water courses.

The second major cause of land degradation is the uncontrolled waterlogging and salinization of the soil in canal-irrigated areas due to seepage from unlined distribution systems in easily permeable soils, and of uncontrolled and often excessive application of water to lands which are not properly drained. Water logging and salinization represent a very serious threat to land resources, because they affect good agricultural soils into which large investments have been made by way of expensive irrigation projects, and which possess high potential for sustained production under conditions of good land and water management.

It is a matter for the most serious concern that most developing countries do not yet seem to be aware of the gravity of the threat which the continued mismanagement of their most basic resources poses to their very survival. It is of prime importance for their future that the situation be brought under control before the point of no return is reached.

Urban growth as seen in the river basin perspective

The problem of urban growth, whether controlled or uncontrolled, poses massive challenges all over the world but particularly in the poor countries. Due to improperly developed communication systems, storages and possibilities for a large part of the population to buy food and fuelwood, the pressure on land and vegetation in the vicinity of urban centres is comparatively much higher in developing countries as compared to industrialized countries.

Urban areas often occupy fertile lowland areas, forcing the local crop production uphills. Urban areas represent pockets in the national economy, where the demands on water may be quite high. Similarly, waste production, both solid and solute, is considerable. In industrialized countries, the urban water demand may create conflicts between domestic, industrial and food production forces.

Within the river basin many subsystems may be identified, socio-ecological as well as human social, which interact in the basin system and call for different solutions.

A multitude of land and water problems are produced in the wake of urban growth, including urban storm runoff, water quality degradation, impacts on ground water, losses of good cultivable land, increasing health hazards, and especially in developing countries, deforestation due to fuelwood harvesting around the city. The character of the composite problem structure changes from area to area according to background and socio-economic conditions.

Water quality degradation

All over the world, water quality deterioration is an increasing problem. In industrialized countries, the pollution of rivers passing through urban areas is considered a large problem, and massive capital investments have gone into treatment plants and other measures to reduce the negative impact of point outlets on fishing and other beneficial uses of the river as a resource. As these water quality problems have been addressed and countermeasures developed, new generations of quality problems have developed as a result of modern agriculture supported by high level use of fertilizers, herbicides and insecticides. Also airborne pollutants are causing serious water quality problems including acidification problems in temperate areas with poor soil buffering capacity.

Integrated Land/Water Management — Means and Approaches

The Seminar, after discussion of these problems in the light of the cases presented, arrived at the following conclusions.

Imperatives of improved resource management

Conservation and optimal management of the fundamental and mutually interacting life-support systems provided by land and water is one of the greatest challenges facing mankind. This challenge is particularly evident in developing countries which are faced with acute problems of poverty, increasing population pressures and land degradation. A sound conservation and management policy offers their only hope of breaking out of the vicious circle in which they find themselves today.

Land use and soil quality may have great influence on water flow and water quality, and vice versa. The situation of serious land degradation in the developing world demands that there should be a clear recognition at all relevant decision-making levels in these countries of the imperatives of good resource management.

Land and water management are so inextricably linked with one another that the conservation and development of both these resources need to be undertaken in an integrated manner at all levels, from the field level to the central government. There exists today a resource illiteracy, particularly among planners. Farmers on the other hand probably understand quite well, but are compelled by everyday subsistence problems.

Two system approaches

Systems analysis is a modern tool in the analysis of systems of great complexity.

Systems analysis could be performed from a purely ecologic perspective in seeing man as part of the *ecosystem*. Several studies exist where the water divide has been used as the ecosystem boundary, and the land and water systems contained have been studied as ecological units. The ecosystem approach can be looked at as a means not only to guide management of complex ecological systems and to solve problems due to misuse and mismanagement of the environment, but also to understand the nature and influences of various interacting forces in response to environmental manipulations by man. The ecosystem modelling is in fact well suited for interdisciplinary work.

The Seminar stressed, however, that such ecosystem analyses have their limitations due to the abstractness of the ecosystem concept as such. It was recalled that many of our concepts refer to a natural stage and may be difficult to apply at stages with strong anthropogenic influences. The river basin is also a hierarchy of subbasins with often vastly different ecosystems, especially in large-scale basins transversing the boundaries of different ecological zones. In cases where river basin development is seen as solely a socio-economic activity involving merely a development of water resources, the ecosystem concept might not necessarily be of specific use for managing the water resources.

In spite of such limitations, the present trends in ecodevelopment thinking, the stress on preservation of land productivity, and the strong link between water and ecosystem functions stresses the need to build up a renewed thinking in this field. System approaches may be helpful also in identifying the structure of many of the above mentioned problems and as a basis for *policy design*.

The complex problems caused by uncontrolled urban growth may also be a field where systems analysis is useful as an aid in the decision-making process for the selection of the appropriate strategies and their consequences. In such applications, the basin may often be seen as composed of urban foci and interconnected subregions with different characteristics in physical as well as socio-economic terms (soil, topography, land use, money vs subsistence economy etc.).

Means to manage conflicts

Legal and administrative tools are generated in response to conflicts — which are generally unavoidable — as a help in managing them. Such tools are frequently proposed in order to create rational systems of water management. They can refer to different levels (international, national, provincial/state, regional/basin or municipal areas) and can be established by legislative powers, agreements, judges, or custom.

The legal and administrative tools, acting as incentives or disincentives for water resources management, can assume different forms: rules on property and water uses, regulations with fines or other penalties in case of violation, water withdrawal charges, pollution charges, user fees, subsidies etc. Land related tools are land banking, consolidation of holdings, and nationalization of land.

Economic sanctions and the price mechanism do play an important role both as regards the orientation as well as intensity of land and water use. Fees, subsidies and market conditions in general may thus decrease or stimulate certain types of land use. At the same time it is clear that these tools are either not systematically applied or controlled by the authorities in most countries, or they are not efficient enough to eliminate unwanted uses of land and water resources. The very unequal economic power of various social groups in Third World countries will e.g. be difficult to adjust for. Demand on the international market level may likewise create considerable pressure on e. g. certain tropical forests. For millions of poor people with acute needs in a subsistence economy, it is also tempting to embark on land-use practices that will give short-term profits although being counterproductive to sustainable production.

In developed countries, market prices of land will generally make a very intensive land-use necessary. Heavy inputs of energy and chemicals are consequently used, and leakages of these chemicals with the passing water are part of the side effects produced.

Obstacles related to problems of scale

Numerous obstacles form disincentives at present and effectively impede the land/water integration needed. Many speakers underlined the problems of scale: both time and geographical scale, and level of development. Increasing scale amplified the obstacles due to limitations in coordination and communication capacity of an organization.

Great difficulties have been witnessed in managing large-scale projects. Without a systematic monitoring and evaluation of established projects, no lessons may be learned to use in future project design and management.

It was stressed that executive decisions of coordinative character are generally taken in the short-term, micro-level perspective, whereas policy-making is related to the meso-scale level and medium term perspective. It tends to be useless, however, unless adequate attention is paid to the general micro-level behaviour and social concepts.

The general development away from hardware — i.e. technical solutions, to software — i.e. solutions involving improved management, has met with considerable difficulties due to lack of capability to implement software solutions. Particular problems emerge when social goals divert from strategic goals formulated in legislation and comprehensive plans, and when bureaucracies work with outdated thinking.

Fundamental human pre-conditions

One has to distinguish between obstacles to the development of policies and plans, and obstacles to the implementation of such policies and plans. Public acceptance generally depends on some degree of *fairness/equity* in the distribution of costs and benefits. To this aim, fairness has to be considered already at the plan development stage. One could talk about equity in terms of equal *opportunities* (e.g. equal costs, prices, access to resources of equal quality etc.), and equity in terms of the *outcome* of a policy. To reach some degree of equity between individuals it is usually necessary that the policy includes directives for a distribution of costs and benefits which is inverse to the existing social conditions. Poor people or regions should consequently pay reduced costs and vice versa etc.

Equity considerations should be seen in some context of efficiency, since they imply various allocations of resources. A problem is that equity and efficiency considerations frequently have a bias towards short term perspectives. In the context of land and water use, it is however, crucial that long term considerations are applied.

Equity considerations are in fundamental ways related to river basin development. In a natural, unregulated situation, people living upstreams have an advantageous position vis-a-vis those living downstream. This applies to quantitative as well as qualitative aspects. However, under the conditions of fairly strict water quality control introduced in many countries, upstream users are affected relatively more than downstream users. For users within one sector, e.g. a certain type of industry, the location downstream may therefore be favourable as compared to upstream locations, particularly if other relative advantages are considered i.e. transport, market nearness etc.

Even if equity is a complex concept, difficult to operationalize, economic and legal tools must be worked out and be consistent with more general policy considerations.

In both developed and developing countries, there is thus a fundamental need of creating a *motivation and understanding* for the need to introduce new policies. *Public participation* may be helpful and necessary in plan/policy development and in securing *public acceptance* for its implementation.

In Third World countries, popular participation is an often used term in the context of rural development, be it in terms of drinking water supply development and management, or production increase in general. Quite frequently it is conceived of as a strategy for saving financial resources and man-power of central agencies. It is supposed to arouse motivation and lead to better functioning of schemes or projects. Generally speaking it is supposed to be democratic, or at least antibureaucratic, and even efficient.

The perspective is certainly different in industrialized countries as compared to developing countries. In the former, it is not so much motivation that is needed for mobilizing the use of resources, but rather a much more keen awareness and knowledge about impacts of current usage patterns. In countries or regions where the policy has been to favour a decrease in production, it has been a question of creating a motivation for the same. In developing countries in general, the top priority is for increases in production. This does not, however, mean an increase in all environments. Indeed, there are radical proposals for changes in land-use patterns, e.g. transforming intensively cultivated lands to pasture or forests.

In both cases, local communities should actually be involved in the planning, decision-making, implementation, maintenance and rehabilitation phases. Furthermore, it is not enough to talk about "communit", but to identify who in the community should/would participate to what extent, under what conditions, etc.

Some Illustrative Cases of River Basin Strategy

The drainage basin constitutes a natural territorial unit for intergrated land and water management. The contributions to the Seminar reflected a multitude of experiences from various parts of the world. Three cases illustrated the river basin development as such.

In the Tisza river, there is a very strong connection between the application of a basinwide strategy and the stage of river basin development. In other words, different strategies are called for at different stages. As the population pressure, agricultural production, urbanization, and industrialization develop in a river basin, so do the water demands and the risks for land and water degradation.

The level of river basin development may be described by a multicriteria index (D). According to the Hungarian experience, basinwide strategy is feasible when D lies in the interval 20—40% and is necessary when D exceeds 40%. Basinwide strategy then calls for integration of water, land use and environmental management.

The Kävlinge river case illustrated four phases of the basin development (land reclamation phase, water export phase, fish damage phase, and phase of multiple river basin integration) and the need for a higher level of flexibility in legislative and administrative controls.

The Thames river case, finally, illustrated the difficulties introduced by numerous small changes in an atomistic system instead of by large changes. easily detected.

Most of the cases presented addressed the river strategy from specific aspects: the land-water interdependence, legal and administrative matters or the problem of urban growth.

Land-water interdependence

The upstream-downstream interdependence along a river was illustrated for the Huang Ho (China), where the parallel needs for erosion control on the one hand, and flood control on the other were two driving forces.

In the Wuding river (China), the land/water interdependence was illustrated by the fact that land conservation is practised to reduce erosion and water conservation is practised to support land use, particularly agriculture.

The need for a normative approach and long-term planning was illustrated for the Han river (China), where an integral view is taken for the whole basin including mitigation of natural flood disasters, promotion of multipurpose benefits of water resources development, and maintenance of productivity both of nature and man.

The control and management of land use and water resources call for an adequate subdivision of land into different categories of landscape elements as was illustrated for the Bhavani river (India).

In the Tana river (Kenya), land conservation is fundamental to reduce the ongoing large-scale sedimentation in a chain of hydropower reservoirs of crucial importance for the energy supply of the nation.

A catchment-based ancient irrigation system was illustrated for Sri Lanka, and presented as the logical response to the challenge posed by its natural environment in the dry zone.

The Mekong river case (SE Asia) stressed the need for maintenance of productivity of resources rather than the conservation of resources as such.

Legal and administrative matters

Some obstacles generated by administrative and legislative arrangements were illustrated for the Colorado river (North America), particularly in the interaction between the upper and lower basin. For the same basin an example was given of the management of conflicts between those interested in high flow and those threatened by high flow and interested in low flow.

The need for new attitudes and a search for a locally adapted philosophy rather than a passive copying of industrialized-country models was illustrated by the Nile river (Africa).

The importance of human irrational factors for reaching given aims was illustrated by the Ganges case (India-Bangladesh), particularly the crucial role of adequate timing of international negotiations.

The success of the cooperative owner approach to coordinated land/water conservation was illustrated by a case from the loess plateau of Huang Ho (China), the Xio Shi Guo brigade.

The quite rapid transfer of water from local common law control to centralized, legislated control emerged from recent English rivers.

Problems of urban growth

There were also a number of cases illustrating the problem of urban growth as seen in the river basin perspective. Four of these cases emerged from industrialized countries whereas one single case illustrated the immense problem in the developing world.

The latter case of the Citarum basin (West Java) illustrated how urbanization forces food production uphill, thereby generating a severe soil erosion problem combined with increased flooding of the river valley bottom. The most important step in controlling this development is education of the hill farmers — former urban citizens forced out from the city in response to economic recession — in adequate soil conservation through use of relevant cultural channels.

The various generations of water quality problems emerging in the wake of urban development was illustrated in the Chicago-Illinois river area (USA), with stress on the importance of seeing river basin problems in a long-term perspective.

Competing rural and urban interests were illustrated for the Hawkesbury river (Australia), stressing the technical management aspects, particularly nutrient control, monitoring programs and water quality standards.

Criteria and Action Components in an Environmentally Sound Management of Land and Water

Based on the discussions organized around the four specific issues approached and the case studies presented, the Seminar arrived at a number of criteria for land/water integration and for legal and administrative tools allowing such integration.

Criteria for land-water integration in environmental management

Water and land are closely interrelated parts of the natural environment. Most of the water in a river has earlier passed land. Most human land uses are water dependent, but influence at the same time water flow, seasonality and quality. This makes the water divide fundamental for the integration of land and water conservation, and the drainage basin an appropriate unit for such integration.

The Seminar stressed that both water and land interventions by man, as a general rule, produce changes in the natural environment. From this aspect, environmental protection as an isolated activity is not meaningful. Instead, environmental management should be seen as inseparably connected with land and water management — it is just another side of the same coin.

Unavoidable environmental changes due to land and water development can in fact be seen as the price to be paid for improved quality of life. These changes can, however, be more or less harmful for a sustained productivity of the resources. A realistic goal is therefore environmentally sound management of land and water, implying the harmonization, in development, of environmental and socio-economic interests in a basin.

14

The role of environmental management should be to maintain both the productivity of the primary resources and their linkages. Coordination of land and water conservation and management should aim at an optimum of the net production on all the basin lands. Land-use patterns must be based on actual levels of fertility as well as the possibilities for compensations of various losses incurred through an intensified land use.

It is important to distinguish between development of a river basin to maximize the availability of water, and the conservation and management of water for maximizing socio-economic benefits in the river basin.

To solve some of the numerous problems from growing urban systems, the drainage basin perspective is particularly helpful in regions with a sloping landscape in facilitating an integrated view on land use and water resources development. The interactions between rural and urban areas must be dealt with in the framework of the carrying capacity of the basin.

Transbasin water transfers may reduce the importance of the drainage basin as a relevant terrestrial unit. The water divide however remains as a natural membrane, in some respects separated from outer environments although passed by import/export flows of water and other goods. The drainage basin for instance retains its importance as a unit within which water and soils come into some sort of integration and, consequently, when determining the carrying capacity of the land and water bodies of the basin.

When analyzing the problems due to urban growth, it may be particularly useful to divide the basin into interacting subregions with different characteristics in terms of background conditions, land use and type of economy and to elaborate on their interconnections. The general goal should be a balanced growth between rural and urban areas.

Systems analysis may be useful as a decision aid tool for selection of appropriate strategies. The fact that the force behind the rural exodus into urban areas is the poor living conditions makes rural development crucial in trying to limit the massive urban growth in many developing countries.

Criteria for legal and administrative tools as incentives

Laws are basic tools for policy implementation. Inadvertent legislative and decision-making structures are important as disincentives to land-water integration. It is therefore fundamental that legal and administrative tools and decision-making structures are based on adequate concepts and due consideration of physical interdependencies of various natural resources and of pertinent environmental factors. For instance, in areas suffering from soil degradation due to firewood harvesting, local hydropower might be an interesting incentive for local watershed management.

The establishment of legal and administrative tools must make sure that every decision-maker is put into a position to take into account all the benefits and all the costs generated by his/her decisions. It is thus necessary to collect relevant information about all significant consequences and to provide a clear reward/penalty structure that will motivate decision-makers to act on that information.

It is fundamental that legal and administrative tools are flexible enough for timely adaptation to changes in water uses and users, to consideration of water use priorities, and to special situations, such as those occurring during revere droughts and floods.

Since on the one hand legal/administrative mechanisms tend to be rigid, and on the other economic mechanisms tend to be much more flexible in directing and motivating appropriate parties to respond to changing circumstances, one should not encourage rules that give precedence to certain uses and users or establish rigid criteria that must be obeyed under all circumstances.

In many countries, particularly developing ones, there is a dual economic system — modern and traditional. It is important to take this fact into consideration when developing legal and administrative mechanisms in water management and development.

The existence of drainage basin as the natural territorial unit for integrated land and water management should be recognized and made use of in the planning and execution of work at all levels, including regions and entire river basins. Each part of a river basin has to develop its full potential in a long-term perspective.

Existing river basin organizations are only water-oriented. The present advocacy of a new strategy of land/water integration involves a new phase and calls for changes in attitudes and institution bulding.

When establishing river basin agencies, clear definitions should be made of the territorial area of jurisdiction and the functional responsibilities vis-à-vis other governmental agencies. River basin agencies should be financially self-supporting when appropriate. Public participation in river basin management should be duly considered and promoted.

Crucial action components for integrated land-water conservation and management

In the selection of a strategy for integrated land/water/environment conservation and management a number of action components form integral parts. The Seminar identified various groups of such components. Crucial components in land/water integration include the following

— the conservation of irreplacable soil resources, affected or threatended, through appropriate measures like afforestation, restoration of pastures, anti-erosion works and drainage etc.

— the conservation of water through appropriate soil conservation and land management practices, through the construction or restoration of innumerable small local storages in mini-watersheds before impounding genuinely surplus and silt-free waters in large scale reservoirs.

— command-area development works (including drainage) in canal-irrigated areas. System management, proper drainage canals are prerequisites for successful on-farm conservation and management. Corrective measures are a matter of high priority in order to attain their full productive potential and to save them from water logging and salinization.

— use of lands — whether for forestry, pastures or cropping — according to their capability, to the needs of society and in a manner which ensures their carrying capacity not to be exceeded and their productivity to be built up over time. At present much marginal lands have been utilized for an intensive use, resulting in low and diminishing returns.

— active involvement of local populations: due to financial, manpower availability, and other constraints at central level, there are faint possibilities to implement the above mentioned measures unless they are described and executed with the direct involvement of populations concerned.

In order to satisfy these various integration components, the seminar identified a number of action components:

Education and training are fundamental components when trying to achieve coordination of land and water management and conservation. This includes formal and informal education and training, changing the cultural and traditional conditions if necessary, extended education and training programs to village level including non-farmers, lower children school (e.g. importance of planting trees). It will take a great deal of effort before an informed public opinion and a strong political will can be generated in each country. Demonstration catchments might be useful tools in farmers' teaching.

Policy making: river basin strategies should be adjusted to regional or geographical scale, to time horizon, and to level of development. A matrix for analysis of obstacles to land and water coordination was presented. Water-land management projects call for long-term policies.

Legal tools should include laws and rules for protection of vegetative cover.

Administrative tools include agencies involved in implementing land and water conservation and management. When these are different agencies, there should be a coordinating body to ensure the necessary coordination, and to monitor and guide various programmes.

Economic components include economic regulations and consideration of economic indicators (pricing system, land evaluation, taxes, fees, charges, subsidies, priorities of the rural areas, increase of the economic capability of the population, investment policy). It was also recommended that explicit and close attention be paid to the relative economic productivity of water in different uses as a guide to optimal allocation of water.

Evaluation/monitoring is needed both for evaluation of existing projects in relation to their original scope, and for the establishment of a system for the permanent control of land degradation.

Technical components, especially stressed in relation to the problem of urban growth, include measures to improve water supply, to manage water demand (reuse, recycling, water saving), and to manage water quality. In the discussion of main obstacles to integrated land/water management the importance of avoiding large scale development projects was stressed. It was also recommended to eliminate or withdraw tillage from areas not suited for agricultural production.

Master plans are useful tools to draw up a realistic plan for the optimal and integrated management of a country's total land and water resources. Such a plan must for obvious reasons be formulated as quickly as possible and implemented within a period of up to 25 years. National governments must in particular assess the various social, educational, legal, administrative, technical and financial measures to be taken to translate these plans into action.

Technical assistance: there is a great scope for international agencies as well as individual developed countries to assist national governments both in education and training programmes and in master planning efforts. It would be eminently desirable for UNEP to accept responsibility for active leadership in the field of coordinated land and water conservation and management including the setting up of a special unit for promoting the ideas and action programmes discussed at the Seminar, and to monitor and evaluate their progress on a regular basis.

Lundqvist, J.; Lohm, U. and Falkenmark, M. (eds.):
Strategies for River Basin Management, pp.019—029
© 1985 D. Reidel Publishing Company

River Basin Development Strategies in the Tisza Valley

David, L., United Nations Environment Programme, PO Box 30552, Nairobi, Kenya

Abstract: An overview of the experiences and perspectives of the river basin development process and related strategies in the valley of the Tisza river, a main tributary of the Danube is presented, covering the period from the 19th to 21st century. Come characters of the region are described and the river basin development and its future perspectives are analyzed. It is pointed out that there is a very strong connection between the application of the basin-wide strategy and the stage of river basin development. During the natural phase of development there is no need to apply the basinwide strategy. But on higher levels of the development the complexity of water and land resources management makes it necessary to apply the basinwide strategy for planning, decision making and implementation of water projects. The environmentally sound management of land and water resources needs also a basinwide river basin strategy. The multi-purpose character of the river basin strategy is also emphasized. Finally a number of recommendations including environmental considerations are summarized.

Introduction

The natural availability of water resources in a river basin, generally does not coincide with the growing water requirements of the socio-economic development and with the constraints of the environment management in the same basin. Consequently, there is a need to establish a continuous balance among natural supplies and human requirements over space, time, quality, quantity, and energy. This balance can be established and maintained by the river basin development strategy which involves a large number of environmentally sound water management and conservation activities. This development is an integrated, planned and comprehensive long-term process, the purpose of which is to make the optimum use of water resources possible on a basinwide scale during socio-economic development. The criteria for optimum use basically depend on the constraints of the socio-economic growth and the environmental management (United Nations 1970).

The development process of river basins can be divided into three consecutive periods: the natural (period I), the developing (II) and the matured or fully developed (III) phases of river basins. The objectives, the means, the strategies for the development of a river basin, the complexity of river basin management and the images of the river basin depend on the stage of river basin development (David 1976).

The purpose of this paper is to present a comprehensive, long-term analysis of the river basin development process and related strategies in the Tisza river basin from the 19th to 21st century. To meet this purpose, the paper is organized in the following way. In the next section some characteristics of the region are described. The river basin development is discussed from a contemporary perspective. This is followed by an outline of the future perspectives and strategies. Finally, conclusions and recommendations are summarized.

Some Characteristics of the Tisza Basin

The eastern part of the Carpathian Basin in Europe is drained by the Tisza river into the Danube. The Hungarian part of the basin is called Tisza valley. The natural hydrological and socio-economic characters of the Tisza valley are summarized by Lászlóffy (1982).

Most of the Hungarian part of the basin is flat, which represents about 30% of the basin's total catchment area of 157,000 km². The other countries sharing the basin and their percentage of the area are: USSR (8%), Czechoslovakia (10%), Romania (46%) and Yugoslavia (5%). The climate is continental in character, and it is rather arid. The annual precipitation in the plains is less than 600mm. The irregularity of precipitation is reflected by its wide variability.

The runoff from the upper watershed attaining 2000 m altitude in some sections, finds its way to the vast plains lying at elevations of 90 to 95 m and consisting of a series of undrained depressions by a dense surface water system. Flood inundated large areas and undrained precipitation provided ample water supply to the wide-spread marshes. This situation prevailed essentially without change up to the beginning of the last century.

The potential water resources of the basin (the multi-annual average runoff) is 25 km³/year at the southern border of Hungary, from which only 2.5 km³/year originates in the Hungarian territory. To completely regulate this volume of water over several years, a storage of 75 km³ is needed, which indicates the wide variability of the runoff.

Socio-economic development necessary for ensuring the desired standards of living for the growing population in the Tisza valley has been made possible by the development of the Tisza valley and the hydraulic engineering work related thereto.

River Basin Development up to Contemporary Situation

Development of the Tisza valley began in the middle of the 19th century. In its natural stage, the basin was a vast marshy area, much of which was covered permanently by water or inundated annually (Fig 1). Over the last 140 years, the development of the basin made the area accessible to cultivation and human settlements, thereby creating opportunities for flourishing economic activities. As a result of this work an area larger than that gained in the Netherlands from the sea, was reclaimed for agricultural use.

The development of the basin began with the strategy to control the floods and to regulate the river. The first flood levees were not high enough and their cross-sections inadequate. They proved uncapable of withstanding the flood waves, which increased as a consequence of flood plain confinement. Ruptures decreased as a result of the levees' gradual strengthening. Due to basin development flood levels continued to rise, until the levees attained the present cross-sectional dimensions to control them (Fig 2).

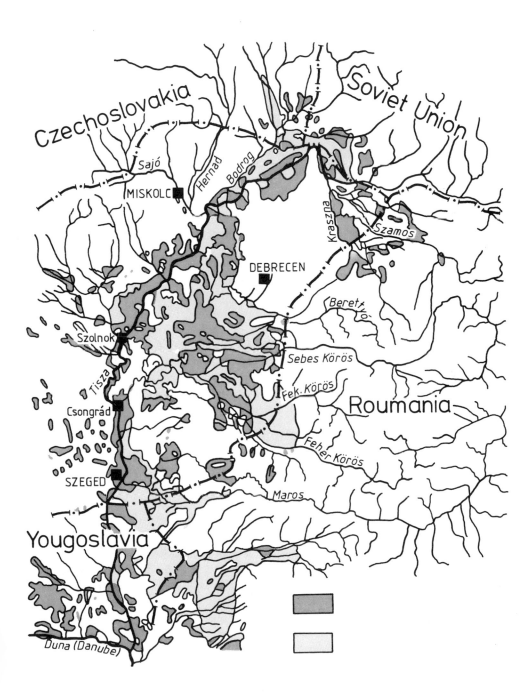

Fig 1 Hydrographic situation in the Tisza river basin before river basin development

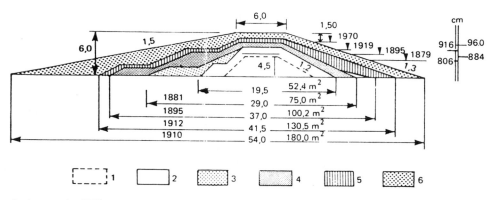

1. levee section 1855;
2. levee section 1879;
3. standard section prescribed after 1881

4. standard section prescribed after 1895;
5. standard section prescribed after 1919;
6. levee section prescribed after 1970

Fig 2 Increase of levee sections along the lower Tisza river.
1) levee section 1855; 2) levee section 1879; 3) standard section prescribed after 1881; 4) standard section prescribed after 1895; 5) standard section prescribed after 1919; 6) levee section prescribed after 1970.

The strategy to construct a levee system against floods was followed by the provision of drainage from the basin to the stagnant surface waters accumulating in the plains. These efforts laid the foundations for agricultural and industrial development, and for the increase in the population density (Fig 3). The marshy area disappeared and a new environment and landscape was created.

Fig 3 Development of water management characteristics in the Hungarian part of the Tisza river basin.

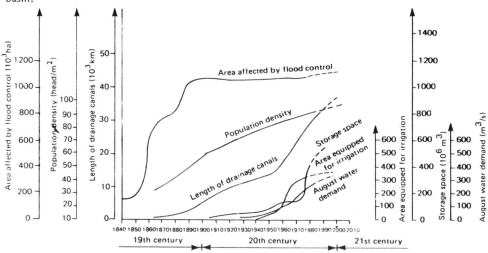

In the 20th century, the major objective and strategy has been to develop the basin-wide runoff regulation (intra and inter basin water transfer and storage) and to regulate the use of water, primarily for agriculture. Increasing attention is being paid to the development of industrial and domestic water supply systems, to navigation and to recreational activity (National Water Authority 1965).

The development of the runoff regulation projects started in connection with the canalization of the rivers. The canalization of the Körös river, one of the main tributaries, started in early 1940's and was practically completed in the 1970's. The first barrage of the Tisza river at Tiszalök was completed in 1954, and the first largescale raw-water transfer canal, called Keleti Föcsatorna (Eastern Main Canal) was completed in 1956. Its transfer capacity is $60 \text{m}^3/\text{s}$ and its length is 100 km. It transfers water from the Tisza to the Hortobágy region and to the Körös river valley for meeting industrial, domestic and irrigation water demands (Fig 4). Later, in 1966, the Nyugati Föcsatorna (Western Main Canal) was constructed in the same region. The second barrage at Kisköre was completed in 1973. Its backwater reservoir has a capacity of 400 million m^3. The water from here is diverted to the Nagykunság and to the Jászság main transfer canals. Their conveying capacity is $120 \text{m}^3/\text{s}$ and their aggregate length is 220 km. These canals were completed in the late 1970's. They supply water to the Körös and the Zagyva river basins. This system forms the Tisza-Körös Water Management sub-system of the Tisza valley. When the development of this sub-system was started, the stage of river basin development was about 20%. Nowadays it is more than 40%.

The stage of river basin development can be measured by a development function which integrates a great number of socio-economic, technical, environmental and hydrological elements of the development. The development function can be considered as a multi-criteria utility function expressing the regional water management "utility" and complexity. It can be composed of a system of indices in which each index expresses a criterion. The development function (D) is a dimensionless figure. Its 0 value expresses the natural phase of development and its 100 value indicates the fully developed phase of the river basin. From practical view points, the following limits can be considered:

$0 \leq D \leq 15$ for the natural phase
$16 \leq D \leq 85$ for the developing phase
$86 \leq D \leq 100$ for the fully developed phase

The D function can be composed according to the characteristics of the investigated river basin and the purpose of the investigation.

It can be determined as follows:

$$D(t) = \sum_{k=1}^{K} W_k \cdot I_k(t)$$

where t = time

I_k = development index where $0 \leq I_k \leq 100$

k = number of development indices

24

$$. \, W_k = \text{weight of } I_k \text{ where } W_k \geq 0 \text{ and } \sum_{k=1}^{K} W_k = 1$$

The I_k expresses an element or a strategy of the river basin development and it is a simple function of different basic (natural, hydrological, social, economic or environmental) factors. All of I_k are dimensionless and most of them are time-dependent. The construction of the development function and the development indeces is described in detail by David et al. (1979).

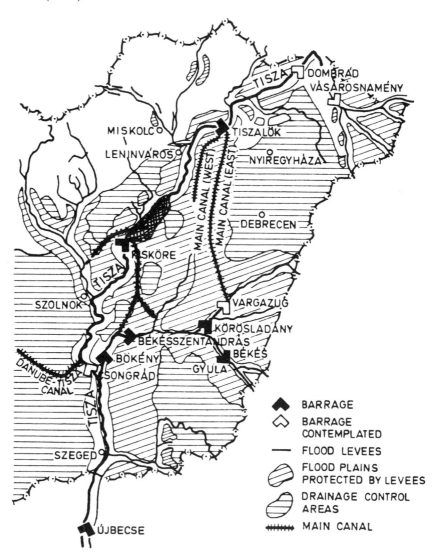

Fig 4 The existing water resources system of Tisza valley.

This development of the mentioned sub-system, as expressed by the 20% increase in the stage of river basin development, has changed completely the natural hydrological situation. This sub-system is able to transfer annually $5 km^3$ of water which is much higher than the $1.3-1.5 km^3$/year natural supply. But this runoff-regulation system also created a new environment along the main canals in the flat and arid region called the Great Hungarian Plains.

Because of the increased use of water resources, the quality of both surface and ground water has deteriorated and attention is now being focused on pollution control. The transfer canals are used for dilution and the drainage canals are used for sewage disposal purposes as well. Due to the gap between the increased domestic water supply and the less developed sewage canalization, the ground water resources as a source for drinking water supply have deteriorated and the ground water level increased in many parts of the region. Therefore, in the recent years the main strategy is to regulate the water resources with respect to the quantity, quality and energy-related aspects to increase the usable water resources.

At present the Tisza valley as a whole is in the developing stage with 38% level of development. About 15 to 20% of the planned runoff regulation has been achieved. The existing water resources system has a flood levee system extending 2800 km, multipurpose transfer, drainage and irrigation canals (2600 km) that form the basis of agricultural activity over 3.2 million ha of land, a number of river barrages and storage reservoirs, and water supply systems and sewage systems for hundreds of settlements and industrial plants. The present level of river basin development is in good harmony with the requirements of the socio-economic development in the region but the environment management aspects need more emphasis.

Future Perspectives and Strategies

The goal of future strategies is to reach the developed stage for the entire valley. According to the socio-economic plans, the realization of this stage is scheduled for completion in the year 2050 at the earliest. Naturally it will depend on the future rate of socio-economic development in the region. Thus, the duration of the present developing stage in the Tisza valley will be about 200 years. Priorities may change with the evolution of time (United Nations 1976).

The main objectives and strategies for the further development of the Tisza valley are the following:

— Regulations of the basin's water resources according to their quantity, quality and engergy-related aspects, including the water transfer from the Danube river;

— Multi-purpose water transfer and supply for consumptive and nonconsumptive users, and water demand control including the control of reuse and the techniques used;

— Water damage prevention (flood control, drainage, waste water treatment and disposal), in conjunction with water resources regulations (storage and water transfer) with emphasis on waste water transfer, treatment and disposal;

- Rational use of natural and socio-economic resources for development (water, land, capital, manpower, energy, etc.);
- Development of international cooperation for water resources regulations (both storage and water transfer) among the five countries in the basin, both on bilateral and multilateral bases. A master plan for developing the water resources of the Tisza river basin has been prepared by these five countries, within the framework of the Council of Mutual Economic Assistance (CMEA), which could serve as one of the inputs for developing such cooperation;
- Environment development by the means of water resources projects, to serve as a balance between the natural and socio-economic development of the valley;
- Flexibility in the choice and application of methods for developing the valley, so as to avoid problems of uncertainties in natural conditions and in the techniques used as, for example, water demand forecasting.

Considering the listed main objectives and strategies of further development, it can be noted that major emphasis is being given to water resources regulation in the future as well. An analysis of the development of water resources regulation in the Tisza valley under the Hungarian conditions, shows that the large and basin scale water resources regulation is a necessary measure of water management after reaching the $D = 40-45\%$ stage of river basin development. To achieve these objectives the application of proper methods and techniques, such as systems analysis and decision — making techniques is needed (Loucks et al. 1981; Giocoechea et al. 1982).

To meet the main strategies of further development of Tisza valley, five alternative long-range water resource systems were planned and ranked on a multicriterion basis (David & Duckstein 1976). One of them could be chosen for development over the next 50 years.

- System I: Danube and Tisza rivers interbasin transfer using a multipurpose canal-reservoir system via a gravity canal in the flatland area, and pumped canal-reservoir system in the Börzsöny Cserhát Mountains.
- System II: Pumped canal-reservoir intrabasin transfer system in the northeastern hilly region of the Tisza valley.
- System III: Flatland water management system, composed of transfer canals and reservoirs of 2—4 m deep, to be developed in the flatland area of the valley.
- System IV: Mountain transfer and reservoir system in the upper Tisza river basin (located outside Hungary).
- System V: Ground water storage system, to be developed in the eastern part of the valley as part of conjunctive surface ground water system utilizing the Tisza river water for both water supply and ground water recharge.

The basic aim of these systems is to develop the natural supply of water resources by comprehensive, environmentally sound runoff regulation, including quantity and quality regulation over space and time by water transfer and storage, while simultaneously trying to fulfill other objectives that reflext economic, environmental, and social aims. Thus, the following planning goals were established: water demand fulfilment, flood protection, drainage and used water disposal, utilization of resources, environmental management and flexibility. The last goal implies that the proposed system should be sufficiently flexible to meet a broad spectrum of future requirements, most of which cannot be foreseen at the present time.

Considering this set of goals, the following twelve criteria (performance indices of measures of effectiveness) have been selected: Yearly cost, probability of water shortage, water quality, energy reuse, recreation, flood protection, land and forest use, manpower impact, environmental management, international cooperation, development possibility, sensitivity. These evaluate how well the system performs with respect to meeting goals and their specifications. Certain factors are evaluated in terms of monetary or marketable criteria, while others can be expressed only qualitatively. Furthermore, a few important factors involving social and environmental elements, such as flood protection, land and forest use, are evaluated in both ways. The costs and expected losses are combined into one criterion, namely the total annual discounted cost. The data reflect the tradeoff between costs and losses, even if the differences between totals are relatively small.

The comparison of the alternative systems were made by the ELECTRE method (David & Dunkstein 1976) and by the Multiattribute Ultility Theory (Keeney et al. 1976). The results imply that System I and II are considerably better then the other systems. The combination of these two systems is proposed for further consideration.

For the implementation of these systems and strategies the multicriteria and multi-regional system's planning should be applied. Interactive analysis of river basin versus water and other project development should be made according to David & Fogel (1982). There is a need to further develop the international cooperation and the multipurpose utilization approach to support the basin wide harmonization of strategies and development plans. Considering that at the higher levels of river basin development the possibilities of conflict situations are increasing more and more emphasis should be given to conflict management and to the development of the legal and institutional framework during further development of the Tisza valley.

Conclusions and Recommendations

An overview of the experiences and perspectives of river basin strategies in the Tisza valley is outlined to contribute to the world-wide evaluation of river basin strategy. When deciding on a river basin strategy and during evaluation there are many aspects that need to be investigated, some of which are emphasized, e.g. the effects of longterm river basin development.

Based on an analysis of a 150 year period, it can be stated that in the Tisza valley the application of the basin-wide approach for choice of a strategy is feasible when $20 < D < 40$ and it is necessary when D is over 40%. It means that there is a very strong connection between the application of basin-wide strategy and the stage of river basin development. During the natural phase and the very beginning of the developing phase, there is no need to apply the basin wide strategy. It is possible, but not necessary. But on higher levels of the developing phase the complexity of the water and land resources management makes it necessary to apply the basin wide strategy for planning, decision-making and implementation of water projects.

During the development of the regional and interregional water management systems, the experiences gained in the Tisza valley underline the importance of the gradual approach. Large and complex basin wide systems can be developed by a step by step method.

The long-term efficiency of the basin-wide strategy can be increased when planning and decision-making are based on multicriterion decision analysis. Local and regional as well as short and long-term requirements of the environmentally sound management of the water and land development projects should be involved in such an analysis.

The multi-purpose character of the river basin strategy is also emphasized. The aspects of water quality in the comprehensive strategy increase during the river basin development. The importance of raw water and sewage water transfer and the basin-wide analysis of water treatment are also needed. The drinking water supply and the sanitation (sewage canalization) should be developed in good harmony. The single-purpose water projects gradually become multi-purpose on a higher level of development. By the river basin strategy a harmonized new environment should be developed in connection with the water projects.

The results and the conclusions of this overview lead to the following recommendations:

1. The river basin strategies should be analysed in connection with the river basin development process.

2. The interactions between the river basin strategies and the growing complexity of water management during the river basin development process need to be harmonized over a long-term basis by considering the projects as an integrated element of the water management system of the basin.

3. The evaluation and the planning of river basin strategies need the application of the multicriterion decision-analysis methods, including the environmental impact and management analysis as an integrated element of the decision-making process.

4. For medium and large river basins there is often a shortage of capital resources. Regional or interregional multipurpose water management systems should therefore be developed gradually and over a long time horizon.

5. Considering that most of the river basin strategies change the natural and social environment to a certain extent, the main objective of any water project should be to develop a new equilibrium between the project and its environment by the harmonization of the original environmental conditions and the project impacts. To achieve this goal the development of environmental monitoring system should be an integrated part of the river basin strategy and it should be initiated from the planning phase.

References

David, L.: River basin development for socio economic growth. In: Proceedings of UN Interregional Seminar on River Basin and Interbasin Development. Vol. 1, United Nations, Natural Resources Water Series, No. 6, New York — Budapest 1976.

David, L.; Duckstein, L.: Multicriterion ranking of alternative long-range water resource systems. Water Resources Bulletin 12, 731—754 (1976)

David, L.; Telegdi, L.; van Straten, G.: A watershed development approach to the eutrophication problem of Lake Balaton, A multiregional and multicriteria model. Collaborative Paper, CP-79-16, IIASA, Laxenburg, Austria 1979.

David, L.; Fogel, M.: Interactive analysis of river basin-water project development. Paper presented to the IV World Congress on Water Resources, Buenos Aires 1982.

Goicoechea, A.; Hansen, D. R.; Duckstein, L.: Multi-objective decision analysis with engineering and business applications. John Wiley and Sons, New York 1982.

Keeney, R. L.; Wood, E. G.; David, L.; Csontos, K.: Evaluating Tisza River Basin Development Plans Using Multiattribute Utility Theory. Collaborative Paper CP-76-3, IIASA, Laxenburg, Austria 1976.

Lászlóffy, W.: Monograph of the Tisza River. Hungarian Academy of Science, Budapest 1982 (in Hungarian).

Loucks, P. D.; Stedinger, J. R.; Haith, D. A.: Water Resource Systems Planning and Analysis. Prentice-Hall, Englewood Cliffs, New Jersey 1981.

National Water Authority: The National Masterplan for the Hungarian Water Resources Development, Budapest 1965 (in Hungarian).

United Nations: Integrated River Basin Development. Publication II. A.4, United Nations, New York 1970.

United Nations: River Basin Development: Policies and Planning, Proc. UN Interregional Seminar on River Basin and Interbasin Development, Natural Resources, Water Series No. 6, New York — Budapest 1976.

Lundqvist, J.; Lohm, U. and Falkenmark, M. (eds.):
Strategies for River Basin Management, pp. 031–040
© 1985 D. Reidel Publishing Company

Ecosystem Modelling of a River Basin

Pantulu, V. R., Environment Unit, Mekong Secretariat, c/o ESCAP, United Nations Building, Bangkok 10200, Thailand

Abstract: An approach adopted in ecosystem modelling of the Nam Pong river basin in northeast Thailand is described. In this context, a river basin is seen as a highly complex ecosystem, comprising numerous interactive and interrelated factors, including human activities. In order to assess the response of the ecosystem to resource management actions, therefore, its functional components and their linkages have to be defined. Furthermore, objectives and instruments of resource management have to be specified, in order to identify management options. The consecutive steps of ecosystem analysis and modelling are illustrated as examplified by the Nam Pong river basin in which a multipurpose reservoir was constructed. The predictive simulation model developed for this river basin illustrates the ecological implications of resource management actions and serves as a decision-making tool.

Introduction

A river basin can be considered to be a functional ecosystem which includes the organisms of a natural community together with their environment. In such an ecosystem the interactions among various components are bewilderingly complex. The major ecosystem fractions involved in a river basin viewed as an ecological unit, can be grouped into three different systems: physical, biological and human or socio economic, each comprising discrete but interactive and interrelated environmental factors (Fig 1). The physical system can be thought of, for purposes of this paper, as comprising the following factors: radiant energy from the sun, the atmosphere, water, and earth including soils and energy influences arising therefrom such as gravity and heat. The biological system comprises aquatic and terrestrial animals and plants, whereas the human or socio-economic system interacts with the physical and biological systems resulting in mutual benefits and adverse effects.

The human system, an interacting force, is composed of man's activities and his social and cultural institutions, all environmentally dependent and which in turn affect the environment. Principal production activities include hunting, fishing, agriculture together with forestry and livestock production, aquaculture (aquatic animal and plant culture; the aquatic parallel of agriculture), the manufacture of goods and products, and the commerce and service industries deriving therefrom. In the human system social institutions embrace health, education and general welfare services, communication including transportation, and cultural and religious traditions. In the context of water resource development, a river basin ecosystem (Fig 1) is characterized by direct feedbacks from human actions. Positive feedbacks go both to the environment (e.g. improvement of available resources

32

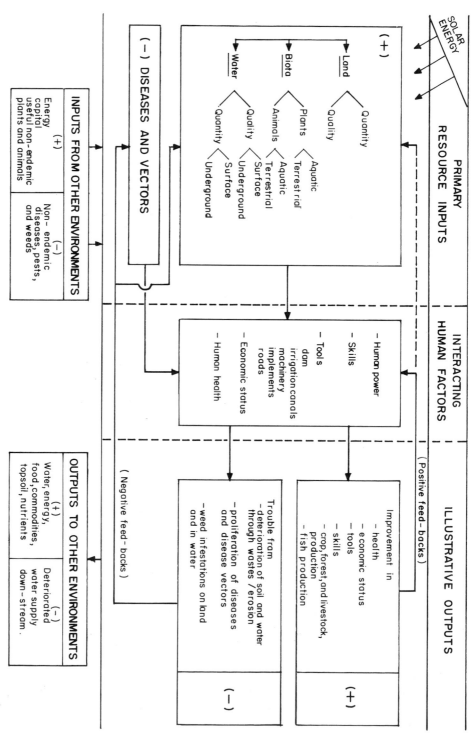

Fig 1 The river basin as an ecological unit

and the development of new ones) and to the interacting factors (e.g. increase in man-power, capabilities and skills, improvement of tools, livestock and so on). Negative feed-backs, however, may go only to the environment, where they become manifest in the de-terioration of resources resulting from such actions as the use of inappropriate technolo-gies, degradation of the environment owing to insidious effects of chemicals and waste products, and proliferation of diseases and disease vectors. With increase in technological inputs and resultant 'developed' state of the system, negative feedbacks can be expected to go directly to the agents of change (human beings) as well (e.g. direct effects of chemical toxicants, occupational diseases and hazards and so on). Thus, the river basin ecosystem must be understood as a naturally evolving complex of environmental components, linked by pathways of energy flows. Both the components and their linkages experience effects of development. The role of environmental managements should therefore be to maintain both the integrity and productivity of the primary resources and their functional linkages.

Sectoral vs. Systems Approach

An attempt was made to test both the sectoral and systems approaches in the context of resource management in one of the sub-basins of the Mekong river, namely the Nam Pong river basin in northeast Thailand, in which a dam was constructed some 15 years ago. Results of this attempt are elaborated in two reports entitled "Environmental manage-ment and water resource development in the Nam Pong basin of northeastern Thailand" (Phase II) (Mekong Committee, 1979b) and "Nam Pong environmental management research project" (Phase III) (Mekong Committee, 1982a). The former relates to a sectoral and the latter to a systems approach in resource management. In Phase II studies of a wide array of sectoral resource management issues such as land use and land capability, fisheries, farm management, reservoir regulation, and socio-economic and health conditions that stem form resource exploitation were examined. It was found that this study did give limited insights to some critical interrelationships which appear to affect behaviour and output of the basin system. However, in this approach, some less obvious, but none-the-less equally important, issues that derive from interactions among components of socio-economic and bio-physical sub-systems were overlooked. Furthermore, issues concerning management by public agencies, such as questions regarding the nature and sequence of interventions, the viability of such interventions in terms of Government policy and so on, further confounded an already complex situation.

In view of the foregoing, a systems approach was attempted to understand adequately the nature and influence of various interacting forces on the behaviour of the basin in response to environmental manipulations by human intervention. The results of this exercise are elaborated in the Phase III report mentioned earlier. Any natural system, such as a river basin, is a highly complex entity. One of the important requirements in systems approach, therefore, is to abstract from this complex reality, key elements to which system behaviour is sensitive and which are amenable to management or manipu-lation by public agencies.

Some Concepts in Systems Approach

In spatial terms the bounds of a river system can be considered to be the river basin. However, in certain situations, where flood control measures and outputs from the system such as nutrients, pesticides or discharge may have effects beyond the specified bounds of the basin, the system may have to be expanded as needed to cover certain other specified but limited functional components. Taking the river as the axis of the natural system, the following six physical subsystems can be identified (Fig 2) in a stylized tropical river system:

1. Upstream watershed system, including upland forests and agriculture;
2. The reservoir sub-system and the drawdown area, the associated lake perimeter population and its activities;
3. The downstream catchment sub-system;
4. The urban and industrial sub-system;
5. The irrigation sub-system, and
6. The estuarine and coastal zone sub-system.

Perhaps the most obvious and important linkages between each of the six sub-systems are the physical flows of water, sedimentation, nutrients and water pollutants (chemical and biological) as shown in Fig 2. Since these are sequential gravity flows from upstream to downstream they create no feedback loops. In general, linkages among sub-systems arise as a result of flows of five classes of attributes namely, material, population, capital, energy, and information, However, it is perhaps more convenient in terms of presentation and analysis to classify these into three classes: physical (material and energy), population, and economic flows. Fig 2 displays possible types of flows between the sub-systems themselves as well as exchanges between each sub-system and the environment outside the basin system.

The complexity of the interrelationships among the six sub-systems is increased with the addition of feedback biological flows and two-way population, and economic flows. The most obvious of these is movement of people across sub-system boundaries. Other feedback loops may result from flows associated with the economic process which comprise products such as fish, crops and livestock, of inputs such as machinery or fertilizer which may be measured in physical or monetary terms. Another potential two-way flow is information on such aspects as production technology or market conditions. For obvious reasons, the urban sub-system appears to be at the centre of economic flows in the basin. Evaluation of the population and economic linkages are of equal importance to the physical flows in assessing resource management issues and, in particular, development environment relationships.

Each sub-system can be further disaggregated into three sets of components (Fig 3) in order to examine functional relationships between sub-subsystems of natural resources and socio-economic sectors which exploit the resources in one way or another. The extent of disaggregation could be virtually infinite, but it is proposed that each subsystem be disaggregated into the following:

1. Natural resource endowment and associated ecological relationships;

Fig 2 Physical subsystems and their interrelationships within a river basin ecosystem and focal points of concern where the system components cross international boundaries

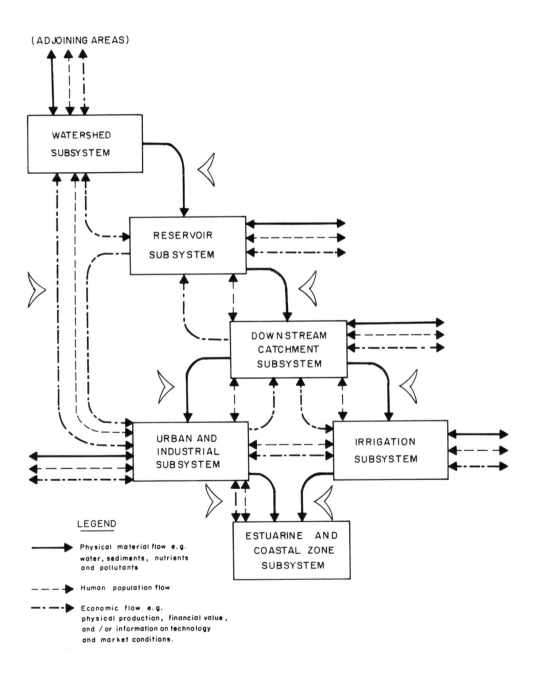

2. A set of primary, secondary and tertiary economic activities measured in terms of energy inputs, employment and area or physical and financial inputs; and

3. The dispersed urban and village centered human populations associated with the above economic activities and their cultural, nutritional, educational and health status.

Having identified the functional components, it is equally important to identify the linkages between components and the crucial areas where management might alter the behaviour of the system in a manner that would result in the achievement of development goals. Such identification should lead to a systematic assessment of critical resource management issues in the basin viewed from a synoptic perspective.

Fig 3 Basic module for disaggregation of a subsystem

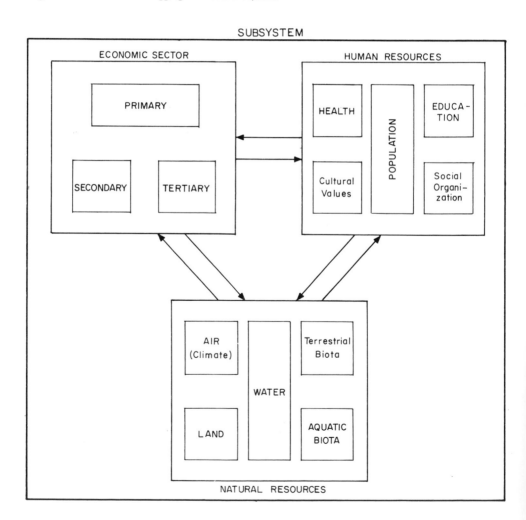

Development Objectives and Performance Indicators

A prerequisite to final specification of the basin system and identification of management options is a listing of the multiple objectives and quantifiable performance indicators by which progress towards goals can be measured. The following may be cited as examples:

1. Perservation of soil capability in the upland areas at a level that will sustain agricultural and/or forestry activities with only limited application to chemicals or other capital-intensive technology.

2. Associated with (1), maintaining a vegetative cover in the upland areas which will at least preserve sediment and flow regimes at current levels.

3. Among the multiple uses of the reservoir, emphasis might be placed on rural development by setting an order of priorities such as: 1. achievement of high and sustainable levels of fish production, 2. flood control, 3. assurance of water supply to urban centres, 4. irrigation, and 5. electric power.

4. Consistent with the rural development and environmental quality focuses of the foregoing three objectives, efforts should be made to stabilize the rural population at sustainable levels.

5. Irrigation in the downstream area should be expanded to the optimum capacity of the water resource.

6. The quality of water discharged downstream from the reservoir at average low flow should be adequate to assure that no prejudicial consequences result to downstream fisheries or riparian irrigation.

With a configuration of objectives such as the above, the "sustained yield" criterion for management of renewable resources and stabilization of rural population become constraints on the management programme. It then remains to be seen, from assessment of policy and the institutional structure, whether such constraints are viable or what constraint must be imposed on public expenditures which establishes the level of effort which can reasonably be expected by each agency, or all agencies together. It should be emphasized that in practice any number of objective sets may evolve with differing budgetary constraints.

Management Instruments

Having specified the basin system and the development objectives which affect resource use and environmental quality, it becomes necessary to examine the elements with the government has at its disposal to manipulate the system in the direction of desired goals. In the first instance the list of possible management interventions will be a function of the objectives. Two types of management instruments may be identified — those which apply to situations where resource use decisions are largely in the hands of primary producers, and those applicable through direct government action with respect to common property resources e.g. the river and reservoir, or state property e.g. forest reserves and infrastructures. In the case of primary producers the instruments will be either regulatory e.g. granting conditional rights to use resources such as licensing of fishermen, control of fishing gear and limitations on catch, or provision of incentives to encourage a particular pattern of use e.g. imparting information, provision of improved productive and social infrastructure, credit facilities or subsidies. Direct public management would apply to such elements as allocation of water in the reservoir between multiple use; fish stocking in the reservoir, logging in forest reserves or provision of health services to control diseases attributable to particular patterns of resource use e.g. liver flukes.

Identification of Key Questions and Information Needs

The problem of methods for identification of critical resource and environmental management issues, and the level of information needed by decision makers may be addressed through a process of conceptual modelling of system behaviour and sensitivity analysis. The primary issue is the physical flow of water, sediments and nurients between subsystems and how this flow affects the socio-economic components in each sub-system. The pivotal point in the basin would appear to be the reservoir. Thus, a number of key questions would revolve around how the flows into and out of the reservoir affect, or are affected by economic activities and social conditions in the upstream watershed, in the downstream area and in the reservoir itself. Illustrative questions in this connection might include the following:

- If, in fact, fish and agricultural production are sensitive to intra- or inter-year variability in the flows, how are such flows influenced by changes in land use in the upstream watershed?
- How is water quality in the reservoir affected by alternative farm management practices in the drawdown zone under alternative water level management programmes?
- What is the realtionship between water quality and fish production?
- What is the relationship between health of the reservoir perimeter population and the pattern of water level change?
- If there is a health-water level relationship, what measures could be taken to render health independent of water level?

On the basis of such an initial set of questions, a listing of potential management interventions can be evolved. Exploration can also be undertaken of how the ramifications of a management action may be transmitted through the system. Fig 4 traces a possible set of chain reactions through a sub-system as a result of a management intervention. Examination of such flow diagrams combined with sensitivity analysis enable exploration of critical areas where further information on system behaviour (key research questions) would be useful in management decisions. Without a systematic assessment of the potential pathways of change and dynamic interrelationships between system components, which is facilitated by the conceptual modelling, some critical questions may be overlooked. Also, there would be an inadequate basis for establishing priorities and the sequence of research, and it would be impossible to establish the precision of information required for management decisions.

Fig 4 Illustrative chain effects from management action in the watershed subsystem

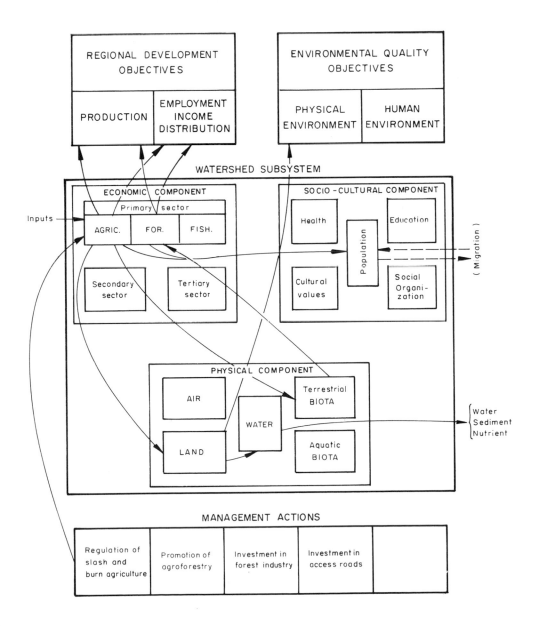

Conclusion

The foregoing concepts and premises were the basis of formulation of the simulation model of the Nam Pong basin in northeast Thailand. This was admittedly an experimental effort with the object of achieving environmentally sound resource management, based on a clear understanding of the dynamic ecological relationships between physical, biological and human components of the river basin system. The process of constructing an integrated simulation model has served, as much as the final product, to make users of the project (managers and affected population) aware of the environmental interdependencies of their actions (Mekong Committee 1982b).

The model itself provided a decision making tool, producing quantitative predictions of the ecosystem's response to management actions. Its particular strengths are that it can be used to illustrate likely consequences and trade offs of a range of possible management actions. Also, it conveys the implication of random events which must be considered by managers, and it presents cohesive indicators of outcomes, which people from a wide variety of disciplines can comprehend (e.g. income trends). Decisions are not formulated by the model. Rather, the model serves as an assessment tool.

References

Mekong Committee: Perspectives and framework for development of Phase III: Nam Pong environmental management research project, northeast Thailand (MKG/R.240). 20 pp. Interim Committee for Coordination of Investigations of the Lower Mekong Basin, Thailand 1979a.

Mekong Committee: Environmental management and water resource development in the Nam Pong basin of northeastern Thailand (MKG/81). Interim Committee for Coordination of Investigations of the Lower Mekong Basin, Bangkok, Thailand 1979b.

Mekong Committee. Nam Pong environmental management research project, final report Phase III (MKG/101). 311 pp. Interim Committee for Coordination of Investigations of the Lower Mekong Basin, Bangkok, Thailand 1982a.

Mekong Committee: Environmental impact assessment. Guidelines for application to tropical river basin development. 123 pp. Interim Committee for Coordination of Investigations of the Lower Mekong Basin, Bangkok, Thailand 1982b.

Lundqvist, J.; Lohm, U. and Falkenmark, M. (eds.):
Strategies for River Basin Management, pp. 041–048
© 1985 D. Reidel Publishing Company

Ecosystem Approaches to River Basin Planning

Reynolds, Peter J., Water Planning & Management Branch, Inland Waters Directorate, Ottawa, Ontario K1A 0E7, Canada

Abstract: Over the past two decades there have been many versions of the "ecosystem approach". Much of the work has been abstract and relatively difficult to implement by decision-makers. Most of these approaches share the following characteristics: a primary focus on ecological phenomena as opposed to engineering, economic, or jurisdictional phenomena; a perception of some self-regulatory capacity on the part of an ecosystem; a recognition of the marked responsiveness of many ecological systems to natural and human activities; and a readiness to strike a pragmatic compromise between detailed reductionistic understanding and more comprehensive, holistic meaning. In the Great Lakes and St. Lawrence River Basin, workers are now trying to implement operational and institutional forms of ecosystem approach. Ten of these initiatives are assessed here, taken from all government levels, directed toward river basin, open lake, coastal, and hinterland components of the basin. This comparison illustrates the range of proposals now under consideration in the basin and identifies their common elements. These features are offered as a definition of the criteria involved in an ecosystem approach. It is suggested that flexible eclectic pragmatism may be the most productive attitude toward Great Lakes environmental problems.

Introduction

A watershed or catchment is an integrated system that transforms precipitation, solar radiation, other environmental variables, labour, and capital in wood products, livestock products, wildlife, recreational and aesthetic satisfactions, and water. The forest management subsystem, the grazing subsystem, the recreation use and development subsystem, and the water management subsystem interact to produce the vegetation, animal, and soil conditions that govern the yield and quality of its products and services. The only level of ecological theory that can effectively guide management of such complex systems is a theory of ecosystems.

The Ecosystem approach discussed in this paper is defined as an anticipatory approach to planning of river basins and general problem solving that relates parts to wholes by alternating "egosystem" and "ecosystem" views. It is based on the knowledge of the operation and interrelationships of systems in Nature and, in consequence, the necessity of ecological behaviour and desirability of adoption of an ethic of respect for other systems of Nature.

Man in Nature

Since every person sees his or her surroundings from a unique point of view, universally acceptable definitions of ecosystem and ecosystem approach must be broad. This does not deny the need for definition. It only means that people can agree on common characteristics of holistic approaches without waiting for universal definitions. The ideas flow directly from abstract concepts to pragmatic initiatives. In the following we discuss these ideas.

Ecosystems are natural or artificial subdivisions of the Biosphere with boundaries arbitrarily defined to suit particular purposes in hand. It is possible to speak of your personal ecosystem (you and the environment on which you depend for air, water, food and friends), the Great Lakes Basin and St. Lawrence River Basin System is a prime example of an ecosystem (interacting communities of living and non-living things in our planetary ecosystem, the Biosphere).

In your lungs there is likely at least one molecule from the breath of every human being that has lived in the past 3000 years and the air around you will be used tomorrow by deer, fish, mosquitos and trees. The same is true of water, sunshine and minerals. Everything in the biosphere is shared. Sharing and interconnectedness are the reasons why the boundaries of ecosystems overlap. People understand this in a vague way. What they have not seen is how to relate to it.

The attitudes, perceptions and behaviour of people and organizations in river basins have been shifting over the past decade from an environmental approach to an ecosystem approach for planning and problem solving. The distinction between environment and ecosystem is comparable to that between house and home. In the former, we see ourselves as outside and separate from the system, in the latter, within and part of such systems. Social, economic and environmental interests are all included in the ecosystem.

Ecological systems are not entirely composed of discrete elements called "structure", "dynamics", and "change", although it has been convenient to break them down in this way. More realistically, ecological systems are interacting assemblages in which any two elements affect the third. One element cannot be modified without changing the other two.

The emergence of an ecosystem approach to planning, research and management in the Great Lakes basin is not accidental. It is the most recent phase in a historical succession of management approaches from egocentric to piece-meal to environmental and now to an ecosystem approach. This succession arose from the growth of population and technology in the Great Lakes basin.

What must be done to practice an ecosystem approach?

— Improve knowledge of the operation and relationships of systems in nature.
— Develop a holistic perspective that takes account of the influences on us of larger systems of which we and our external environments are parts.
— Act in ways that are ecological (take account of that knowledge and perspective), anticipatory (forestalling events that would bring later regret), and ethical (showing respect for other systems of nature comparable to our respect for other persons).

In the past there has been too much focus on hierachical structures and intra-organizational concerns.

Some examples of the evolution of management styles may help to clarify what is meant by ecosystem approach and to show the extent to which it is now in development:

1. Organic waste. First it was dumped wherever convenient — best of all in streams or lakes. Next, because of downstream problems, we developed energy-consumptive sewage treatment systems. The ecosystem approach focuses on energy and material recovery from sewage.

2. Eutrophication. First, it was ignored. When the odours became too strong, nutrient-rich effluents were diverted downstream. Then phosphorus was removed from sewage effluents. The ecosystem approach promoted reduction of use through low-phosphate detergents and more efficient use of fertilizers combined with nutriet recycling.

3. Oxides related to acid rain. At first the problem was not recognized. When problems arose locally, the solution was to build taller smoke-stacks. Then came removal of acids by scrubbing. Now, the ecosystem approach advocates energy conservation and the recycling of sulphur.

4. Water diversions and consumptive use. The first rule was: divert, the more the better. Then the scale was increased to meet new shortages, encouraging export as a commodity. On the basis of the ecosystem approach the new rule is to export water sparingly — and only to recipients that practice an ecosystem approach.

5. Cancer. People were never indifferent to cancer; however, it is still commonly viewed in terms of single causes. For the ecosystem approach real cures must be based on the knowledge that cancer is 80–90% environmental and with many contributing causes.

6. Toxic chemicals. At first, chemicals were used indiscriminantly. Then they were dealt with one by one with regulations after-the-fact as in the case of pesticides. With an ecosystem approach the rule is to design with nature, particularly for long-lived compounds.

7. Energy shortages. Successive solutions were to ignore the problem then to increase the energy supply and expand the grid with pricing to encourage greater use. The ecosystem approach encourages conservational pricing with inverse rate schedules to discourage greater use.

8. Traffic congestion. Successive solutions have been to curse; to build more roads and super-highways; and to improve public transport and with staggered hours. The ecosystem approach encourages decentralization through new forms of transportation and communication.

9. Pests. At first it was "run for your life". Then came broad spectrum pesticides. Next it was selective, degradable poisons. The ecosystem approach calls for integrated pest management.

Ecosystem Approach to the Great Lakes and St. Lawrence River Basin

Ecosystem approaches have been initiated independently for a range of problems in the Great Lakes and the St. Lawrence River basins; from terrestrial to aquatic; from urban to rural; from large to small scale; from single-stress response to complicated multi-stress syndromes. The institutional framework which exists to implement these management proposals is also extremely variable, and often inadequate to deal with the expanded management horizons.

Traditionally narrow management interests and mandates have been stretched to include and coordinate, or at least cooperative with many general planning functions; public involvement, regulation of land use, mechanisms for increasing program funding, and more comprehensive administrative organization.

In this paper, ecosystem approaches are brought together to Great Lakes basin management problems, similarities and differences are identified, and at times, insufficiencies in these approaches, and recommendations are made for some general features of comprehensive ecosystem approaches are outlined.

The ecosystem approaches considered here have been developed or adapted with respect to various purposes such as: international Great Lakes water quality management, international Great Lakes fisheries management; Great Lakes basin urban planning; Ontario provincial fisheries management; environmental planning and management in Ontario; "ecological" planning in general, and environmental information systems in Canada. The approaches (Fig 1) are:

- The International Joint Commission/Great Lakes Research Advisory Board (1978), The Ecosystem Approach, IJC/EA,
- Leman & Leman (1976), The Great Lakes Megalopolis, GLM,
- Great Lakes Fishery Commission (1979), Rehabilitating Great Lakes Ecosystems, GLER;
- Federal-Provincial Strategic Planning for Ontario Fisheries, SPOF;
- International Joint Commission/Science Advisory Board/Ryder (1979), Ecosystem Objectives for the Laurential Great Lakes Based on the Fundamental Requirements of the Lake Trout, Lake Trout Indicators for Management, SAB/LT;
- Ian McHarg's environmental planning method (1969) as demonstrated in the Toronto General Water-front Planning Study (1974-present), CWPS;
- International Joint Commission/International Reference Group on Great Lakes Pollution From Land Use Activities (1972–1978), PLUARG;

Environmentally Sensitive Area Planning in Ontario, ESA Planning:

- Great Lakes Basin Eutrophication Models, GLEM;
- Rapport & Friend (1979), Rapport & Regier (1980), The Stress-Response Environmental Information System, S-RESS.

Other ecosystem approaches not addressed here include the epidermiological (Swain 1981), island biogeography (Magnuson 1976), adaptive management (Holling 1978), trophody-

CRITERIA	APPROACH									
	I.J.C.'s Ecosystem Approach	Great Lakes Megalopolis	GLER (Rehabili-tation)	SPOF (Ontario Fisheries)	Lake Trout, Ecosystem Indicator	McHarg's CWPS Method	PLUARG (Land Use Pollution)	ESA Planning	GLEM (G. Lakes Models)	S-RESS (Environ. Statistics)
Primary emphasis is on ecological phenomena	●	○	●	●	●	●	●	●	●	●
Boundaries reflect ecological integrity	●	■	●	●	●	■	●	●	●	●
Mapping, monitoring and modelling are used to assess ecological state and processes	■	○	●	●	●	■	●	●	■	●
Ecological self-regulation is considered	●	○	●	●	●	○	●	●	●	●
Ecological responsiveness is considered	●	●	●	●	●	●	●	●	●	●

Fig 1 Ten ecosystem approaches to the Great Lakes with different levels of ecological criteria for an "ecosystem approach".

● = Criterion is used,
■ = Criterion is partially or inconsistantly used,
○ = Criterion is not considered.

namic models (Stewart 1980; Stewart et al. 1981), and electic ecosystem approaches to regional planning and design (Dorney 1976; 1977; Dorney et al. 1981). All of these and others as well, are currently being adapted or developed to deal with some Great Lakes issues.

Principles of Conservation

In a broad yet pragmatic policy orientation ecosystem approaches should relate to the following general principles:

1. The ecosystem should be maintained in a desirable state such that
   — consumptive and nonconsumptive values could be maximized on a continuing basis,
   — present and future options are ensured, and
   — risk of irreversible change of long-term adverse effects as a result of use is minimized.

2. Management decisions should include a safety factor to allow for the fact that knowledge is limited and institutions are imperfect.

3. Measures to conserve a wild living resource should be formulated and applied so as to avoid wasteful use of other resources.

4. Survey or monitoring analysis, and assessment should precede planning use and accompany actual use of wild living resources. The result should be made available promptly for critical public review.

Ecosystem Approach to River Basin

The River Basin is an ecosystem subject to multiple use. The geomorphologist, the hydrologist, the social scientist, and the climatologist can all contribute to an ecosystem approach to river basins, but, because these specialists are so bound up with other aspects of the ecosystem, it becomes necessary to have an integrated approach. The River Basin provides a natural boundary to evaluate the ecosystems contained therein. In most ecosystems there is no actual membrane separating system from environment. The boundary in most cases is imaginary and is located at the convenience of the observer. Once the observer has made his boundary definitions, the ecosystem may be visualized as being connecting to the surrounding biosphere by a system of inputs and outputs. Inputs and outputs may be in the form of radiant energy, water, gases, chemicals, or organic materials moved through the ecosystem boundary by meteorological, geological, or biological processes.

The ecosystem approaches have bridged the disciplines of natural and human ecology to form an integrated or transdisciplinary approach to river basins and lakes. The general features exhibited in ecosystem approaches for the Great Lakes and St. Lawrence River Basin appear to be the following:

1. Concepts and methods of appropriate
   - spatial and temporal scope,
   - scale and detail,

2. Analytically, sufficient specification of the ecosystem with respect to:
   - external natural processes and components,
   - internal natural processes and components,
   - ecosystem man-nature interactions;

3. Holistically, adequate comprehension of
   - internal evolutionary/successional process,
   - responsiveness/adaptiveness to external factors,
   - recovery sequences after relaxation of external stress;

4. Information set includes
   - maps of spatial ecosystem features,
   - monitored time series of diagnostic features,
   - models of relational/causal processes,
   - management cost studies;

5. Management objectives involve compromises with respect to
   - material well-being/progress,
   - cultural opportunity/equity,
   - environmental harmony/ethics;

6. Political-administrative processes
   - include the active involvement of a broad representation of users of the ecosystem,
   - implement a range of legal-economic mechanisms designed to reduce user stresses,
   - regularly and critically review the planning, regulatory, and budgeting functions of all agencies involved in the management of the ecosystem, and

7. Disciplines/professions include
    — ecologists/geographers/economists/political scientists ...,
    — engineers/planners/environmental-resource managers.

It can be concluded that it would be neither possible nor desirable to specify a unique, unitary, unified universal "ecosystem" approach to River Basins. The seven features above tend to complement each other, as do their sub-features. Any practical attempt to be effective will need to deal with some of those criteria in a dialectial way, or seek some pragmatic compromise. The River Basin is a challenging ecosystem.

Many questions remain unanswered or partially answered, e.g.,

— Can one organization represent all interests in a river basin?
— What are the major constraints or obstacles preventing implementation of the ecosystem approach?
— Can research objectives be made compatible with ecosystem objectives?
— Can cost-benefit analysis be made to fit in with the ecosystem approach?
    (e.g. soil erosion, impaired fish habitat, degraded water quality are all items having intangible costs and difficult to balance in economic terms).
— Incremental or piece-meal project planning dominates most river basins. What does it take to get a stronger political voice in decision-making?
— What are the difficulties in organizing baseline data for river basins?
— What are the experiences in sustaining productivity through the ecosystem approach?

References

Dorney, R S.: Biophysical and cultural-historic land classification and mapping for Canadian urban and urbanizing land. In: Proc. Workshop on Ecol. Land Class. in Urban Areas. Can. Comm. on Ecol. Land Class, Toronto, Ontario 1976.

Dorney, R. S.: Planning for environmental quality in Canada: Perspectives for the future. Unpubl. paper presented to the Canadian Institute of Planners Annual Meeting, June 1977. Toronto, Ontario 1977.

Dorney, R. S.; Eagles, P. F. J.; Evered, B.; Hoffman, D. W.: Ecosystem planning, analysis and design in Ontario as applied to environmentally sensitive areas. Unpubl. paper presented to the Amer. Assoc. for the Adv. of Sci. Annual Meeting, January 1981. Toronto, Ontario 1981.

Great Lakes Basin Commission: Post PLUARG evaluation of Great Lakes water quality management studies and programs. Vols. 1 and 2. Sullivan, R A. C.; Sanders, P. A.; Sonzogni, W. C.: US Environmental Protection Agency, Report No. EPA-905/9-80-006-A, Chicago, Illinois 1980.

Great Lakes Fishery Commission: Rehabilitation Great Lakes ecosystems. In: Francis, G. R.; Magnuson, J. J.; Regier, H A.; Talhelm, D R (eds.). Great Lakes Fishery Commission Tech. Rep. 37. Ann Arbor, Michigan 1979.

Heidtke, T. M.: Modelling the Great Lakes system: update of existing models. Great Lakes Basin Commission. Ann Arbor, Michigan 1979.

48

Holling, C. S. (ed.): Adaptive environmental assessment and management. Inter. Inst. for Applied Syst. Anal., sponsored by the United Nations Environmental Program. John Wiley and Sons, Chichester and Toronto 1978.

International Joint Commission: Pollution in the Great Lakes basin from land use activities. International Joint Commission Report to the Governments of the US and Canada. Windsor, Ontario 1980.

International Joint Commission/Great Lakes Advisory Board: The ecosystem approach. Special Report to the International Joint Commission. Windsor, Ontario 1978.

International Joint Commission/International Reference Group on Great Lakes Pollution From Land Use Activities: Environmental management strategy for the Great Lakes. Final report to the International Joint Commission. Windsor, Ontario 1978.

International Joint Commission/Science Advisory Board: Anticipatory planning for the Great Lakes. Vol. 1. Summary. Workshop sponsored by the International Joint Commission through its societal Aspects Committee. Windsor, Ontario 1979.

International Joint Commission/Science Advisory Board/Ryder, R. A.: Ecosystem objectives for the Laurential Great Lakes based on the fundamental requirements of the Lake Trout. (1979 unpublished)

Leman, A. B.; Leman, I. A.: Great Lakes Megalopolis from civilization to ecumenziation. Ministry of State for Urban Affairs, Canada 1976.

Magnuson, J. J.: Managing with exotics: a game of chance. Trans. Amer. Fish. Soc. 195, 1—9 (1976)

McHarg, I. L.: Design with Nature. The Natural History Press, Garden City, New York 1969.

Rapport, D. J.; Friend, A. M : Towards a comprehensive framework for environmental statistics: A stress-response approach. Cat. #11-510 — Statistics Canada, Ottawa 1979.

Rapport, D. J.; Regier, A H.: An ecological approach to environmental information. Ambio 9, 22—27 (1980)

Swain, W.: An ecosystem approach to the toxicology of residueforming xenobiotic organic substances in the Great Lakes. US National Academy of Science 1981.

Lundqvist, J.; Lohm, U. and Falkenmark, M. (eds.):
Strategies for River Basin Management, pp. 049–061
© *1985 D. Reidel Publishing Company*

Methodology for Monitoring and Evaluation of Integrated Land and Water Development

Biswas, Asit K., Biswas & Associates, and President, International Society for Ecological Modelling, 76 Woodstock Close, Oxford, England

Abstract: In spite of an urgent need to consider land and water development in combination, fragmentation in education and training and existing institutional structures form important constraints. Intense past disappointments with water projects in many countries have now created a new readiness to persuade planners and decision makers to consider an integrated approach to land, water and other related resources in river basin development. The author proposes that with the massive investments in water development projects in developing countries, systematic monitoring and evaluation should become an integrated part of the management process. The information produced has to be timely, cost-effective, relevant and correct, and free from professional biases. Monitoring and evaluation have to comprise both planning, design and construction; operation and maintenance; agricultural production; and achievement of socio-economic objectives.

Introduction

While nearly all water resources development projects in developing countries are multipurpose in nature, the major purpose without any doubt has been generally to provide irrigation water to increase agricultural production. The two other important purposes often considered simultaneaously are hydroelectric generation and flood control.

Timely, reliable and well-managed water supply and its effective use is a most crucial requirement for the modern high-yielding agricultural production. This is clearly indicated by the fact that even though only 20% of the world's agricultural land is irrigated at present, they account for 40% of the global agricultural production (IDRC 1979). It is quite clear that without reliable water control, the world food problem cannot be resolved.

Since the main emphasis of water development projects in developing countries is to increase agricultural production, it is no longer adequate to consider water development in isolation. Increasing agricultural production requires simultaneous optimal use of resources like water, land, and energy, and availability of other inputs like good quality seeds, credit, and extension services. Furthermore, increasing overall agricultural production cannot be the exclusive goal: the issue of equity — who receive the benefits and by how much — should be a major consideration. Equally important is the consideration how the benefits are affecting the socio-economic conditions of the basin area: how the lifestyle and quality of life of people are changing.

Even though there is an urgent need to consider land and water development simultaneously, it is not happening at present. Fragmentation in education and training and existing institutional structures in countries are two important constraints that need to be overcome if an integrated approach is to be adopted successfully and widely.

Lack of such an integrated approach has contributed to nearly half of the world's irrigated area being afflicted with some degree of salinity, alkalinity or waterlogging. It was estimated during the UN Water Conference that by 1990, out of 92 million ha of irrigated land in developing market economies of Africa, Asia and Latin America, 45 million ha would require improvement at an estimated cost of more than $ 22,000 million at 1975 prices (Biswas 1978). Futhermore, water resources development projects generally do not appear to have contributed to equity. An analysis of the experiences of the United States Agency for International Development (US AID) indicates that irrigation is "at best a reaffirmation of the existing social and economic distribution of assets, but more often, it will tend to exacerbate differences in both income and social prestige" (Steinberg 1983). Often estimates of cropping patterns and intensities, average yields, farm prices, employment and income generation, and availability of credit and inputs like pesticides, fertilizers and seeds, extension services and marketing facilities have turned out to be pious hopes rather than reality. In addition, environmental and health costs of irrigation projects have been substantial (Biswas 1982a, 1982b).

There is however a good possibility that some countries may be willing to adopt an integrated approach because of the intense disappointments with the results of certain irrigation projects undertaken during the past decades. For example, a review of the irrigated agriculture projects in the Sahel by the Club du Sahel and CILSS (1980) concluded that the area under modern irrigation doubled during the period 1960 to 1979, but "generally speaking, during the past few years, the development of new areas has barely surpassed the surface (area) of older ones which had to be abandoned". A major conclusion of a Workshop on "Aid to Irrigation", convened by the Development Assistance Committee of OECD (1983), was not only an expression of general dissatisfaction with the performance of large-scale irrigation projects in developing countries but also the radical suggestion that for some areas of Africa, "irrigation should not be generally promoted until existing schemes were shown to be productive and until well-tested technology and comprehensive plans have been prepared."

Such reasons for pessimism can, hopefully, now be used constructively to persuade planners and decision-makers to consider an integrated approach to land, water and other related resources to river basin development.

One of the main reasons for the past failure of water development projects to meet the approved objectives is due to the lack of adequate monitoring and evaluation, and the failure by the management to use monitoring and evaluation successfully as a management tool. It is not unusual to find that monitoring and evaluation are completely neglected.

Monitoring and Evaluation as Tools in the Water Development Process

In the context of the present discussion, monitoring is defined as continuous or periodic surveillance over the implementation of an activity (and its various components) to ensure that input deliveries, work schedules, targetted outputs and other required actions are proceeding according to the plan. Since the purpose of monitoring is to achieve efficient and effective project performance, it is an integral part of the management information system and is an internal activity. Evaluation is defined as a process which attempts to determine as systematically and objectively as possible the relevance, effectiveness and impact of activities in the light of their objectives. It is a learning and action-oriented management tool and an organization process for improving activities still in progress and future planning, programming and decision-making.

It is essential that monitoring and evaluation become an integral part of the management process to ensure future stream of benefits to occur at the appropriate time-scale to the right target group. It can be equally argued that for this purpose to be achieved, monitoring and evaluation projects have to meet some principal requirements (Fig 1):

i)   timeliness;
ii) cost-effectiveness;
iii) maximum coverage;
iv) minimum measurement error;
v)   minimum sampling error; and
vi) bias-free.

Information has to reach the dicision maker at the right time

Management decisions generally have a time dimension, and timeliness of certain decisions could be more important than others. Thus, monitoring and evaluation (M&E) information needs to reach decision makers on time so that the contents of the information sup-

Fig 1 Requirements for monitoring and evaluation.

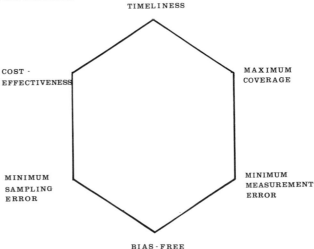

plied serve as a fundamental basis on which rational decisions can be taken. In other words, M&E information needs to be converted into action as shown in Fig 2. Management success depends not only on the timeliness of the information but also quality, extent and form of the information channelled into the decision making process. Problem arises because even if required M&E information is available, it is often not channelled into the decision making process since it is either in a diffused or inappropriate form or could not be obtained and analysed within the time frame by which decisions have to be made (Biswas 1976).

If M&E information from the project does not reach the decision makers on time, it is possible that one or more of the following consequences, which are not mutually exclusive, may occur:

— wrong decision may be taken;
— decision taken may not be optimal in terms of agreed objectives;
— no decision may be taken when one is essential;
— decision taken may result in irreversible damages; or
— decision taken may unnecessarily increase the cost of the project and/or time required for completion.

It is therefore essential that M & E information from the project should reach the people who need it on time and on a continuing basis.

The process has to be cost-effective

Since financial resources, expertise, man-power and equipments available in developing countries are invariably limited, there is always pressure for the M&E system to be effective. This may essentially mean sensible trade-offs between the depth and context of information, as well as between amount, relevance and accuracy.

Information collection, analysis and processing require time, money and expertise. Thus, as a general rule it should be remembered that the value of information collected should exceed the cost of obtaining that information.

Fig 2 Relationship between water project, monitoring and evaluation, and decisions.

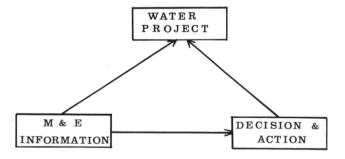

For most projects, from a decision making viewpoint and at any specific point in time, value of information generally increases with increasing extent and accuracy of information available. However, for most decisions, value of information generally reaches almost a plateau after a certain point, beyond which the value may increase only very, very marginally as shown in Fig 3. The cost of obtaining information, however, continues to increase with increasing coverage and accuracy. Thus, it can be argued that the shaded area in Fig 3 is the cost-effective zone, beyond which the cost of obtaining information will rapidly exceed its intrinsic value. Exactly where within the shaded area a decision has to be made will depend on the specific projects concerned, but the trade-off consideration will very often be a value judgement.

Minimum sampling error — Since it is generally neither possible nor desirable to monitor developments in the entire project universe, sample surveys are essential. M & E of irrigation projects will invariably cover not one issue but several, and it should be remembered that what may be considered a suitable sample size for one issue or discipline may be too large or too small for another. For example, for analyses of rainfall, one rain gauge per km^2 will be considered by hydrologists to be a very dense network, and not necessary unless for very specific purposes, but similar sample sizes would be totally unacceptable to statisticians or sociologists. Thus, based on ultimate use of the information to be collected and practical considerations, sample sizes have to be decided carefully.

Minimum measurement error — In contrast to anthropologists and sociologists who tend to prefer maximum coverage, engineers and physical scientists are more concerned with the accuracy of measurements and data collected. A good M & E system naturally should have small sampling error and minimum measurement error. In any event, there is always a trade-off between coverage, sampling, precision, and time needed for analysis. The final decision that has to be taken on all these aspects would have to be case specific.

Fig 3 Relationship between cost, coverage and accuracy of information.

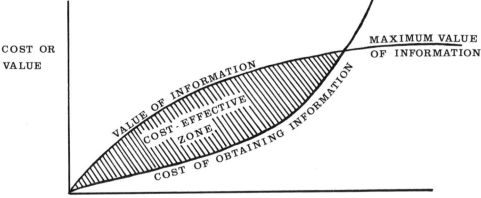

COVERAGE AND ACCURACY
OF INFORMATION

For water development projects, measurement error is a real problem, especially when benefits occuring to small farmers and landless labourers are considered. They are often illiterate and thus may have considerable difficulty with explicit numerical quantifications. Accordingly, they may be somewhat vague or imprecise about changes, especially when they are of the order of less than 15—20%. The questionnaire designers should be aware of this problem and ensure that questions that cannot be answered are not asked. For many purposes it is adequate to get "more/less" responses.

One of the main measurement problems for socio-economic data of existing water development projects is the virtual lack of baseline information. Since the original status of socio-economic conditions in the river basin when development started is often unknown, it is difficult to make quantitative measurements of improvements of these conditions which can be attributed to the project. Thus, a base-line survey of socio-economic conditions of the basin area, before the development process is initiated, is absolutely essential for future evaluations.

The information should be free from professional biases

M & E of water development projects often suffer from biases, which stem from the general tendency to concentrate on some specific issues at the cost of other equally important issues. Biases commonly observed may be concerned with the following:

— major structures and canals but not watercourses;
— head of watercourse but not tail;
— water entering watercourses but not losses;
— irrigation but not drainage:
— roads along canals and watercourses but not fields where access may not be easy or comfortable;
— visible structures but not people;
— contact with rural elite but not farmers;
— large farmers but not small;
— farmers but not landless labourers;
— men but not women;
— senior project staff but not junior;
— users of services but not non-users;
— review during healthier, better-fed dry season when climate is pleasant but not during food-scarce, unhealthy and unpleasant wet season;
— emphasis on the visible but neglect of social relationships;
— extrapolation of trends over the entire project area on the basis of a glimpse at one specific point and/or period of time; and
— evaluation soon after completion but not ten years later when projects is fully developed and conditions have stabilized.

There is also a tendency in irrigation projects which are multifacetted, to introduce biases in terms of one's own discipline. Thus, M & E carried out by indisciplinary people often tend to emphasize areas that are of primary interest to them. The problem emphasis of different disciplines in the area of irrigation, and the standard solutions proposed could be the ones suggested in Tab 1.

Tab 1 Problem emphasis and biases of different disciplines in the area of water development

Discipline	Problem Emphasis	Solution
Administrators	Poor co-ordination	New organization with administrator as co-ordinator
Agricultural Economists	Agricultural prices and marketing	Improve marketing and prices
	Lack of credit	Provide credit
	Risks of production	Reduce risks
Agriculture Extensionists	Farmers unaware of good agricultural and water management practices	More extension service to farmers
Biologists	Inundation impacts on flora and fauna	Reduce impacts by changing scale or location
Economists	Inefficient water use	Water pricing
	Low return on capital and underutilization of potential	More investment
Engineers, Agricultural	Poor land levelling	Level land
	Poor maintenance or lack of field channels	Improve situation
Engineers, Civil	Inadequate structural development	Construct more/better structures
	Poor operation and maintenance (O & M)	Provide more funds for O & M
Engineers, Drainage	Salinity and waterlogging	Construct comprehensive drainage system
Environmentalists	Too much damage to the environment and ecosystems	Stop construction or reduce scale of development
Lawyers	Central-Provincial relations or international implications	Resolve potential legal problems
Political Scientists	Inequitable distribution of agricultural production and water	Change power structure
Sociologists	Inequity and conflict over agricultural production and water distribution	People's participation and conflict resolution
	Rehabilitation of displaced people	Preparation and implementation of better plan for rehabilitation

It should be realized that in the real world an issue is an issue, and it only becomes technical, economic, or sociological depending on a person's discipline, experience and ways and means of approaching it. Thus, for M & E of water projects, it is preferable to use interdisciplinary individuals, who may specialize in one discipline but are flexible, observant, sensitive, eclectric and are capable of intermixing and questioning inventively. Since in reality, it may be difficult to find such interdisciplinary individuals, one may have to depend on who is available. This can be done by carefully selecting a multidisciplinary team, which can offset biases by juxtaposition of insights of various disciplines. However, it should be noted that the choice of a multidisciplinary team does not necessarily produce a real multidisciplinary approach and effort.

Four Levels to Monitor and Evaluate

Complexity of the problems

Water projects are complex to monitor and evaluate, since a large number of specific and specialized tasks have to be performed, both concurrently and sequentially, in a coordinated manner, by a variety of professionals, with decisions being made which may have direct impacts on the projects by local, regional, national and international institutions. In addition, all the project benefits and costs, both direct and indirect, are not confined to the project boundary: some of them could occur far from the area. Thus, it is not easy to define an area which could be said to contain all project impacts.

Time dimension of impacts is another complicating factor. Some impacts are immediate, and thus can be identified during the implementation phase or soon thereafter. Some other impacts, however, could be slow to develop, and thus may not be easy or even possible to monitor meaningfully in the early stages. For example, some unanticipated changes in ecosystem and the environment could easily take more than a decade of operation of the project, before they could even be identified and thus their monitoring could begin. Salinity development in irrigation projects, under certain circumstances, could take 15—20 years, but in others it could take only 2—3 years, depending on physical conditions, drainage provided and effectiveness of operation and maintenance procedures. Thus, water projects need continual M & E, even when the projects appear to be functioning most efficiently for several years. The time dimension also makes intercomparison of impacts of different irrigation projects a difficult task.

The different levels

From monitoring and evaluation viewpoints, water development projects can be logically arranged into following four interrelated levels:

i) planning, design and construction of physical facilities,
ii) operation and maintenance, especially of irrigation, drainage and hydropower generation facilities,

iii) agricultural production, and

iv) achievement of socio-economic objectives.

Of these four levels, probably the easiest one to handle in terms of monitoring and evaluation is the first one — planning, design and construction of physical activities. It is also a discrete phase, which is completed once the construction of physical facilities is over. In contrast, the other three levels require continual monitoring and evaluation during the project life to ensure that the system is operating at the desired efficiency, and that the objectives of the project are being continued to be met.

Planning, design and construction of physical facilities

This is one level of activity where some forms of monitoring and evaluation have always been a standard practice. It has been a common practice among engineers and surveyors to monitor:

i) progress of planning and design on schedule and within financial limits;

ii) use of equipments and construction materials;

iii) construction of structures according to previously designed plans;

iv) project costs do not exceed budget estimates; and

v) construction proceeding on schedule.

Thus, normally for the engineering and technical aspects, there would normally be a technical inspection and cost-accounting system already integrated within the project in some fashion. What may be necessary is to review the existing system or the system proposed to see if some further improvements can be made to make it more efficient.

There are areas, however, where monitoring is essential at this level, but are seldom done — except in an anecdotal fashion. Information on the following aspects is recommended.

Employment: Employment creation during the planning, design, construction and operation phases of water development projects is rarely considered to be an important criterion, and is seldom explicitly considered in most developing countries in order that the potential could be maximized. If the projects are designed to use labour-intensive methods from the beginning, the poor — especially large number of unskilled workers, including women — can benefit from them best (Biswas 1980). Employment of small subcontractors should be encouraged as much as possible to stimulate local business development.

It is necessary to monitor the wages being paid to men and women to ensure that women receive the same wages as men for identical amount of work. It is equally necessary to ensure that children are not employed in contravention of agreed international labour conventions.

Equipments: Classified by number and cost, by type, and whether manufactured nationally or imported.

Materials used: Classified by type of materials, by units used and cost, and wheter produced nationally or imported.

Farmers' participation in project planning: Very seldom do farmers participate or are consulted on project planning and design, including important issues like canal alignment, which is a very important consideration in terms of equity. Information should be collected on the extent of participation during this phase, type of people who participated, e. g. large or small farmers, landless labourers, etc., the process through which the participation took place, positive and negative aspects of the participation process and the extent of women's participation.

Participation of local authorities in project planning: Participation of local authorities in project planning, especially if they are elected, is essential for the long-term sustainability of any project. Was the project imposed on the local people by the central government or was it seen as essential by local authorities?

Operation and maintenance of water control facilities

Operation and maintenance (O & M) is one of the most underestimated aspects of water projects in developing countries. And yet, if the benefits from such projects are to occur on time and to the specific target groups, it is essential that O & M be carried out efficiently to ensure that water supply is reliable, electricity generating facilities are properly maintained, farmers in the tailend receive their regular quota of water, and drainage system is functioning properly so that salinity and waterlogging problems do not occur. A review of past water projects will indicate that most project agencies are generally not ready to undertake O&M work when the construction phase is completed. Until recently, O&M was accorded low priority, at least when judged by the actual performances, by both governments and international institutions or donor agencies. Thus, not surprisingly, funds available for O & M are mostly inadequate, and often maintenance efforts continue to be postponed, until a major crisis appears and it can no longer be postponed. During this period, the efficiency of the projects continues to decline, and during crisis situation, generally the problem faced is more complex to resolve technically and more funds have to be expended than had the maintenance works been carried out on a regular basis.

Another problem pertains to the attitude of the technical staff. Design and construction phases of water systems are considered to be glamorous, and thus not only do the best staff prefer to work in such areas, but also their superiors prefer to assign them to those tasks. O & M assignments are seldom considered to be desirable (Hotes 1983), and thus are often staffed by inexperienced and/or below-average calibre staff.

Primarily as a result of the above two factors, efficiency of irrigation systems a decade after construction is mostly very low: around 20–40%. This means that 60–80% of water abstracted from the rivers do not reach the agricultural fields (Biswas 1983). Similarly hydroelectric generation declines with time because of poor maintenance, and often the blame for this is attributed to the lack of rain or water in the reservoir.

Another major problem worth noting is the fact that poor though O & M is for irrigation, it is even worse for drainage. Poor drainage contributes to salinity and waterlogging development, but since such problems take some time to develop in most cases, the magnitude and extent of the problems are seldom realized until they become serious. This is especially true where M & E system does not exist or is not functioning properly.

Agricultural production

The fundamental objective of any irrigation project is to provide efficient water control in order to increase agricultural income. Efficient water control, referred to at the previous level, by itself is not the sufficient condition to maximize agricultural production, which simultaneously requires other essential inputs as seeds, fertilizers, pesticides, machineries, energy, as well as extension, credit and marketing facilities. It is equally important to ensure that irrigation water and the factors mentioned are available to the farmers in an integrated and timely basis. For M & E at this level, all the factors mentioned — with the exception of irrigation water which has already been considered in the previous level — need to be considered.

Information needs to be collected at critical times for each cropping season, which can then be used to provide better co-ordination between the different organizations responsible for the various inputs and services. At the end of the cropping season, an overall performance review of the season needs to be carried out. This review will be help-ful in preparing an integrated, and more improved plan for the subsequent cropping season.

In many irrigation projects, M & E of agricultural production may require the maxi-mum effort when compared with the other three levels mentioned in this section.

Achievement of socio-economic objectives

The fundamental objective of irrigation is to increase agricultural production, which will not only increase availability of food for people, but also directly contribute to increased income generation of both farmers and non-farmers. Increased productivity and rise in farm income could go a long way to achieve the socio-economic objectives to the project.

It is, therefore, absolutely essential to monitor the impacts of the project on the pro-posed beneficiaries. For example, it is quite possible that an irrigation project may enhance the employment and income potential of landless labourers on farms and in neighbouring towns due to intensified agricultural activities. Equally, it could replace overall employ-ment potential by undue emphasis on mechanisation, which could make the life of land-less labourers far worse than the pre-project level. Similarly, it may be possible that the income of small farmers and landless labourers increases significantly due to the project, thus making more equitable income distribution in the area. Alternately, the benefits could accrue primarily to the large farmers at the cost of small ones, and thus make in-come distribution even more skewed than ever before. Depending on specific irrigation projects, both alternatives have been observed in the past.

It is equally important to monitor the impact of increased income on some quality of life indicators. For example, is the increased income improving the quality of life of the people in the project area, e.g. better literacy rate, improved health services, provision of clean water and sanitation, etc., or is it being primarily used for conspicuous consump-tion, as has been observed in certain instances.

From a management viewpoint, it is essential that M & E be carried out continually so that decision-markers are aware of the developments in order that appropriate policies may be formulated and implemented on time to reverse undesirable trends. To this end, both intended and unanticipated impacts should be monitored.

As mentioned earlier, time factor is very important for this type of evaluation since some of these results may not materialize within 9–12 years. Socio-economic monitoring need, not to be carried out as frequently as monitoring of operation and maintenance of water control facilities or agricultural production. Key variables could be monitored annually, and others could be surveyed once in 2–5 years.

Concluding Remarks

During the past three decades, water resources development projects in developing countries have received massive amount of financial support (OECD 1982). Irrigation is the largest subsector of the agricultural and rural development sector of the World Bank and the IDA lending (Hotes 1983). During the period March 1948, when the very first loan was made to a developing country for irrigation and power ($ 13.5 million for Chile), and June 1982, World Bank's agricultural lending has amounted to $ 26.7 billion, of which more than $ 10 billion was for 285 irrigation projects. The total project costs have been around 2.5 times the amount of the loans.

With such massive investments in water development projects in developing countries, it is essential that systematic monitoring and evaluation of the projects be carried out in order that:

i)   timely corrective actions can be taken for maximising project impacts, and thus achieving the project objectives;
ii)   goal achievements can be determined;
iii) lessons can be learnt for more effective project design and management;
iv)   project assumptions can be verified; and
v)   overall project impacts can be analysed.

Because of these considerations and massive investments in irrigation projects, monitoring and evaluation issues have assumed greater importance in recent years than ever before. Because of general scarcity of investment funds, it is essential that the funds available be used as efficiently as possible to maximise development efforts, which without continual monitoring and evaluation can clearly not be achieved.

References

Biswas, Asit K.: Major Water Problems Facing the World. International Journal for Water Resources Development, 1, 1, 1—14 (1983)

Biswas, Asit K.: Environment and Sustainable Water Development. In: Water for Human Consumption, pp. 375—392, Tycooly International Publishing Ltd., Dublin 1982a.

Biswas, Asit K.: Health Impacts of Hydropower Development. Key-Note Address, "Health Impacts of Different Sources of Energy", pp. 527—537, International Atomic Energy Agency, Vienna 1982b.

Biswas, Asit K. (ed.): Water Development and Management. Proceedings of the United Nations Water Conference, Mar del Plata, Argentina, March 1977. Pt. 3, pp.907—942, Pergamon Press, Oxford 1978.

Biswas, Asit K.: Systems Approach to Water Management, 429 pp. McGraw-Hill Book Co., New York 1976.

Biswas, Asit K.: Socio-Economic Considerations in Water Resources Planning. Water Resources Bulletin, 9, 4, 746—754 (1973)

Biswas, Asit K.; Durie, R. W.: Sociological Aspects of Water Development. Water Resources Bulletin, 7, 6, 1137—1143 (1971)

Club du Sahel & CILSS: The Development of Irrigated Agriculture in the Sahel. 33 p. Club du Sahel 1980.

Hotes, Frederick L.: World Bank Irrigation Experience. International Journal for Water Resources Development, 1, 1, 65—76 (1983)

IDRC: Opportunity for Increase of World Food Production from the Irrigated Lands of Developing Countries. Report to the Technical Advisory Committee of the Consulative Group of International Agricultural Research, IDRC, Ottawa 1979.

OECD: Recent Trends in Aid to Irrigation. Working Document OCD/82.20, 24 p., OECD, Paris 1982.

OECD Development Assistance Committee: Major Conclusions of DAC Workshop on Aid to Irrigation. Report OCD/83.2, 13 p., OECD, Paris 1983.

Steinberg, David I.: Irrigation and AID's Experience: A Consideration Based on Evaluations. 85 p. + appendices, Agency for International Development, Washington, DC 1983.

Lundqvist, J.; Lohm, U. and Falkenmark, M. (eds.):
Strategies for River Basin Management, pp.063–070
© 1985 D. Reidel Publishing Company

2 COMPLEMENTARITY OF LAND AND WATER CONSERVATION

2.1 Importance of soil conservation in the context of river basin realities

Problems Related to Coordinated Control and Management of Land and
Water Resources — Some Perceptions Derived from the Indian Experience

Vohra, B. B., Govt. of India Advisory Board on Energy, Sardar Patel Bhawan, New Delhi
110003, India

Abstract: Developing countries, faced with the twin problems of poverty and rapidly growing populations, must give the highest priority to the optimum mangement of their natural resources of land and
water, which constitute their basic life-support systems. The management of these resources must
necessarily be attempted in an integrated manner, so closely and inextricably are they inter-linked
with each other. However, since the soil is an irreplacable and non-renewable resource and since proper
soil and land management automatically results in the conservation of a great deal of water, it would
be good strategy to pay special attention to this subject, particularly because it has suffered from
serious neglect in the past. The afforestation of denuded water-sheds and waste-lands, the provision of
adequate drainage in canal-irrigated areas and the greater use of ground water for irrigation must accordingly form an important part of all river basin development programmes.

Background

A consideration of the concrete problems which have emerged in the field of land and
water management in India would yield valuable insights into their relevance to developmental goals as well as the possible methods of solving them. Such an approach would
also bring out the close inter-connection between the management of land resources on
the one hand and of water resources on the other and demonstrate why it is inescapably
necessary to undertake the integrated management of both on a complete watershed basis
if infructuous and even counter-productive expenditure is to be avoided.

The Indian situation demands that the conservation and optimum utilisation of the
country's vast land and water resources should be achieved as early as possible if any headway is to be made in the fight against poverty. Great urgency attaches this matter because
nearly half the country's total population of over 730 million is already below the line of
absolute poverty. However, we are posed to reach the 1000 million mark in another 15 to
20 years and the only hope of sustaining such a large population at anything like a human
level lies in making the best possible use of the country's life-support systems, represented
essentially by its land and water resources.

Priority to Land Resources

It is well to remember that while water is a replenishable resource, which is gifted to us annually, the soil is, for all purposes a non-renewable and irreplacable resource and there-fore deserves greater care and attention than water. It accordingly makes sense to consider the problems of land management first and only thereafter those of water management — if any still remains which requires attention. Such an approach will in fact show that these problems are more or less identical and that if we are able to look after the land properly the problems of water management get automatically solved in the process.

The problem of land management in India is truly formidable in size and scope. According to official statistics (India 1980, p. 343), out of a total area of 266 million hectare (mh) which have a potential for production, as many as 175 mh suffer from de-gradation caused for the most part by soil erosion but also by water logging and salinisa-tion. This means that, on an average, at least 2 out of every 3 hectars of land in India are today not in good health. However, it is also known that at least half of the sick lands are almost completely unproductive (Vohra 1980). In other words, around one third of our lands is almost completely unproductive, another one-third is partially productive and it is only the remaining one-third which is in good health. It should be quite obvious, even to a superficial observer that there is no earthly chance of eradicating poverty and surviving as a self-respecting nation so long as such a state of affairs persists.

Threats due to Deforestation

The most serious threat to the soil is posed by deforestation, denudation and soil erosion which together account for over 150 mh of sick lands, and of which over 50 % are under cultivation. It is impossible to calculate the damage caused to the economy by these threats. Apart from the progressive loss of productivity which occurs on eroding soils — whether productivity of trees or grasses or crops, according to the use to which the land is put — the displacement of the top soil also causes the premature siltation of costly and often irreplaceable reservoirs which represent valuable potential for irrigation, for flood control and for hydel generation. Again, displaced soil raises the beds of rivers, thus reducing their water-carrying capacity and causing floods. Yet again, the excessive run-off of rain water along denuded slopes reduces its percolation into soil and sub-soil strata and results in the loss to the sea, often after causing floods, of a resource which would otherwise have been available for round-the-year use as ground water. Floods and droughts are thus two sides of the same coin of poor land management and conservation of the soil auto-matically results in the conservation of a great deal of water also.

This simple hydrological fact has, however, yet to be fully grasped by the country's decision makers and planners who continue to suffer from what has been very aptly described as "resource illiteracy". Thus, disproportionately large sums continue to be pre-empted for big irrigation projects in the mistaken belief that these offer a panacea for all our ills, even as acute problems of deforestation, denudation, soil erosion and sedimenta-tion remain comparatively neglected for want of adequate awareness of the terrible toll they take of the economy.

Canal Irrigation — A Mixed Blessing

Water-logging and salinisation, which constitute the second major threat to the soil, have already claimed 13 mh and threaten many more, unless adequate preventive steps are taken in time. The lands affected are for the most part situated in canal-irrigated areas and have suffered basically because of the absence of adequate drainage. These lands represent a most valuable asset on which very high investments have been made by way of costly irrigation projects and must not therefore be allowed to deteriorate, let alone go out of production altogether. The situation calls for provision, as a matter of the highest priority, of adequate water distribution as well as drainage systems in all canal-irrigated areas, as prevention is always better than cure. For on no account must we permit improvident canal irrigation to turn into a curse what was meant to be a blessing.

However, even this threat is not yet viewed seriously enough, with the result that command area development and drainage works languish for want of resources even as huge amounts are spent on new irrigation works. Unfortunately, irrigation establishments are still thinking in terms of gigantic new projects — including some for the linking of river systems and the creation of a "National Water Grid". The conservation and optimal management of land resources does not enter into their calculations to any large extent. The mistake of course lies in treating canal irrigation as if it was an end in itself, instead of only a means to greater productivity of the soil.

Options for Improvements in Land Productivity

The utterly lop-sided resource management situation in which we find ourselves today can be corrected only if we learn to consider how the productivity of the soil can be improved under different land use conditions, and to utilise our limited resources to the best advantage of the community, after taking into full account the pros and cons of the various options before us. Let us consider what these options are.

Canal irrigated areas, which account for around 50% of the total net irrigated area of around 45 mh, offer the greatest scope for a quick increase in production. According to the Sixth Plan document, thanks to the absence of field channels, proper land-shaping and drainage systems over vast areas, the productivity of such lands is on the average, only around 30% to 40% of what may be reasonably expected from them. Very high priority, therefore needs to be given to the completion of command area and drainage works in such lands even if this involves the slowing down or postponement of new irrigation projects. For it obviously makes better economic sense to quickly put existing irrigation potential to optimal use and to save precious land from damage by water-logging and salinisation than to create additional capacity, which under existing policies, is bound to remain under-utilised as well as pose a grave hazard to the soil.

The 20 odd mh of agricultural lands, which are served by ground water, face comparatively few problems. This is so because of the intrinsic nature of this resource, which lends itself beautifully to development by individual farmers at little expense and in record time. However, the very attractiveness of this resource creates the danger of its overexploitation. This can result in an abnormal lowering of water tables, and in areas where there are saline aquifers in juxta-position with fresh water aquifers — as is often the case in coastal regions — in the saline infection of the latter. The situation calls for systematic investigations into the nature and recharge characteristics of ground water aquifers and the imposition of necessary controls over withdrawals wherever these are called for. At the same time, however, the use of ground water, wherever it is available in plenty must be encouraged by all possible means such as the consolidation of holdings, the electrification of pump-sets, the provision of loans and the extension of competent technical advice to the farmer. But, more than anything else, the replenishment of ground water must be maximised by taking comprehensive afforestation and soil and water conservation measures in the watersheds. However, as already noticed, such measures need to be taken in any case for many other reasons also.

Advantages of Ground Water Utilization

In considering a rational strategy for the extension of irrigation in the interests of increased production, due account must be taken of the many unique natural advantages which ground water enjoys over surface water as a source of irrigation. These advantages flow from the fact that no expenditure has to be incurred on either the storage or the conveyance of ground water, no land needs to be acquired for either the reservoir or canal systems, and no seepage losses — which often amount to 50 % of the water released at the reservoir — have to be incurred. Ground water is also not susceptible to evaporation losses either during storage or transit. Again, surface water schemes have very long gestation periods — often running into decades — which are responsible for heavy cost escalations, whereas tubewells can be installed and commissioned within a matter of weeks, if not days. Yet again, land shaping for ground water use can be carried out by the farmer himself without reference to the lie of surrounding lands, whereas in canal irrigated areas such work must be done on the basis of entire outlet commands.

Problems of water distribution and drainage, which cause water-logging and salinisation in canal irrigated areas, are also almost non-existent in areas served by ground water because there is no seepage from canals, no interference by the latter with the natural drainage of the areas, and above all, because the farmer can easily plan his own distribution system and is careful not to apply excess water to the land. But more than anything else, it is the ease and certainty with which the farmer can use water for irrigation exactly when he wants to and to the extent he wants to — without the intervention of a big and often corrupt bureaucracy — that makes this resource so useful.

It therefore stands to reason that wherever it is felt necessary to supply water to non-irrigated lands, the possibility of using ground water should be explored to the full before turning to the surface water option. While planning new surface irrigation projects which are considered unavoidable, due account must also be obviously taken of the full cost of items like command area development, drainage and the rehabilitation of oustees from submerged lands — in the past such costs have often been under-estimated for cosmetic reasons.

Conservation Measures in Rainfed Areas

Let us now consider the problems of the country's 100 odd mh of rainfed agricultural lands. As already noticed, a large proportion of these lands suffer from soil erosion. Common sense demands that those rainfed lands which have good soils and do not lie on slopes which are too steep must be saved from the further loss of top soil by suitable engineering works such as bench terracing and bunding, and thereby made fit for sustained cultivation. Such works will however prove to be infructuous if the higher slopes, which are often under forests or pastures, are not simultaneously treated for soil conservation. Such lands must also be provided with irrigation facilities to the maximum extent possible. Where ground water is not readily available, recourse should be had to the construction of small storages at suitable local sites before exploring the possibility of importing water from a distance, and building big canal systems. Rainfed lands which are subject to wind erosion must be protected by wind brakes and shelter belts.

However, while considering the optimum utilisation of rainfed agricultural lands, due account should be taken of the fact that a significant proportion of such lands are intrinsically unsuitable for sustained cultivation on account of either the steepness of their slopes or the shallowness of their soils or both. Such marginal lands which come under cultivation only because of growing population pressures, should be progressively taken out of cultivation altogether and placed under permanent vegetal cover — of suitable trees or grasses — for economic as well as ecological reasons. At a rough guess, such lands may be around 40 mh in extent.

Reduction of Acreage under Cultivation for Increase in Production

It is certain that if such a resource-oriented approach is adopted towards the management of agricultural lands, India would be able to ultimately achieve far greater production from around 100 mh of properly managed and irrigated soils than it is presently able to do from around 143 mh of rather indifferently managed soils, and at the same time achieve greater ecological stability. It is important to remember in this connection that while we are able to produce only around 140 million tons of foodgrains per annum from almost as many hectares, China produces over 300 tons from only around 100 mh, thanks mainly to higher levels of double cropping as well as of per hectare yields, both made possible by better land and water management.

We may now turn to the problems of management of forests and pastures both of which have suffered great damage as a result of over-exploitation. While the 12 mh officially classified as pastures are in fact almost devoid of grasses, about half of the 68 mh classified as forest lands are also without much tree cover, thus making a total of about 46 mh which need to be placed under permanent vegetal cover. It is, however, to be remembered that another 40 mh of land, described in official parlance as "Culturable waste lands and fallows", are also lying almost completely unproductive on account of soil degradation. These lands are obviously far more suitable for being placed under trees or grasses in the interests of both conservation and productivity than for being reclaimed for agriculture. The restoration of permanent vegetal cover has, therefore, to be carried out in any case over around 86 mh of non-agricultural lands classified as forests, pastures and waste lands.

If we add to this figure the 40 mh or so of marginal rainfed lands which may ultimately be reverted from agriculture, we arrive at a total of around 126 mh which need to be put to better use in the non-agricultural sector. Incidentally our only hope of saving the 34 mh or so of good natural forests that are still left with us, lies in tackling these vast unutilised and underutilised areas in as short a time as possible. For the country is losing around 1.5 mh of forests every year, against which it is currently able to create only around .5 mh of fresh plantations.

Economic Considerations

It is easy to imagine what a transformation of the national economy would take place as a result of optimal land utilisation. What would be the financial cost of such an effort? A very broad idea of the investments required can be obtained from the fact that apart from the 175 mh which already stand degraded, at least another 25 mh will have to be treated for proper land and water utilisation between now and the year 2000 AD, considering the scale on which denudation is continuing to take place and the large additional areas which are planned to be brought under canal irrigation. This will make a grand total of at least 200 mh. Even if we assume an extremely modest average outlay of only $ 500 per ha and even if we keep cost escalations out of consideration, the total bill will not be less than $ 100 billion. This is a truly mind-boggling figure. However, it will become even bigger when we add to it the cost of the big irrigation projects which are already in hand and of the innumerable small ones which will have to be built to serve local needs — this cost may well be of the order of $ 50 billion. It is going to be a very difficult job indeed to find resources of this order even if the expenditure is to be spread over 20 to 25 years.

The only way of finding adequate funds for the inescapable task of making the best use of the country's life-support systems will lie in the reordering of national priorities so that expenditure on un-productive activities is reduced to the minimum. It is necessary to point out in this connection that observers of the global environmental situation have also come to the conclusion that the world's environmental problems cannot be tackled successfully unless there is a very significant reduction in the expenditures of around $ 700 billion which the international community incurs every year on its armed forces.

Policy Changes Necessary

However, the paucity of funds is not something which needs bother us unduly just now, for before we can spend money on the requisite scale to save the land from further degradation and to put it to optimal use, it will be necessary to carry out far-reaching policy changes to re-orient the thinking of departmental agencies which have got used to working in water-tight compartments with very limited objectives, and to build entire new cadres capable of planning and executing works on the required scale in a time-bound manner. All this will take time.

Participation of local communities

An aspect which will need serious consideration is whether we should think in terms of employing only departmental agencies for executing afforestation, land levelling, drainage and other land improvement works. We have to remember in this connection that departmental agencies have not only proved to be often inefficient and corrupt and, therefore, expensive but that they are also quite un-prepared, as of now, to take up huge new responsibilities with any sense of confidence. The new jobs are in fact so enormous in size and scope that they can hardly be achieved without the active cooperation and indeed the participation of local communities who have a direct stake in the better utilisation of the resources on which they depend for their living. Fortunately, a great deal of the work involved — whether afforestation or land levelling and bunding or the construction of field channels and drains is fairly non-technical in nature and can be easily performed, with a little guidance and training, by the farmers themselves, and at a fraction of the cost incurred by departmental agencies. The participation of the people is also, of course, inescapably necessary from the point of view of ensuring the proper maintenance and protection of the new works and plantations.

A new ethos

The Government agencies and departments concerned with various land management programmes will, therefore, have to acquire an entirely new ethos if they are to measure up to the task ahead. They will have to give up their present attitudes of distrust, distance and naughtiness towards the people and learn to work with and for them in a spirit of cooperation, mutual confidence and comraderie. Such a change will in turn be possible only if the present bias for over-centralisation of authority is given up and field organisations are given much greater powers and flexibility than they enjoy today, so that these in turn may be shared by them with local communities in the interests of quick and cost-effective implementation.

All this, however, is easier said than done. In the last analysis, these exceedingly formidable problems can be successfully tackled only if there is a much greater awareness of their relevance to the common good and of the urgency which attaches to them. It is only after such general awareness has been created that an informed and articulate public opinion may be expected to emerge and, in turn to generate the strong political will which is needed to place resource management on a sound footing. It is indeed only after a strong political will has come into being that it will be possible to carry out the far-reaching administrative and organisational reforms required to put matters right. These will include the creation of an integrated Ministry for Land and Water at the national level in place of the present Ministry of Irrigation and its being made squarely responsible for the conservation and optimal management of these twin resources. It is only after this basic reform has been carried out that it will be reasonable to expect adequately structured and empowered river basin authorities to be set up to take a synoptic view of the total resource situation in each basin and to re-order priorities in the best interets of the community. Such authorities will incidentally be ideally situated for also tackling such connected matters as the control of water pollution, promotion of navigation, the generation of hydel power and the planning of urban expansion on correct lines.

Future Has to Be Built now!

Such developments are, however, still very, very far away and it is no use imagining that they will be easy to bring about. A real-life example will show how exceedingly difficult it is to bring about changes in thinking in this field. It was as far back as in 1972 that a suggestion was first made that the absence of a national agency with responsibility for the proper management of the country's land resources should be rectified as a matter of priority. (Please see the author's "A charter for the land" of 1972, printed in Econ. & Pol. Weekly 1973, see also Vohra 1980). Although this suggestion received the blessing of the Government at the highest level in 1973—74, it was not till 1984 that such a body was actually set up and even then with only a limited and watered-down brief. The way things are, it may well be another decade before our land (and, therefore, also our water) resources begin to receive the kind of attention they deserve. By then, however, the degradation and loss of the non-renewable resources of the soil may have reached and passed the point of no return, at least in relation to the needs of a greatly increased population. Something of this kind does in fact seem to have already happened in certain African countries.

In these cirumstances, the responsibility of those who can see the writing on the wall is quite clear. They must obviously do everything in their power to create public interest in the vital issues at stake and not rest till policy makers and politicians have been shaken out of their present attitudes of unawareness and complacency. This task is no doubt going to be long and difficult, but it is definitely worth doing.

In the words of the poet:

What do we ask for them?
Not for pity's pence nor pursy affluence,
But that we may be given the chance to be men.

References

Govt. of India Planning Commission: Sixth Five Year Plan 1980—85, New Delhi 1980.

Vohra, B. B.: A Charter for the Land. The Economic and Political Weekly (March 31, 1973)

Vohra, B. B.: A Policy for Land and Water. Sardar Patel Memorial Lectures, 22nd and 23rd of December 1980. (The full text is available in Vohra, B. B. 1982, Land and Water Management Problems in India. Training Volume 8. Dept. of Personnel & Adm. Reforms, Ministry of Home Affairs, New Delhi.)

Vohra, B. B.: A Charter for the Land. The Economic and Political Weekly (March 31, 1973)

Lundqvist, J.; Lohm, U. and Falkenmark, M. (eds.):
Strategies for River Basin Management, pp. 071–079
© *1985 D. Reidel Publishing Company*

Seeking a New Balance in the Deteriorating Upper Citarum Basin

van Bronckhorst, B., Provincial Planning Office, Bappeda, West Java, Indonesia

Abstract: The Citarum basin in West Java has been increasingly strained by rapid population increase and uncontrolled urban growth of Bandung, second city of Indonesia. Development has come to a point of overpopulation of the watershed. In the wake of economic recession urban dwellers are moving out and becoming farmers on the mountain slopes, bringing ever larger slope surfaces under cultivation. The article describes the emerging problems, especially heavy hillside erosion and severe valley inundations during heavy rains. It discusses the problems of integrated regional planning to replace sectorial planning in specialized sectors. In the efforts of creating a better understanding of nature supportive activities among people, the traditional puppet theatre, wayang, is being used as a powerful intermediary in the communication process.

Introduction

A lot of knowledge has accumulated in the world about river basins and their proper management. Most of it is only recurrent in seminars, and does not reach out to the people who are dependent on river basin ecologies ... what is processed by clever analysis remains far away from the ones whose daily life is mainly survival.

The problem how to involve people in deciding what is to be done, in working out the system, and later in operating it well, has often been raised. Usually, however, scientists and engineers stick to the hardware part, while expecting the required complementary software part to develop on its own. (Usually it does, but it takes long time and the emerging socio-economic forms not often match with expected outcomes of a project!)

When the problem is too big, on the other hand, and too complicated to handle, when money lacks and the time perspective extends far beyond any political horizon, most development will depend entirely on socio-economic — and thus political — processes!

It West Java, the most important river basin is that of the Citarum (Fig 1). Its modern development started early in the 20th century when the Province's swampy northern plains were drained and turned into a large rice areal. In the thirties, proposals were put forward for even more extensive works, doubling the irrigated area and constructing three major dams to control the water discharge and generate electricity.

It took a quarter of a century before the first part of these proposals were brought to realisation, and another twenty-five years to come to the second and third stages of the plan. Meanwhile the Indonesian population has more than doubled and strong urbanisation tendencies changed the entire situation. Now the water in the Citarum is hardly

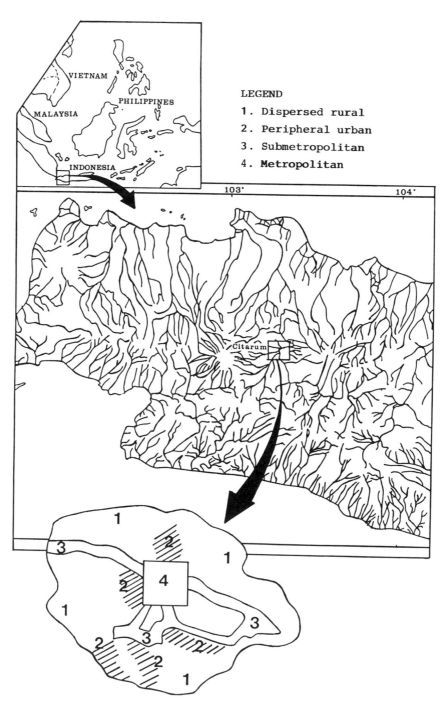

Fig 1 West Java with Citarum basin. Detail of the settlement pattern of Bandung

sufficient to meet with the ever rising demand for electrical energy, and the need for rice field irrigation competes with industry's new claims on land and water.

Ever increasing complex sites confuse the planners, and conflicting requirements lead to deadlocks in negotiations. In the meantime processes of growth go on. The major catchment area of the river, the upper Citarum basin, became densely populated, containing the second city of Indonesia: Bandung.

The article summarizes the problems of this basin. At present it appears that development planning has to integrate and bring to a new level what once has started as sectorial planning by various specialised government agencies on one hand, and the actual processes of growth in rapidly modernising Indonesia on the other. The need is now for regional planning and implementation. The article also stresses the problem of getting people involved and the need for intermediaries in the communication process.

Problems

Well endowed with precipitation

Indonesia is generously endowed with precipitation: only a very few meteorological stations register less than 1,000 mm per annum. Rains are more intense in the mountainous regions — the record in one region is 7,000 mm.

However, the seasonal variation is of more importance than yearly totals. Rains are of several types, ranging from monsoons to equatorial double rain periods, with local peculiarities caused by standing temperature differences. West Java may experience five to seven 'wet' months a year, while some rain might fall every month. Locally, the variation is high. A mountain city like Bogor has a rain pattern with a yearly average of 4,000 mm, and a montly average rainging between 250 mm and 400 mm. Bogor is the thunderstorm capital of the world, with an average of 350 storms a year. Bandung registers 1,900 mm annually, with a montly range between 50 and 250 mm. Elsewhere on the Bandung complex, rainfall might be as high as 4,500 mm per annum.

Bandung — a city of random growth

Bandung's explosive growth — up to 10 % in the recent past — has stretched the city far beyond the current administrative boundaries. The growth rate of Bandung itself has dropped to about two percent, but the peripheral population, which is economically oriented on Bandung, keeps rising. The present city population is 1.5 million, with another million in the vicintiy. The highest population density is in an area of south Bandung, with 30,000 per km^2. The lowest density is 500 people per km^2 in the upper-class neighbourhoods.

The environmental load must be examined carefully, as the city has already outgrown the natural limits of the Cikapundung delta and is expanding east- and westward along the slope of the Lembang escarpment, and southward into the Citarum river alluvium.

Construction was creeping northward, uphill towards Lambang, until the Ministry of Population and Environment put a stop to that. Not only was it damaging the natural water storage, but it was increasing erosion and rainwater runoff, which is already too high. Bandung often experiences flooding as water runoff cannot be readily channeled to the natural drainage system of the Citarum river. The slum areas in the lowlands are invariably flooded as water overflows the limited drainage system of the city.

There are too many problems that need too many points of view to find solutions. Bandung and Bandung Raya have been the subject of planning for the last three or four decades. Plans have always lagged behind reality. A far-reaching and inspiring vision is needed to carry Bandung Raya from its clogged present to an open future. People are finally realizing that something radical must be done. Community voices are calling for another approach. Recently, a book, Bandoeng Tempo Deoleo dan Masa Depan, was published. After the appearance of that book, the regional newspaper Pikiran Rakyat organised a day of a discussion on the future of the city. Many people active in the city development attended. The newpaper summarized the meeting as follows:

> "The city is in a crisis! A crisis resulting form random growth. An excess of people conflicts with the city layout, leading to unrestricted illegal building. People compete for places to stay."

> "How can this city be helped! All kinds of theories from abroad have already been offered and discussed, and lead to even greater confusion!"

Recession forces urban dwellers up in the mountain slopes

The upper Citarum basin provides, in a concentrated form, an example of the complications of development. From the point of view of scientific analysis the situation is clear; there are simply too many people living in that watershed.

Geology and hydrology have strongly influenced all patterns of settlement on West Java. Every deviation from these natural patterns immediately will lead to costly constructions and installations, so it is obvious that concentration occurred along least cost profiles! On West Java people live concentrated for various reasons, one being the limited accessibility of all parts of the province, which inevitably leads to economic differences. And then there is the scarcity of good land, with proper irrigation and drainage. Within the upper Citarum basin, therefore, a strange pattern of urban development has evolved, looking on the map like a collar around the floodplain of the river in the altiplane, following the old pathways that roughly coincide with the prehistoric lake's borders.

With economic fluctuations, seasonally and periodic, the need for labour in the urbanised area is always changing. When money wage is scarce, people move back to the mountain slopes and start to plant. Associated with the present depression it can be observed now how ever larger surfaces of the slopes, even where they are rather steep, are being brought under cultivation, with negative consequences for the ecological balance. But people must live, and therefore eat. For ages the earth of Java has provided that food, being continuously and gently watered through the fans of ancient eruptions from the many volcanoes that also act as buffers in the water cycle. Yet, in conditions nearing overload the natural processes cannot maintain themselves, and in several places on Java the result of such situation can be observed as leached and entirely deteriorated soils.

Yet there are better methods that have been tested out in the world, such as terracing and agro-forestry. In some areas of the province these methods are also tried out and with success! Recently, we have tried a similar experimental development on the hillslopes south of Bandung city: terracing and agro-forestry.

Two major economies in the basin

A matrix setting physical and social geography against way of life distinguishes 12 subtypes (Fig 2). These reflect the complexity of life in West Java more accurately than the

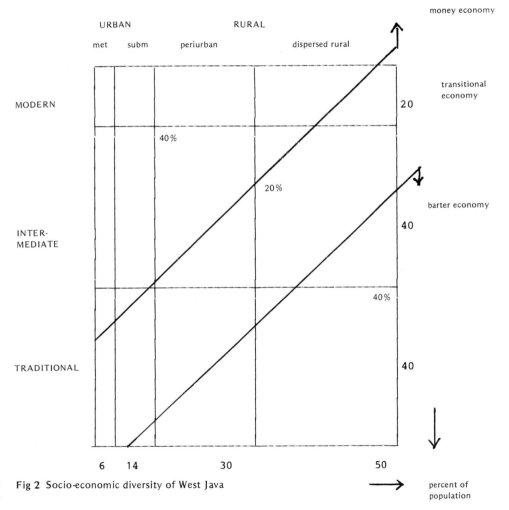

Fig 2 Socio-economic diversity of West Java

usual urban/rural dichotomy, even when combined with the formal/informal classifica-tion. The latter can be misleading, as it only refers to whether an enterprise of self-employed individual has been registered with the government.

A line drawn diagonally across the matrix separates two major types of economic interaction: predominately money-based, and predominately barter-based. This boundary delineates the people in transition, the migrants in transit from the traditional life to the modern. Here we find the deepest poverty in Java, where people lack not only money, but the surrounding nature environment and support of their families that formerly made their poverty bearable.

40 percent of the West Javanese live is a money economy. Another 40 percent still live in a barter economy. The remaining 20 percent live in the limbo between these two economic regimes — truly 'the poorest of the poor'.

Measures Taken against Flooding

Heavy rains keep flooding Bandung

During January 1984, newspapers in Bandung carried stories of urban flooding almost daily. Heavy rains — ten times the ten year average for the first two weeks of January — had caused widespread flooding in the Bandung plateau. Over 5,000 houses were inundated and about 10,000 hectares of paddyfields were under water.

This was a serious threat, as the paddyfields would be needed to produce 250,000 tons of rice.

Past observation has shown that both the length and intensity of the rainy season varies over the years. For centuries, the Javanese have used a time reckoning which incor-porates a thirteen year periodicity to guide their planting and harvesting cycles. It is said that the same periodic variation can be derived by analysing the currently available figures. Perhaps the heavy flooding south of Bandung was simply a manifestation of a long term climatic variation cycle. At any rate, the existing drainage systems could not handle this enormous surge of water.

Two problems: peak flows and sediment load

For a long time observers had to rely on disparate sources of information, using explanatory models based on Western thinking. Statistics were not fully satisfactory, neither was the unit-hydrograph method. All they could do was admit that runoff due to tropical storms was sometimes so high that the river simply could not carry the water away quickly enough to prevent flooding.

Following models that had proven their worth in other countries, they studied the possibilities of somehow retaining the water, or regulating the rivers. A plan now exists to normalize the river with dikes, and to build a large number of smaller reservoirs in the upper courses for the Citarum system's feeder rivers. Costs are estimated at several billions of rupiah, and execution at a decade or more.

But no one knows if it will work. It certainly will not solve the problem of the high sediment load of the inflowing water. The forestry department suggests augmenting the reforestation process, claiming this could reduce runoff and the quantity of suspended solids.

Problems of regional planning

A systems approach is the only feasible way of examining this complex problem, especially for a planner. Dr. Emil Salim, Minister of Population and Environment, has said: "We need planners who are not generalists, but integralists". He means Indonesia needs planners who combine the contributions of several disciplines into a workable whole, creating operational models from which viable, long term solutions will emerge.

Most of the solutions found by commonly used methods of analysis may not be wrong in themselves, but they do not fit the peculiarities of the situation of the Bandung plateau complex. The unit-hydrograph method is one example. It can be applied successfully to certain conditions, but, as one foreign consultant to the Water Resources Research Directorate of the Public Works Ministry warned: "It might not apply to conditions in Indonesia".

Appeal to Protect and Develop the Soil

Land use changes are badly needed

Many questions come to mind. Can land use be limited or revised? Where would the people go? Can we do anything at a reasonable cost? How long would such solutions be valid?

Bandung's settlement patterns, with its strong metropolitanization, is at the root of the problems. Concentration in this kind of natural environment leads to conditions of overload. But deconcentration, efforts to spread the population over the valley, have so far not succeeded, because of the bad overland connections. This is understandable, given the difficulty of building roads in the terrain. Meanwhile, the phenomena mentioned earlier in this paper indicate that something is going wrong. But then, these signs are usually taken as starting points for other drives towards quick solutions — large scale solutions for large scale problems caused by large scale mistakes.

Most people have little awareness about the world they live in. That is a fact everywhere on earth, not only in the Bandung plateau complex. Everyone sees only what is relevant to their own lives. People cannot be faulted for what they are doing. After all, it does not differ from what their parents and grandparents did. No one can know from intuition how to change habits.

The means must be discovered to create new understanding for people. But this does not mean asking everyone to pay attention to the deteriorating environment. Everyone from the slum dweller to the farmer in the hills knows it is deteriorating. They have eyes to see and ears to hear — all can understand.

But they still must live. When asked whether it is right to take wood from the remaining forest, the answer is: "No!" Asked why they do it, they say: "Well mister, you gotta eat".

No solution to environmental deterioration can be found without overcoming the shortage of income opportunities. Until that problem is solved, man can do nothing but rob nature. The great stumbling block is that people know little of other opportunities. The farmers on the hillsides do not know about methods for making level terraces, so they accept poor land and poor yields.

Neither do city dwellers know how to take care of water in their wells, and what to do with wastewater. They need education.

The matrix in Fig 3 is a representation of the problems and priorities of regional planning. The two planning authorities, BAPPENAS and BAPPEDA, have different criteria for defining development priorities. BAPPENAS takes a geographical perspective, dividing West Java into regions, whereas BAPPENAS wants to focus project money on measures to facilitate resettlement.

BAPPEDA feels the need of solving pressing problems in the whole province, and so takes a subject view. Among the priorities of BAPPENAS are those of the sector Land and Water; the province being subdivided into river basins — or major watersheds — which need specific attention according to their natural condition and settlement and employment/economic situation.

The development matrix seeks to come to grips with the problems of West Java by using subject categories to break the province down into cells, each incorporating a subject and a geographical area. Employment in the Galunggung disaster area, for example, is a more urgent problem than transportation, while in the Greater Bandung area, the reverse is true. The matrix will help planners visualize the development priorities, and so target the available resources more effectively.

REGIONS / SUBJECTS	Bandung Raya	Peripheral Recencies	Gallunggung Disaster Zone
Land and Water	Soil Conservation Water management	Upland conservation and river basin development	Restoration of damaged watersheds
Overland Communications	Commuter flows Extension of road and rail	Access roads and rail Fishing harbours	Reconstruction of roads and bridges
Settlement and Employment	Dispersal of settlement to relieve city centres	Village reconstruction and non-farm employment	Rebuilding of villages and generation of new employment

Fig 3 Regional Development Matrix for West Java

Needed: a new Understanding of Nature Supportive Activities

It is already understood, however, that planning alone will not lead to good results without people's involvement. What is needed is a new vision that can be shared by all the groups that seek a living in that upper Citarum basin. What is needed is a new understanding that converts present ecologically neglective behaviour into nature supportive activities, so that we may hope to divert the present danger course to a more safe one.

With this we are back to the initial question of this paper: how to convey the messages of scientific and engineering knowledge to the millions that inhabit the upper Citarum basin? (Or, any river basin!). And: how to organise for feedback from socio-economic realities to scientific investigation and modeling? I believe we need intermediaries; people that can understand the languages spoken on either side, and can translate the messages in comprehendable terms.

Since ancient times, most societies have indeed thrived on their informal communication systems that are based on art. Supported by these systems, built up from all sorts of art, like theater, story telling, plays, ceremonies, communities could absorb new understanding and adapt themselves to newly emerging conditions. Science once was part of it all, before it took off and established as a separate entity in society. With scientific approaches also standardisation entered and its associated training methods led to an impressive extension of knowhow and its application all over the world. But it could go on only as long as there was enough new space to work in. Where resources were limited, even locally, standard methods only seemed to lead to skewness in development and to the emergence of unbridgeable gaps between rich and poor layers in society.

Messages channeled through puppet theatre

Luckily, in Indonesia many of such traditional communication systems still exist, among them the important tradition of puppet theatre, wayang, that has always been an informal means of communication and learning. It appeared possible to revive this tradition for the purpose of environmentally sound development in West Java. The driving force behind this activity is a new puppet play — wayang lakon — specially created for telling people what happens to nature, and what can be done about it. Brought in the traditional setting and performed by the ancient characters, the message is conveyed in a very strong way.

It is one good example of how artists can be mediators between knowledge and daily usage, how they can help their people to absorb and adopt more sensible behaviour, better suited to the emerging circumstances. Wayang performances are played almost every night in many villages and now the new play is among those: the Tangkal Rahayu Rat, meaning the Tree of Peace and Welfare for the whole Earth. It is an appeal to recover nature and care for trees to protect and develop the soil. The performance is based on best agricultural and ecological knowledge available, and was ideated by one of the formost West Javanese puppet players — dalang — Abah Sunarya. Young and old see and hear, and experience in their own language, what is told by the personalities familiar to them all.

Lundqvist, J.; Lohm, U. and Falkenmark, M. (eds.):
Strategies for River Basin Management, pp. 081–089
© *1985 D. Reidel Publishing Company*

Methods of Land and Water Conservation in the Wuding River Basin

Gong Shiyang and Mou Jinze, Ministry of Water Resources and Electric Power,
PO Box 2905, Beijing, China

Abstract: The work of harnessing the Wuding river has lasted for almost 30 years. The natural conditions of its subregions differ from each other calling for unique strategy and the selection of regionally adapted land and water conservation measures. The wind drift sand area primarily needs protection from wind erosion, measures to transform sand dunes into cultivable flat land, and reservoirs to store flood water. The river source area needs check dams in the deep gullies to trap sediment, improve vegetation cover, and river dams for sediment retention and water storage. The gullied hilly loess area, finally, consists of many small gully watersheds and is therefore the most difficult to manage.

General Characteristics of the Wuding River Basin

Flowing through the loess plateau, the Wuding river constitutes one of the heavy sediment-laden tributaries of the Huanghe (Yellow River). The river basin, with an area of 30.260 km², is situated in the arid and semi-arid region. According to geomorphological conditions, the basin may be subdivided into three subregions, i.e. (1) wind-drift sand area; (2) river source area; and (3) gullied-hilly loess area (Fig 1).

The main features of water resource and sediment of the river are as follows: River runoff is scarce and unevenly distributed in both space and time. Annual precipitation in the Wuding river basin varies from 350 to 540 mm in different parts of the basin. As a result, annual runoff averages only $1,530 \times 10^6$ m³, corresponding to 4,245 m³ per hectare of arable land and 1,084 m³ per capita in the basin. The runoff is however rather different for different subregions.

In addition, owing to the influence of monsoon from south-east, more than 60 % of annual total precipitation occurs in a period from July to September. Consequently, except for the winddrift sand area, the runoff in this period in the other two subregions comprises more than 50 % of the annual total (Tab 1). It is furthermore concentrated in several flood events, offering difficulties to its utilization.

Most part of the river flow abounds in sediment. Rain falling in July, August and September is mainly in the form of storm. In general, one heavy rainfall may amount to 10 %–40 % of the annual total and the short duration intensity may reach 2 mm/min or more (Gong Shiyang & Jian Deqi 1979), causing severe soil erosion in the loess-overlain terrain of the basin, and making the river heavily sediment-laden. Observational data indicate that the average modulus of sediment yield of the basin is around 8,400 t/km² year. This corresponds to an annual average sediment yield of 254×10^6 m³, out of which 87.7 % occurs in the period of July to September. The maximum monthly average sediment content of runoff recorded in this period amounted to 456 kg/m³ and that of flood in general varies from 700 to 1000 kg/m³.

Fig 1 Map of the Wuding river basin.

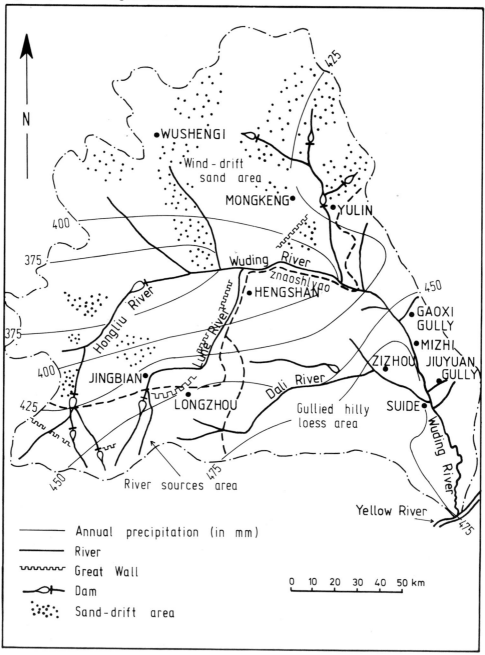

Region	Area	Annual average runoff 10^6 m^3		Depth of runoff	Amount of runoff per ha	Amount of runoff per capita
	(km^2)	total	% in July to Sept	(mm)	(m^3)	(m^3)
Total river basin	30,260	1,530	40.2	50.6	4,245	1,084
Wind-drift sand area	16,446	959	32.3	58.3	16,635	3,098
River source area	3,454	109	50.8	31.6	3,960	1,234
Gullied-hilly area	10,360	462	53.8	44.9	1,725	453

Tab 1 Distribution of runoff and sediment in basin

Strategy and Measures in the Individual Subregions

As the natural conditions of the river basin subregions are remarkably different from each other, unique strategy and measures have been selected for each subregion selectively in order to arrive at optimum utilization of water and sediment resources.

Wind-drift sand area

Located at the northern and western parts of the river basin, this area is mainly covered by movable, stable and semi-stable sand dunes. Between dunes there are varied-sized loess farmlands and pastures. Dunes and a thick sand layer allow rapid infiltration and offer a giant underground reservoir for storing precipitation and regulating runoff. Soil erosion by water is therefore slight. While the watertable about 2 m beneath surface ground water is abundant and easy to exploit. Surface runoff from this area amounts to 959×10^6 m^3, comprising 63% of the total of the river basin, whereas the cultivated area is only 3.5% of that in the whole area. In addition, the population density (18.8 persons/km^2) is relatively smaller than that of the other subregions, so that the water resources are not only enough for the area but can also be supplied to other subregions of the river basin.

The main problem in this area is the severe wind erosion, which results in drift of sand dunes, damaging and covering farmlands and towns. The main objects of harnessing and development are stabilizing sand dunes, properly reforming the land into farmlands and pasturelands to protect and increase the utilization of soil resources, and in the meantime, to exploit ground water, detain and store runoff of the river for development of local irrigation and supply of water to other water deficient areas in middle and lower reaches of the Wuding river. Measures adopted are mainly as follows:

— Establishing anti-wind forest belts around the boundaries of the area as well as the farmlands, and planting grass on the bare land surface in order to stabilize sand dunes. By the end of 1981, the area of forest belts and afforestation lands had amounted to 241,000 hectares which is 15% of the total area. According to measurements, average annual wind velocity at some places had been decreased about 21 % and drift velocity of sand dunes had been decreased from 7–8 m/year to 1 m/year; the phenomenon of sand dunes in the western part drifting southwards had disappeared, and black shells with a thickness of about 4 mm have been formed on the surface of some shifting sand dunes.

—   Diverting water from reservoirs, lakes and canals in the upland area to sluice the uneven surface of the sand dunes and allow the formation of flat land, at the same time utilizing the sediment in summer floods for warping the bare lands to create fertile farmlands of meadow. Not only more land may be utilized for cultivation but also the shift of sand dunes could be halted. By the end of 1975, 27,000 ha of arable land had been created by means of this method, and the unit grain production of some farmlands had reached 4,600 kg/ha (Fig 2).

—   Building reservoirs in the river to regulate runoff and enlarge the irrigated area. Six reservoirs with a capacity of more than 10×10^6 m^3 each have been built so far, achieving a total capacity of 327×10^6 m^3 or adequate for irrigating farmlands of 26,000 hectares. In order to meet the requirement of the water supply for irrigation in alluvial plain and terraces in the main stem of the Wuding river below the con- fluence of the Luhe, three reservoirs are planned to be built. Among these the Wang-

Fig 2 Scheme of engineering works in diverting water and sluicing sand to create arable land.

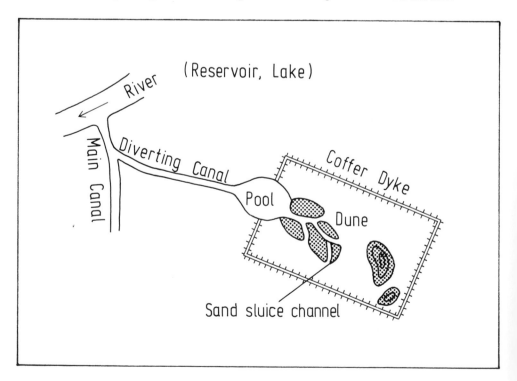

kedu reservoir with a capacity of 272×10^6 m^3 could be used to provide water for irrigation of 17,000 ha. From these and other new reservoirs the requirement of water supply for irrigation in the alluvial plain and terraces on the main stem of Wuding river may primarily be fulfilled. Due to slight erosion by water, average annual sediment yield in the wind-drift sand area is only about 912 t/km^2. The annual capacity loss of existing reservoirs is less than 1%, but as viewed in a long run, the sluicing of sediments from reservoirs should be properly considered in order to preserve a usable capacity. As mentioned above, 60% of total annual sediment comes in July to September with only 30% of the total water. Therefore, for the proposed Wangkedu reservoir, after 22 years of operation as a storage reservoir, the annual operation scheme will be changed to sluice sediments by operating at a lower water level during July to September, such that a capacity of 150×10^6 m^3 may be preserved for seasonal regulation in a long term perspective and the water supply for irrigation could be maintained at a confidence level of 50%.

River source area

This area is located in the south-western part of the basin and consists of loess and plateau basins which constitute the main farmlands in this area. The channel is in general cut down by more than 50 m. Gully erosion, bank sliding and slumping are extremely active processes, continuously reducing the area of farmlands. Average annual sediment yield is a high as 16,000 t/km^2. Surface runoff in this area is rather insufficient with an average annual runoff of only 31.5 mm. 50% of annual runoff occurs in July to September, carrying about 90% of the total sediment of a year. In consequence, mud flows with hyperconcentration of sediment of more than 1000 kg/m^3 take place frequently.

It is very difficult to utilize the water resources by means of ordinary methods. The ground water surface lies generally 20 m below the ground surface, and may locally be extremely low. Therefore, main objects for harnessing and development in this area are reduction of soil and water losses, especially control of gully erosion, sliding and slumping to protect cultivated land (the sediment yield caused by these erosion modes is about 80–90% of the total yield). Measures of utilization in this area are mainly as follows:

1. In the gullies, check dams are built for trapping sediment to form new arable land, and flood water and sediment are diverted into the basin for warping and irrigation. Longzhou Production Brigade is a typical example (Fig 3). The land area of this Brigade is 90 km^2, which is divided into two parts, gullied hilly land and the basin. The density of population is 42 persons/km^2. Since 1956, apart from planting trees and grasses to increase vegetation cover for reducing loss of soil and water, 33 check dams have been built in the gullies, which have trapped sediment and formed arable land of 32 ha (may later be increased to 47 ha). In addition, the remaining flood water has been diverted into the basin for warping and irrigation, reforming 127 ha sandy land into fertile farmlands (may be further enlarged to 200 ha). Some reservoirs have been built on main gullies to store the relatively clear water during the non-flood season. Since most sediment had already

86

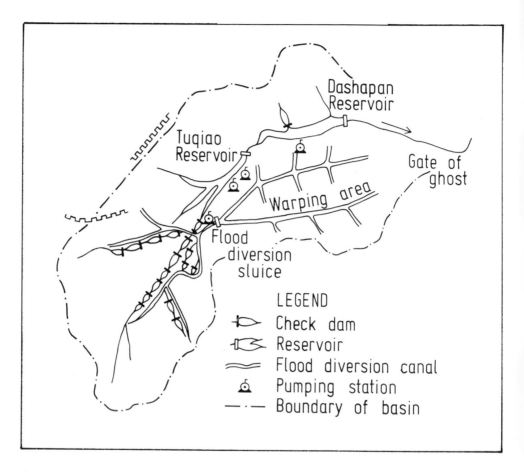

Fig 3 Plan of engineering at Longzhou Production Brigade.

been detained and diverted by check dams and flood diversion canals, only small amounts of sediment were deposited in the reservoirs. Pumping stations built around the reservoirs had further developed irrigation for about 400 ha. Since 1970, all the water and sediment resources in this Brigade have been used in this way, and the grain production has been increased twofold and economic income by three times.

2. On the river, a series of dams are built for sediment retention, water storage, irrigation and power development. The upstream of Hongliu river with a drainage basin of 1332 km^2 and a population density of 30 persons/km^2 may be considered as an example. Xinqiao reservoir was first built in 1959 for irrigation with a storage capacity of 200 × 10^6 m^3. Silting in the reservoir developed rapidly since the gully erosion of uplands and bank slumping had not been controlled. Later, 16 reservoirs of different sizes in a group of check dams were built in succession on the tributaries above Xinqiao reservoir. A lot of sediment was trapped, and the elevations of these gullies and river bed was raised respecti-

vely. Above the dam site the river bed level in a 40 km reach was raised about 40 m. Consequently, gully erosion and bank slumping reduced significantly due to the lifted datum plain.

As a result, sediment yield has been remarkably decreased. The amount of annual sediment deposition in Xinqiao reservoir decreased gradually from 14.5×10^6 m^3 to 1.45×10^6 m^3, and then to 0.29×10^6 m^3 in 1974. Moreover, aggradation of reservoirs in the wind-drift sand area downstream the Xinqiao reservoir was also avoided. So far, in the region upstreams the Xinqiao reservoir, irrigated land of 2,100 ha and an area of more than 1,000 ha silted up by check dams have been developed as local farmlands of cereal production. According to the observed data, the proportion of water resources utilized in the area concerned has reached 27.4%. Another 52% of surface runoff has been turned the ground water, raising the ground water table near the reservoir by 5 to 20 m. Water inflow to the reservoirs on the lower reach of the Hongliu river have been increased. Irrigation water is supplied to an area of 6,200 ha by pumping, and hydropower with an installed capacity of 2,800 kW, Fig 4.

Gullied-hilly loess area

This area is located on the south-east of the basin. The ground surface is mainly covered by loess, and dissected into criss-cross ravines and gullies, with hills in the form of ridges and mounds. Here, ground surface with a slope of more than 25° constitutes above 60% of total area, and the gullies have a depth in general of about 50—100 m and has a population density of 70—150 persons/km^2. With an index of cultivation as high as 60%, hydraulic as well as gravitational erosion have been very vivid producing an average annual sediment yield of 17,800 t/km^2, which ranks highest in the whole basin. Average annual runoff is only 44.9 mm, corresponding to 453 m^3 of water for each person. This is equivalent to only 1/3 and 1/7 respectively of the amount in the other two sub-areas mentioned above. Ground water is scarce and lies deeply.

All this makes this area the most difficult one in the whole river basin for developing and utilizing water resources. Main development measures include the utilization of clear water during nonflood season. During flood season flood water and sediment have to be rised as much as possible, in order to build a number of bases for cereal production. Along with the continuous increase of crop yields, part of the sloping farmlands should be gradually recreated by afforestation or grassing, in order to make the land use more rational and increase economical benefits.

According to natural conditions, this area in fact consists of many small gully watersheds. Each of them may be considered as an independent unit from the point of view of water and sediment utilization and land usage. Measures for development of these small watersheds may include: (1) Planting trees and grasses on wild grazing lands, steep slopes and abandoned farmlands to minimize the soil loss and increase economical income; (2) Terracing horizontally the slope farmlands and silting up lands by check dams in gullies, to establish bases of cereal production, storing and utilizing precipitation, runoff and sediment and building small reservoirs or ponds for irrigation.

In this respect, Gaoxigou Productivity Brigade might be an excellent illustration. In this Brigade with an area of 4 km^2 and a population density of 110 persons/km^2, harnessing work began in 1958. A number of horizontal terraces, check dams, small reservoirs and pumping stations were established for irrigation, and more sloping farmlands were stored by afforestation and grassing. By the end of 1979, as compared to 1958, the amount of

Fig 4 Plan of Hongliu river reservoir.

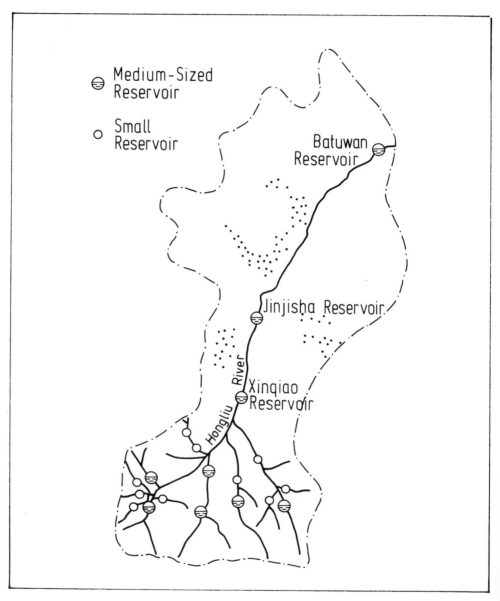

croplands had been decreased from 167 hectares to 67 hectares (in which 90% could be irrigated), i.e. decreased by 60%, yet the total output of grain had increased twofold. The area of woodlands and grasslands had increased by 4.4 and 3 times respectively, and the economic income was increased by 5.3 times.

Concluding Remarks

Work of harnessing the Wuding river has lasted for almost 30 years and certain achievements have been obtained in the field of water and sediment utilization. The utilization rate of water resources approaches 14.1%, and the amount of sediment transport into the Yellow river has decreased by almost 50%. The total output of grain in the whole basin has increased by 2.2 times. Obviously, for such an arid and semi-arid region with a population density of 43.5 persons/km^2, the achievements obtained have not been so easy. The strategy and measures undertaken in harnessing the river have proved not only suitable for the local conditions but also effective in forming a good base for further development.

It should be pointed out, however, that harnessing the Wuding river still remains a hard job. Especially the gullied hilly loess area, which constitutes one third of the whole basin, is difficult to manage. Further investigation and practice will be indispensable in order to make the further progress needed in the development of the Wuding river.

References

Gong Shiyan; Jiang Deqi: Soil Erosion and Its Control in Small Watersheds of the Loess Plateau. Scientia Sinica XXLL (1979)

Lundqvist, J.; Lohm, U. and Falkenmark, M. (eds.):
Strategies for River Basin Management, pp. 091–097
© *1985 D. Reidel Publishing Company*

Soil and Water Conservation on the Chinese Loess Plateau —
The Example of the Xio Shi Guo Brigade

Gustafsson, J.-E., Dept. of Land Improvement and Drainage, Royal Institute of
Technology, Fack, S-10044 Stockholm 70, Sweden

Abstract: This paper reviews briefly the comprehensive planning of land and water resources in the
Yellow River basin. The soil and water conservation work in the Xio Shi Guo Brigade on the Chinese
Loess Plateau is presented. The paper suggests that a cooperative ownership approach is an essential
criterion for the control of land and water resources under conditions of population pressures.

The Yellow River Basin

There are few areas in the world, which better illustrate the necessity of a comprehensive
approach to planning an mangement of land and water resources than the Yellow River
basin (Fig 1). In older periods the Yellow river was also named "China's Sorrow" due to
the many dyke breaks occuring in its lower canalized section.

The Yellow River is the world's most silt-laden river. It carries an average of 1600 million tons per year, which compares with a wall with a square metre section wound around

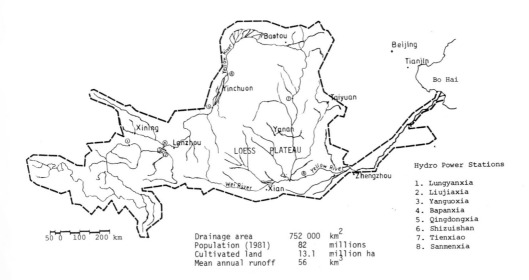

Drainage area	752 000	km^2	
Population (1981)	82	millions	
Cultivated land	13.1	million ha	
Mean annual runoff	56	km^3	

Hydro Power Stations

1. Lungyanxia
2. Liujiaxia
3. Yanguoxia
4. Bapanxia
5. Qingdongxia
6. Shizuishan
7. Tienxiao
8. Sanmenxia

Fig 1 The Yellow River basin

the world 27 times. About 400 million tons of sediment are deposited in the lower section, where the river crosses the gentle plain area before entering the Bohai Sea. The sedimentation has meant that the river bed is on average 4—7 metres above the surrounding plain. If adequate control and regulation work is not undertaken, it is estimated that the theoretically flooded area would be 15 million ha, i.e. approximately the area of the entire Shandong Province (Todd & Eliassen 1940).

Big floods in combination with a high silt content have caused the Yellow River to change its lower course 26 times of which nine have been serious ones (602 BC, 132 BC, 11 AD, 1048, 1060, 1194, 1288, 1855, 1938). In 1938 the river changed its course to the southern side of the Shandong peninsula. Chiang Kaishek deliberately destroyed the dykes at Huayuankou to defend his troops from the Japanese invasion army. An area of 5.4 million ha was flooded, 890 000 Chinese people were killed and 12.5 million people were affected by the flood. It took nine years to repair the dykes before the river got its present outlet.

Since 1949 there has been no further serious dyke break. By reinforcing the dykes and dams, building flood retention basins in the lower reach, constructing reservoirs in the whole basin and dredging and widening the river course, dyke breaks have been avoided. The 155 large and medium-sized reservoirs built have a total storage capacity of 56 300 million m³ or similar to the average annual flow of the river (Gustafsson 1984).

The main reason for the accumulation of silt in the lower reach and in reservoirs is the extensive areas of barren loess land along the middle reach of the Yellow River (International Symposium on River Sedimentation, 1980). This means that soil conservation is the only longterm solution to the complicated Yellow River problems.

Already in the 1950 the Yellow River Commission was established and was given the overall planning responsibility of the entire basin. It clearly recognized the importance of the coordinated control and management of land and water resources. However, due to various reasons the implementation of this principle was weakened during the 1960s and most of the 1970s. But since the end of the 1970s planning authorities in the basin have reconsidered soil conservation as the basic measure for a successful water conservation.

The Loess Plateau

The erosion problem

The Loess Plateau region embraces an area of 53 million ha of which the yellow loess earth covers 32 million ha. It is the unfavorable combination of a very erodable soil and very erosive rainfalls, which makes the soil erosion so serious on the loess plateau. The forest cover is only 5 % and the plateau has an uneven topography over an area of 43 million ha.

Studies in small basins have shown, that practically all the eroded soil is transported out of the basins and into the tributaries of the Yellow River. The average annual silt removal is more than 20 000 tons per km² in the central gully hilly area of the plateau.

At Sanmenxia the average annual silt transport in the Yellow River is 37.6 kg per m³ with a maximum recorded value of 666 kg per m³.

About half of the annual Yellow River silt load or 800 million tons is derived from the Shaanxi part of the loess plateau. The gully hilly area in Shaanxi, located basically in Yanan Prefecture, covers 6.5 million ha. The precipitation ranges from 400 mm in the north to 700 mm in the south. Around 60 % of the annual precipitation occurs in July—September. The area is covered by a thick yellow earth. Sometimes the thickness exceeds 100 m. The bare round-shaped loess hills dominate the landscape and make it very beautiful. The hills are separated by deep gullies, which sometimes are widened to valleys.

Because of the population pressure in the area, the hilltops and the steep slopes are usually cultivated. On the hilltop the farmers sow winter wheat, on the slope millet and some buckwheat and at the gully bottom or in the valley maize. In the north and central parts the length of the gullies are 5—7 km per km² and the average soil loss 15 000—30 000 tons per km². Further to the south the length of the gullies has decreased to 3—4 km per km² and the average soil loss to 10 000—15 000 tons per km². The average annual soil loss in some forest steppe areas is 100—1 000 tons per km².

Integrated soil and water conservation on the brigade scale

The Xio Shi Guo Brigade is situated within the gully hilly area about 40 km northeast of Yanan City. The patient soil and water conservation works in this brigade illustrate that much environmental, economical and social improvements could be attained, if the reliance on the comprehensive planning and management of land and water resources is adhered to (Gustafsson 1984).

All the land is located along a big gully and five smaller gullies, which alleviate the conservation works. The brigade has 154 inhabitants, living in 34 households. The labour force numbers 59 people. There are 40 cattle and 142 goats and sheep. Some basic facts about the land use is given in Tab 1.

The brigade started to build the first siltation dams in 1956. At that time the number of households was only 1/3 of the present. The peasants were very poor and each person had a grain ration of only 100 kg. In 20 years this brigade has put in 68 000 workdays in

Tab 1 Land use in the Xio Shi Guo Brigade in 1982, hectares.

Arable area	28
of which	
— irrigated area	4
— new dam land	5
— total terraced land	3
— terraced land on hills and gully slopes	2
Orchards	6
Forests	16
Grazing area and other land	58
Total brigade area	108

Fig 2 Maize growing on new dam land in the Xio Shi Guo Brigade in the gully hilly area Yanan County, Shaanxi

soil and water conservation works, moving 300 000 m³ of earth to a total investment of 42 000 Yuan (approximatively 20 000 US dollars). The government has supported this work with 10 000 Yuan.

In the beginning the first dams were destroyed by heavy rainstorms, because no spillways were built. But today all the gullies are controlled thanks to the comprehensive planning and management applied. The brigade has used labour to build 33 dams and create 5 ha of new dam land. There are series of dams in each gully, which are needed in order to control the storm floods (Fig 2).

The dam land is suited for maize or sorghum cultivation. Agriculture, animal husbundry and forestry have been planned and managed together. Trees have been planted and grasses sown, which reduce the risk for dam collapses during rainstroms. The grass gives fodder and could be used for fertilizing purposes. The planting of fruit trees has diversified the economy. The combined result of all these measures has been, that all the flood water is conserved within the gully and if it is not used by plants, it will become ground water.

The Xio Shi Guo Brigade first learnt the technique of building a siltation dam by sending a person to a soil conservation workshop. Such workshops are arranged every year by the Yanan County Soil and Water Conservation Bureau.

Soil conservation will indirectly conserve the precious water. The brigade has constructed three small and shallow waterponds behind three earthdams. Throughout the

Land type	Crop	Average productivity
Plain land and new dam land	Maize*	7500
Hillslope land	Millet	1500
Hilltop land	Winter wheat	1500

* New maize varieties have helped to increase the yield.

Tab 2 Average productivity of different types of cultivated land in the Xio Shi Guo Brigade, kg per ha.

year there is a continuous flow through the ponds. They are used for irrigation, and they are also stocked with grasseating fish species. The fish production is enough to supply the brigade with its own fish demand. The small fish seedlings are bought from the Yanan County Soil and Water Conservation Bureau, which has a 7 ha dam raising area just outside the city.

The brigade has also dug a ground water well with a depth of 10 m and a diameter at the bottom of 3 m. This well could supply pumped irrigation water to 1/3 ha of dam land each time. During the irrigation season the well is recharged with ground water within one day. They plan to dig some more wells in order to control the irrigation and drainage needs of the dam land. Finally the families get their drinking water from four natural springs.

The constructing of high yielding new dam land has created a favorable situation for a better land use. The average productivity on different kinds of cultivated land is given in Tab 2. The Xio Shi Guo Brigade previously had 60 ha of arable land of which 33 ha was hillslope land. Today the arable area is only 28 ha. The rest has been converted into forests, orchards and grazing areas. The brigade's immediate future task is to level a hilltop of 3 ha with the purpose of conserving water for winter wheat cultivation.

The various soil and water conservation works have substantially increased the productivity of the land and the standard of living (Tab 3). The total grain yield has increased from 46 tons in 1971 to 81 tons in 1981. The year 1980 was even better, when 103 tons were harvested. In 1981 each brigade member had a grain distribution of 409 kg as compared to 255 kg in 1971. The distributed collective income was 212 Yuan in 1981 as compared to 56 Yuan in 1971.

Tab 3 Total grain production and the distribution of grain and collective income in the Xio Shi Guo Brigade.

	1971	1975	1980	1981
Total grain yields, tons	46	78,5	103	81
Grain sold to the state, tons	–	–	16	13,5
Agricultural grain tax, tons	–	–	–	1,5
Distributed grain per capita, kg	255	286	498	409
Distributed collective income per capita, Yuan	56	102	183	212

Advantages of the cooperative ownership approach

The experience of Xio Shi Guo Brigade is successful as almost all soil and water has been conserved within the gullies. The conservation has been a time-consuming, labourious and patient work. To a high degree the peasants have learnt the necessity of the comprehensive planning and management of land and water resources by their own experience. As it is shown in Tab 4 Xio Shi Guo Brigade appears not to be an isolated case. In 1981 2.5 million ha or 25% of the Shaanxi loess plateau area was under soil conservation control.

Nowadays it is generally acknowledged that the organization aspect of rural development in Thirld World countries has been strongly neglected (Johnston & Clark 1982). In the Chinese case all the land and water at the local level has been planned and managed by cooperatives.

A cooperative ownership approach towards land and water resources development has had many advantages. First, in the Xio Shi Guo Brigade the arable area per household is only 0.8 ha. It is not likely that a rational control and management of resources could be achieved by a system of small peasant land ownership under such circumstances of land scarcity. Only the cooperative ownership approach could give a solution to the problem of scale economy in agricultural production. Second, the cooperative system of investment allocation and income distribution has transferred the element of risk in agricultural production from the individual peasant to the cooperative unit. Thereby the low-income peasants' aversion to the introduction of new technology and modern agricultural inputs has been greatly reduced. Third, the cooperative village framework has provided the needed linkage between governmental authorities centralized planning and economic guidance and the local level decentralized planning and management. Fourth, the co-operative organization has furnished the village level with a functional democracy, which has promoted the peasants grass-root participation in planning, implementation and management.

In conclusion, the present study arrives at the following hypotheses:

1. small-scale peasant land ownership is not consistent with a comprehensive control and management of land and water resources under conditions of population pressure.

2. only a cooperative ownership approach is consistent with a comprehensive control and management of land and water resources under conditions of population pressure.

Terraces on slope land	350 000
Terraces on plateau land	210 000
New dam land	43 000
Land overflooded with silty water	12 000
Soil conservation forests	1 000 000
Artificial grassland	300 000
"Fenced off" hilly forest land	180 000
New irrigated land	420 000
Total soil conservation area	2 515 000

Tab 4 Area under soil conservation control on the Shaanxi loess plateau at the end of 1981, hectares.

In China, the cooperative ownership approach provided the most important judicious and administrative tool for a comprehensive planning and management of land and water resources.

References

Gustafsson J.-E.: Water Resources Development in the People's Republic of China. Trita-kut 1035, Royal Institute of Technology, Stockholm 1984.

Johnston, B. E.; Clark, W. C.: Redesigning Rural Development: A Strategic Perspective. Baltimore 1982.

International Symposium on River Sedimentation. Beijing 24—29 March 1980. Proceedings. Guanhua Press and China National Publishing Industry Trading Cooperation.

Todd, O. J.; Eliassen, S.: The Yellow River Problem. American Society of Civil Engineers Transactions 105, 346—453 (1940)

Lundqvist, J.; Lohm, U. and Falkenmark, M. (eds.):
Strategies for River Basin Management, pp. 099–113
© 1985 D. Reidel Publishing Company

Catchment Ecosystems and Village Tank Cascades in the Dry Zone of Sri Lanka A Time-Tested System of Land and Water Resource Management

Madduma Bandara, C. M., Dept. of Geography, University of Peradeniya, Peradeniya, Sri Lanka

Abstract: An ancient irrigation technology based on natural drainage basins, namely, 'tank-cascade' systems, in a tropical monsoonal environment is analysed on the basis of available information. An attempt has been made to identify the causes that led this system to persist over long periods of history. It is thought that some equilibrium of the cascade system with nature and the prevailing social organization that supported to maintain it, constituted the main raison d'etre. In this context, particular attention is drawn to the possible existence of a resource management system in which necessity for co-ordination of different components of the cascade was built into the system so that a self regulating mechanism became eventually established.

Introduction

The Dry Zone of Sri Lanka has been the home of one of the ancient 'hydraulic civilizations' (Leach 1959, Wittfogel 1957) in monsoon Asia. The origin, development and decay of this civilization is a saga in the annals of man's experience in harnessing water resources for his sustenance, well-being and aestheticism. It is remarkable that, most of the irrigation systems which supported this civilization do still survive inspite of the vast changes that have taken place in their natural, socio-cultural and economic environments. These irrigation systems, particularly the small village tank systems provide an interesting field of investigation which has some bearing on the river basin strategies for long-term carrying capacity of life supporting systems.

The objective of this paper is to analyse the salient features of the traditional irrigation systems, particularly those related to tank cascades in the Dry Zone of Sri Lanka, with a view to understand the principles of resource management that supported these irrigation technologies to perpetuate over long periods of history, often running back to twenty centuries. The most conspicuous characteristic of these traditional irrigation systems is that they are organized within natural drainage basins in the Dry Zone landscape. Thus in certain ways these miniature catchment systems may be taken to represent a microcosm of larger river basin systems in Sri Lanka and possibly in other parts of the world. Therefore, it is hoped that an understanding of the nature and management of 'thank-cascade systems' would provide useful insights into river basin based planning of land and water resources.

Dry Zone of Sri Lanka

Conventionally, Sri Lanka is divided into two major climatic zones, namely the 'Dry Zone' and the 'Wet Zone' (Fig 1). However, on the world climatic mosaic almost the entire Island could be considered as a single macro-climatic region in a general sense. Thus in terms of annual rainfall most part of the so-called 'Dry Zone' recieve twice as much rain as the world average of some 857 mm. On the other hand, the marked seasonality of rainfall with its bimodal regime creates serious water shortages for agriculture during certain

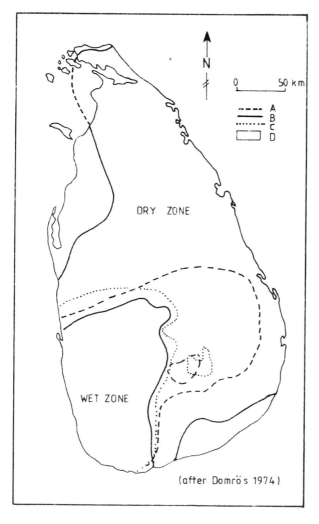

(after Domrös 1974)

Fig 1 The dry zone and wet zone A 75 inches (= 1904 mm) isohyet of the mean annual rainfall; B 20 inches (= 508 mm) isohyet of mean SW monsoonal rainfall September; C boundary line of effective dry period; D areas with less than 50 inches (=1270 mm) mean annual rainfall also described as 'Arid Zone'

periods of the year. It is to meet this natural challenge that a system of irrigation reservoirs began to emerge in this area from the early periods of history.

The precise demarcation of the Dry Zone of Sri Lanka has also posed several difficulties to the climatologists, hydrologists and ecologists. At the most elementary level, it was defined as that part of the island receiving 1270–1905 mm of rainfall. When it became clear that this did not tally with certain other ecological manifestations, different definitions were attempted by different scholars (Fig 1). Some writers (Kayane 1983) discovered that the distribution of irrigation tanks provides a reasonable indication of the areal spread of the Dry Zone. In general, the 'Dry Zone' covers over 60% of the land area of Sri Lanka.

Distribution and Density of Irrigation Tanks

The elaborate nature of the ancient irrigation works in the Dry Zone of Sri Lanka began to be rediscovered during the process of topographical mapping of the interior of the Island by the British in the latter part of last century (Brohier 1934). Cook (1951) referred to the density and distribution of tanks in Sri Lanka and made some attempts to relate their spatial patterns to climatological and ecological factors. Thus Cook thought that 'the seventy five inch (1270 mm) isohyet cuts off nearly all the tanks'.

The map showing the distribution of ancient irrigation tanks (Fig 2) indicates a greater preponderance of these reservoirs in the region of the ancient kingdom of Rajarata centered around Anuradhapura and in the principality of Rohana in the southeastern parts of the island. A closer inspection of the topographical maps reveals that in some localities the density of tanks exceeds 15 for an area of mere 10 km^2. Although a large number of these tanks are operational at present, a considerable proportion lies in an abandoned state under forest cover. It is generally believed that most of these tanks have fallen into disuse after the collapse of the Rajarata Civilization around the 12th century.

In certain parts of the country these minor irrigation tanks are so numerous that some scholars hesitate to believe that they were operational during the same periods of history. The close juxtaposition of most tanks compelled others to believe that these tanks were not used entirely for irrigation but for storage purposes and to support animals and to maintain higher water table levels in their vicinity. Some scholars who made attempts to determine the population of ancient Sri Lanka on the basis of tank densities were confronted with the problem of arriving at too high estimates. Thus, it is obvious that there is hardly any consensus regarding the nature and water resource management functions of ancient irrigation tank systems.

The register of Irrigation Projects (1975) lists some 3119 minor irrigation tanks spread out in different parts of the Dry Zone. It is estimated that the total number of tanks in use and in abandoned state exceeds 12,000 (Madduma Bandara 1984). The irrigable extent under minor irrigation works in Sri Lanka comes closer to 200,000 ha, and constitutes nearly 40% of the total irrigable extent under both major and minor schemes.

102

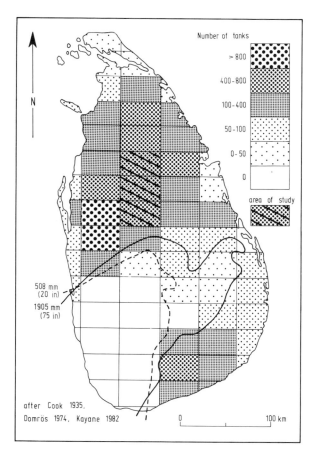

Fig 2 The relative number of tanks (in use and abandoned). The rectangle indicates the different sheets under which Sri Lanka is mapped on the scale of one inch to a mile

Number of tanks

> 800

400 - 800

100 - 400

50 - 100

0 - 50

0

area of study

508 mm (20 in)

1905 mm (75 in)

after Cook 1935,
Domrös 1974, Kayane 1982

0   100 km

Catchment Ecosystems

The peneplained landsurface of the Dry Zone of Sri Lanka with its characteristic undulating topography and its relatively dry and seasonal climatic regime displays an array of stream catchments of varying sizes and shapes. These catchments contain ephemeral rivulets that carry some water only during the rainy season. The crystalline hard rock base of the Dry Zone is a relatively poor aquifer and the occurrence of ground-water is often restricted to pockets of alluvium which are often influenced by the presence of tanks (Madduma Bandara 1982). The water table in these areas fluctuates seasonally and reaches ground surface in the valley bottoms during wet spells of weather. The watersheds of catchments are usually marked by outcrops of resistant crystalline rocks and the slopes extending from them towards the ephemeral stream display soil catenas with well drained coarse grained reddish brown soils on upper segments and finely grained humic soils in the valley bottoms (Panabokke 1959).

The dry zone stream catchments which are not entirely cleared for cultivation or other purposes, remain under a cover of deciduous dry zone forest. The plant species in these forests which are generelly regenerated woodlands due to shifting cultivation show characteristics developed to withstand the seasonal moisture stress. The fissured barks, deciduous habits and species with thorny characteristics are obvious indications of this natural adjustment. Inspite of the high mean air temperature (23–30 °C) throughout the year, the march of seasons is discernible in the vegetative cover in consonance with the seasonality of rainfall. Irrigation tanks on the other hand have developed their own aquatic ecosystems with a rich variety of aquatic plants, freshwater fish and bird life.

It is within the framework of the natural stream catchments described above, that the ancient hydraulic civilization grew with its three salient components; i.e. the tank, paddy field and the *dagaba* (dagaba is a buddhist monument and a place of worship). The tank and the paddy field occupied the valley, while the *dagaba* occupied the higher ground where the rock outcrops and inselbergs were converted into works of art, places of worship and of spritual retreat.

Land-Use Zoning Related to Water Flow

Some recent studies indicate that there is a high degree of morphological regularity in the spatial organization of land use in a *purana* (old) village system centred around an irrigation tank (Tennakone 1974). Five landuse zones in a purana village system are identified, namely, the tank, old field, field blocks, parkland and the forest. The main axis running through all these zones is represented by the ephemeral stream which enters the tank and passes through the paddy fields. The forest cover, with its characteristic fauna and flora, where it still exists, serves to separate one settlement from another. In other words, the original tank settlement with its necessary appurtenances displayed a fine adjustment of man's activities to nature, providing a stable background for its long-term persistence. As Brohier (1934) puts it 'in Ceylon time cannot stale, nor can usage wither; 'tank' would appear to be synonymous with 'village' implying thereby that, one section of the ancient population was composed of a number of agricultural republics each of which had a tank and a paddy field below it'.

'Cascades' of Village Tanks

One of the traditional land and water management systems which has obviously developed on the basis of catchment ecosystems is the system of 'tank chains' or 'tank cascades'. A cascade is a connected series of tanks organized within a micro-catchment of the Dry Zone landscape, storing, conveying and utilizing water from an ephemeral rivulet (Fig 3). As indicated by topographical maps, system of tank cascades had been a widespread

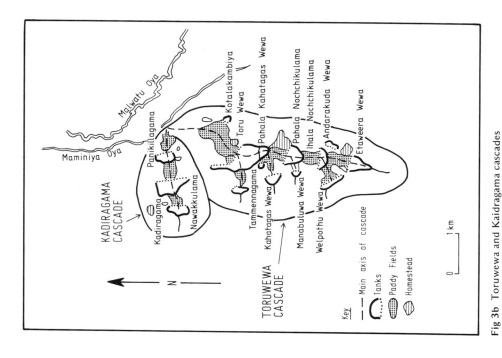

Fig 3b Toruwewa and Kaidragama cascades

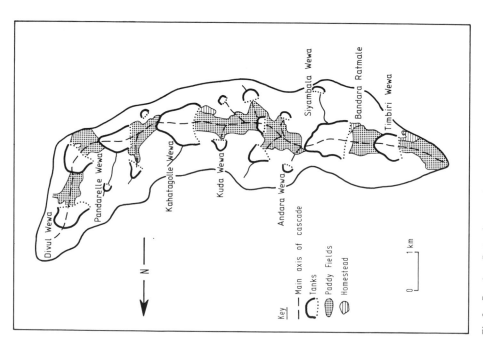

Fig 3a Bandara Ratmale cascade

phenomenon of the Dry Zone landscape in the past. In most areas these cascades are still operational, inspite of the diminishing forest cover and the changes in the traditional land and water management practices.

A study of the nature and operation of these cascades systems could be attempted in two ways, namely, by (a) examining a large sample of cascades and (b) by examining in detail the organization and operation of a few selected cascades.

In order to obtain a statistical understanding of the cascade systems, measurements were made on all the cascades that appear on two topographical maps of Sri Lanka (i.e. Anuradhapura and Medawachchiya). These two maps cover an area of about 1225 km^2 representing a region of high tank densities. For purposes of measurement 'a cascade' was defined as a series of tanks located in succession one below the other. At least two currently functional tanks were considered necessary to make a cascade. Thus according to the stream ordering terminology of fluvial geomorphologists tank systems selected for investigation may be considered as 'first or second order' cascades. There were of course a few exceptions where higher orders had to be included in the sample due to the complicated nature of some tank systems. On the other hand cascades which were affected by major irrigation canal systems have been eliminated.

Some results of the cascade measurement exercise are given in Tab 1. This shows that there were some 127 cascades within an area of 1225 km^2 giving an average density of about one cascade for every 12 km^2. The density becomes still higher when the area affected by major irrigation canals is excluded.

Tab 1 also shows that over 85 % of all the operational tanks were located within cascade systems. The tanks actually remaining outside the cascades are often found to be small *olagam* (tanks without settlements) draining directly into higher order dry zone streams, or those belonging to cascades of higher orders. Fig 4 shows the relationship between catchment areas of cascades and the irrigation tanks within them.

It is possible that when the hydraulic civilization began to collapse around the 12th century, and the kingdoms began to shift, the remaining population gradually abandoned *olagam* and concentrated their activities under the bigger tanks in the cascade systems.

Tab 1 Some statistical aspects of tank cascades in the Nuwara Kalawiya region

Extent of study area	= 1225 km^2
Number of cascades in the study area (excluding those affected by major irrigation structures)	= 127
Mean size of the catchment area of a cascade	= 6.42 km^2
Total number of operational tanks in the study area (*excluding* those affected by major irrigation structures)	= 575
Total number of operational tanks located within the cascades	= 496
Average number of operational tanks per cascade	= 3.90
The percent of operational tanks within cascades (out of the total in the study area)	= 86.26

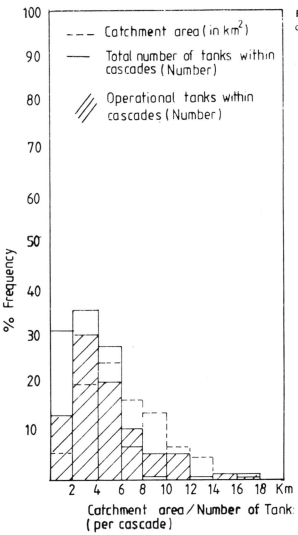

Fig 4 Frequency distribution of cascade characteristics

These main tanks would have obviously had a greater chance of survival. However, when there was a growth of village population, and demand for land increased, *olagam* were brought back into operation. In certain ways this represents the flexibility and the re-silience of tank cascade systems in a relatively hostile environment.

Village Irrigation Tanks

The total number of village tanks (including those falling within major irrigation areas) in operation in the area covered by the two topographical maps stood at 929 by 1972. There

Number of tanks (for which information is available)	= 43
Average catchment area of a tank	= 1.49 km^2
Average capacity of a tank	= 136,700 m^3 (= 13.67 HM*)
Average irrigable extent of a tank	= 15.84 ha
Average number of share holders	= 38.97

*HM = Hectare meters

Tab 2 Village tank characteristics in the Toruwewa Tulana

were also some 340 tanks in abandoned state. This gives an average density of about one tank (either in use or in abandoned state) for every square kilometre in the study area. The number of abandoned tanks and the forest cover has progressively diminished in recent years. This is clearly visible from a comparison of aerial photographs taken in 1956 and in 1981. It is interesting to note that nearly 73 % of all tanks were in operation by 1972, leaving only 27 % in abandoned state.

In order to obtain an understanding of the nature of village tanks and their command areas, an attempt had been made to analyse some statistics collected by the Water Resources Board (1968) for one part of the study area, namely, Toruwe Tulana. A summary of this information is listed in Tab 2.

Tab 2 shows that an average village tank has a catchment area of about 1.49 km^2 (including the catchment areas of tanks above), and a capacity of about 136,700 m^3 (13.67 HM). Thus an average village tank can store only about 92 mm of water from rainfall that is received on its catchment. This works out to 6—7 % of the mean annual rainfall in the area. It should be noted, however, that there can be several tanks within the catchments of some tanks located downstream. In such cases each tank can collect 6—7% of the rainfall from its own catchment. Therefore, in an average cascade system (i.e. with say four tanks) the maximum amount of rain water that could be retained would be around 20—30% of catchment rainfall.

Fig 5 shows the frequency distribution of catchment areas, capacities and the command areas of tanks in the Toruwe Tulana. The skewed distributions of these parameters indicate the predominance of smaller tanks. Nevertheless, if we consider a hypothetical case of an average tank which fills once a year and its total irrigable extent is cultivated once a year, the tank can supply about 863 mm of water per hectare, assuming of course that the effects of losses due to evaporation and seepage are counterbalanced by incremental replacement due to continuing rainfall and incoming drainage water from the catchment area.

Although these calculations are somewhat hypothetical, it is interesting to compare them with water use in the major irrigation projects, where, under common field conventions ex-sluice duties range between 914—1524 mm of water per hectare, for maha cultivation (Chambers, 1974). In this respect water use under the village tanks appears to be more sparing than in the major irrigation areas. On the other hand, it shows the irrationality of planning to expand further the irrigable areas under village irrigation tanks. It is now well known that even in a year of abundant rainfall, around 40 % of the irrigable area remains uncultivated under minor tanks. This of course if often attributed to fragmentation of paddy lands (WRB Survey 1968).

Toruwewa and Kadiragama Cascades

In order to study the detailed operation of cascade systems, two cascades in close proximity to each other were selected from the area of study (Fig 3). Out of these two, Toruwewa cascade is a system of 12 village tanks draining an area of 9.55 km². At least eight of these tanks are considered as *olagam*. Some basic information on 11 tanks in this cascade are given in Tab 3.

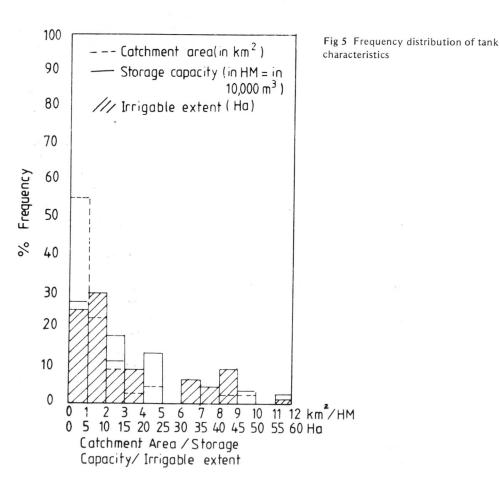

Fig 5 Frequency distribution of tank characteristics

As can be seen from Tab 3, Toruwewa is the largest tank in the system into which water from seven other tanks drain. The three tanks above Toruwewa are located on the main axis of the cascade, while five smaller tanks are located on its either side. Land under these small tanks are normally owned by farmers under the main tanks. For example both Welpotuwewa and Andarakudawewa fields are cultivated by farmers from Etaweerawewa.

The Kadiragama cascade is a system of three tanks located in direct succession, one below the other. Compared with Toruwewa, Kadiragama is a much smaller cascade covering an area of only 2.97 km². The details of the Kadiragama cascade are given in Tab 4.

In terms of capacity and the area irrigated, Kadiragama is the biggest tank in the cascade. However, in terms of the number of share holders, Panikillagama (the lowest tank) is the largest village. In fact, it was suggested by farmers of Panikillagama to increase the capacity of their tank and to extend its irrigable area. This has not been attempted recently, because any such attempt would have submerged the lower part of the paddy fields at Kadiragama (WRB Survey 1968). It is evident that the Panikillagama tank should have had a bigger capacity in the olden days.

Tab 3 Characteristics of tanks in the Toruwewa cascade

	Catchment area (km²)	Storage capacity (in 10,000 m³/HM*)	Total irrigable area (ha)	Number of share holders
Toruwewa	7.40	43.38	51.59	80
P/Nochchikulama	3.78	19.75	—	—
I/Nochchikulama	2.82	8.51	10.31	41
Manabuluwa	0.26	5.61	5.66	14
Etaweerawewa	1.52	12.34	—	—
Tammennagama	0,82	12.34	4.45	21
I/Kagatagaswewa	0.82	2.22	2.42	18
P/Kagatagaswewa	0.84	2.02	2.43	14
Welpotuwewa	0.38	88.14	13.15	24
Kotalakembiya	0.36	2.46	2.02	3
Andarakudawewa	0.12	2.71	2.53	3

Source: WRB Survey, 1968   * HM = Hectare meters

Tab 4 Characteristics of tanks in the Kadiragama cascade

	Catchment area (km²)	Storage capacity (HM/10,000 m³)	Irrigable area (ha)	No. Share holders
Nawakkulama	0.49	7.40	3.84	24
Kadiragama	1.78	24.69	14.77	39
Panikillagama	2.51	7.40	9.51	51

Source: WRB Survey 1968)

Discussion

In each cascade system, water received from rainfall appears to have been ingeniously utilized with minimum wastage. Thus water that is used for irrigation under one tank is passed on to the next tank *(Vel Pahu Watura)* through a drainage line for reuse. Where a paddy field is very extensive, small anicuts *(Amunu)* were used to tap this water from the drainage channels *(Wagala)*. There are five anicuts in operation today in paddy lands below the Toruwewa Tank. Thus even at a time of abundant rains, hardly any water escapes through the final outlet of the cascade system. Cascade technology conforms clearly with the famous dictum of King Parakramabahu (1153) who declared that "not a single drop of water received from rain should be allowed to escape into the sea without utilizing for human benefit". This statement is often quoted by scholars to highlight the priority accorded to water resources development by ancient kings. It may be stated that the cascades of tanks do infact provide an excellent technical and organizational expression of Parakramabahu Philosophy.

The preliminary observations indicate that the size and capacity of tanks increase in a downstream direction. Thus larger tanks located near the outlet of a cascade can store excess water draining from the tanks located in the upper reaches. This became clearly evident during the recent high rainfall spell (1983) that led to floods in most part of the Dry Zone. On the contrary, in an area fed by major irrigation canals it was observed that during rainy weather much water flows out through drainage lines due mainly to the absence of village tank-type storage systems.

The irrigation network under each tank in the cascade appears to have been adjusted over the years for economising the use of tank water. There are at least two sluices for an average-size tank, one for use when the water level in the tank is high *(goda horowwa)* and the other *(mada horowwa)* when water level is low. The two main irrigation ditches that originate from the sluices run usually along the two sides of the old paddy field which is divided into long strips of paddy tracts (Leach 1959). The drainage channel that appears towards the centre of the lower portion of the paddy fields collects all the water that pass through the paddy fields and carry them to the next tank.

Coordination Necessary

It is obvious that a considerable degree of coordination and inter-dependence should have been necessary for different tanks in the cascade system to operate together. For example, during a time of unusually heavy rainfall a breach of the bund of the highest tank in the cascade can lead to a breach in the second tank and so on. In other words it can progressively collapse as a set of dominoes. Therefore, the people living in a downstream tank settlement should have had a serious interest in the safety of upstream tanks. There is evidence to suggest that it was customary for farmers in the chain-tanks to pool their resources to repair a breached bund.

On the other hand in a year of low rains, farmers have traditionally resorted to *bethma* or *Irawilla* form cultivation. Under these systems only a portion of the old paddy fields is chosen for cultivation on a common proportional ownership basis. In a situation of serious water shortage, a threatened crop could be saved if the tank above is in a position to release some water. In view of the highly localised nature of rainfall in these areas it is possible for different tanks in the cascade to receive different amounts of water from rainfall.

Similarly, any attempt to raise the bund or the spill level of a tank to increase its storage capacity is bound to submerge the paddy fields belonging to farmers in the village above. This problem could be observed in the Ihala and Pahala Nochchikulama tanks and in the Panikillagama tank. On the other hand any attempts to extend the paddyfields of one tank towards the upper water-spread area *(jala gilma)* of the tanks below would always lead to some conflict of interests. The old practice of tank-bed cultivation *(tāvalu)* was of course there, but it was only a seasonal phenomenon. The lack of an understanding of the cascade principle often led land development planners to encourage expansion of irrigable areas and to support the increase in storage capacity of tanks in a haphazard and piecemeal manner in response to popular demands.

The integrated management of a cascade system of irrigation tanks would not have been possible without a strong support from the prevailing social organization in the villages. This is a highly complex and often controversial area covered by many scholars (Geiger 1960; Pieris 1956; Leach 1961; Paranavitana 1960; Farmer 1957; Gunawardhana 1979). An interesting account of the early social setup in the Dry Zone is presented by Ievers (1899) in his Manual of the North Central Province.

The most conspicuous aspect of the rural society, is that the farming communities in the tank settlements belonged to a hierarchy of social groups called 'castes', often occupying separate villages. The great majority of the people belonged to the cultivator *(govigama)* caste. As recorded by Ievers (1899), out of some 945 sinhalese villages in the North Central Province 65 % belonged to this group which represented the mainstream of the rural society. It is unlikely that the picture has changed significantly during more recent times. Field investigations for the present study indicate that the position of a tank in a cascade was however, not obviously related to the caste structure. It was not unusual to find the so-called low-caste villages on the top end of a cascade system. Nevertheless, it could be observed that tanks with larger capacities and with larger command areas often located in lower parts of the cascade systems are predominantly occupied by farmers belonging to higher caste groups.

Although cultivators formed the majority of rural people, the apex of the hydraulic society was apparently represented by the almost defunct and numerically insignificant group called the *'Vannihuru'* who claim ancestry to Wijaya — the first king of Sri Lanka. They performed most of the rituals related to water and apparently possessed the specialised knowledge and experience of water resources management. This group may possibly be the last in the line of *Kulinas* whose extermination by the invading south Indian forces that contributed to the collapse of the Rajarata hydraulic civilization (Paranavitana 1960).

Ecologically, the cascade system appears to be a logical response to the challenges posed by the natural environment in the Dry Zone. Water is the most scarce resource in this environment and its rational use formed the basis of all dominant human activities. In an environmental setting where ground-water resources are relatively limited, it is natural that all the attention is focussed on surface waters. The management of surface water resources in a cascade is accomplished within the framework of the natural drainage basin. In a climate of seasonal water shortage and occasional water surplus, the development of small village tank systems appears to have been the best answer. Once a system of storage reservoirs is established, a host of microclimatic, hydrological, ecological and socio-cultural factors became the determinants of its success. The preservation of the forest cover in the watershed areas, thus became part of the water conservation strategy.

It would be interesting to study the nutrient flows from the forested watershed to the tank and from there to the paddy fields and then to the next tank within a cascade system. The shifting cultivation had significantly affected the forests, but it was carried out systematically, so as to allow adequate time for natural regeneration. The tanks provided a variety of freshwater fish and it was the custom to share the catch according to the ownership of land in the old field. Tanks provided water for animals (mainly buffaloes used for draft power and dairy cattle) and for domestic purposes. Tanks also provided a variety of aquatic flowers (lotus, water lilly, etc.) for temples located near the rock outcrops which formed the least useful segments of rural lands.

References

Brohier, R. L.: Ancient Irrigation Works in Ceylon. Government Press, Ceylon 1934.
Chambers, R.: Water Management and Paddy Production in the Dry Zone of Sri Lanka. ARTI, Colombo 1974.
Cook, E. K.: Ceylon. Macmillan, London 1951.
Domroes, M.: The Agro-climate of Ceylon. Geoecological Research. Franz Steiner Verlag, Wiesbaden 1974.
Farmer, B. H.: Pioneer Peasant Colonization in Ceylon, A study in Asian Agrarian Problems. Oxford University Press, London 1957.
Geiger, W.: Culture of Ceylon in Medieaeval Times. Otto Harrassowitz, Wiesbaden 1960.
Gunawardhana, R. A. L. H.: Irrigation and Hydraulic Society in Early Medieval Ceylon. Past and Present 53 (1979)
Ievers, R. W.: Manual of the North Central Province, Ceylon. Ceylon Government Printer, Colombo 1899.
Kayane, I.: Evapotranspiration and Water Balance in Sri Lanka. In: Yoshino et al., Climate, Water and Agriculture in Sri Lanka, Tsukuba 1983.
Leach, E. R.: Hydraulic Society of Ceylon. Past and Present 15, 2–26 (1959)
Leach, E. R.: Pul Eliya: A Village in Ceylon. Cambridge 1961.

Madduma Bandara, C. M.: Some Aspects of the Behaviour of the Groundwater Table in the Vicinity of some Major Irrigation Reservoirs in the Dry Zone of Sri Lanka. In: Beiträge zur Hydrologie, Kirchzarten, FR Germany 1982.

Madduma Bandara, C. M.: Green Revolution and Water Demand — Irrigation and Groundwater in Sri Lanka and Tamilnadu. In: T. Bayliss Smith & Sudhir Wanmali (Eds.), Unterstanding Green Revolutions. Cambridge University Press, Cambridge 1984.

Panabokke, C. R.: A study of soils in the Dry Zone of Ceylon. Soil Science, 87 (1959)

Paranavitana, S.: The Withdrawal of the Sinhalese from the Ancient Capitals. University of Ceylon, History of Ceylon, Vol. 1, Peradeniya 1960.

Pieris, R.: Sinhalese Social Organization: The Kandyan Period. Ceylon University Press Board, 1956.

Tennakone, M. U. A.: Spatial Organization of Agriculture in the Traditional Rural Settlements in the Mahaweli Development Area. Staff Studies, 4, 2 Central Bank of Ceylon, Colombo 1974.

Water Resources Board (WRB): A Survey of the Water Resources Utilization and Cultivation Habits and Practices by the NCP Peasant Cultivator in the Toruwe Tulana of Anuradhapura District, Colombo 1968.

Wittfogel, K. A.: Oriental Despotism: A Comparative Study of Total Power. Yale University Press, New Haven 1957.

Lundqvist, J.; Lohm, U. and Falkenmark, M. (eds.):
Strategies for River Basin Management, pp. 115–122
© *1985 D. Reidel Publishing Company*

Control and Management of Land Use and Water Resources of Bhavani River Basin in Tamil Nadu, India

Sivanappan, R. K., Water Technology Centre, Tamil Nadu Agricultural University TNAU, Coimbatore 641003, India

Abstract: The problem prospects of Bhavani River Basin have been identified in all aspects. The various agencies involved in the development process are tackling the problems in the basin starting from prevention of erosion, reducing siltation in the reservoirs and better utilisation of land and water to obtain maximum production. Since the water availability is limited in the State, many alternate suggestions are given for the project to increase the water use efficiency in the basin. If the land and water resources are managed properly, the wealth of the basin could be increased considerably.

Introduction

Bhavani river is one of the important tributaries of the River Cauvery in south India (Fig 1). The river originates in the adjoining Silent Valley in Kerala State, passes through Nilgiris hills and joins Cauvery at a place called Bhavani in Tamil Nadu. The total drainage area (basin) is 6730 km². The tributaries for the river are Pykara, Siruvani, Kundah, Coonoor and Moyar. The basin receives rain from southwest and northeast monsoons. The lower Bhavani reservoir constructed across the river, provides a storage capacity of 906 million m³ to irrigate about 84,000 ha. In addition, there are many small reservoirs constructed in hills to store about 473 million m³ which are used for hydro power generation which is about 50% of the hydro power production in the State.

The basin elevation ranges from 166 m above sea level in the plains to 2634 m in hills. The entire basin can be broadly divided into four physiographic regions namely western ghat, eastern ghat, plateau region and Bhavani valley.

The index and river basin maps are appended (Fig 1 and 2).

Drainage

Bhavani and Moyar are the main rivers draining the basin. The Lower Bhavani reservoir was created by constructing a dam across the river Bhavani just below its confluence with Moyar in Satyamangalam taluk. From the reservoir, a 200 km long Lower Bhavani main canal takes off with a capacity of 64 m³/sec. It has three major and 196 minor distributaries and several direct sluices. The total length of all the canals put together is about 1130 km. Apart from the Lower Bhavani Project, Kundah-Pykara Hydro Electric complexes have series of reservoirs and small reservoirs across the tributaries have also been constructed for irrigation and for drinking water purposes.

116

Fig 1 Index map of Bhavani river basin

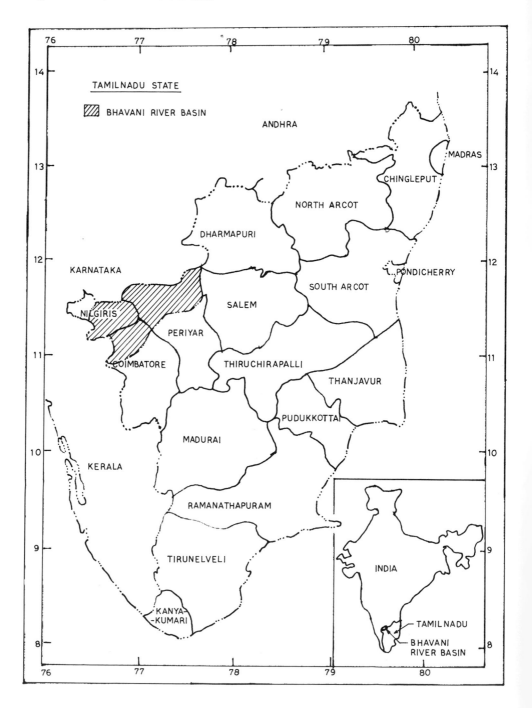

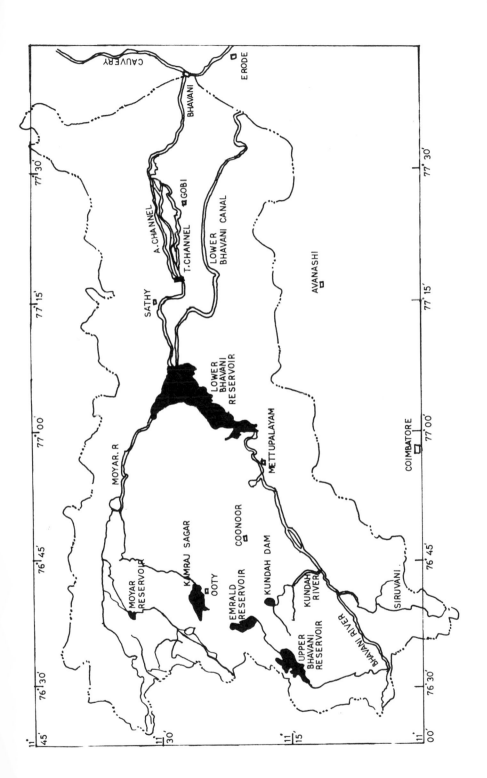

Fig 2 Bhavani river basin

Climate and geology

The basin has a monsoonic climate as it lies within the tropical monsoon belt of Asia. The monsoon period is from June to December and January—May forms no-monsoon season of the basin. During monsoon, the basin benefits from the south west (June-September) and the north east (October—December) monsoon. They cause heavy stream flow in rivers, recharge acquifers and fill up reservoirs and tanks. The temperature of the basin varies largely. It may be observed that the summer season lies between March—May (maximum 41 °C in plains and 17 °C in hills) and the winter season is between December and February (maximum 25 °C — minimum 8 °C). The average rainfall of the basin varies from 700 to 4600 mm. The western parts of the hill receive heavy rain and the eastern part and plains receive less rain. The annual average evaporation rate above 1500 m, ranges from 1000 to 1500 mm and below 1500 m, it is between 1500 to 2000 mm. Geologically the basin is underlined by three major rock units for archean age, the charnockite, mixed gneisses and granities. The charnockite rocks occupy major area of the basin while the mixed gneisses form a good source for ground water exploitation.

Soils and crops

The typical soils of the basin are: red, black, alluvial, mixed red and black and lateritic. A soil survey of all the erosion hazard areas of the basin has been carried out and documented. The basin is predominantly composed of mollisols, entisols, alfisols and inceptisols. The soil series, their orders, irrigability classes, hydrologic soil grouping and their distribution have been studied and documented. Vegetation of the basin was surveyed and every vegetation has its own individuality and different from others. They differ from each other in their vertical and horizontal composition and distribution of leaf canopy. Plant cover and land use should be well managed for the optimum yield of water with minimum sediment. Ecology of vegetation has a direct effect on climate, rainfall runoff and other parameters of the environment. Agriculture is practiced widely in the basin. Rainfed agriculture is common in hills while dry farming is practiced in the plains. Irrigated agriculture has spread to a large area after the construction of Lower Bhavani Project and a wide variety of crops are cultivated.

Erosion

Rainfall records which range from 30 to 100 years in the basin are quite adequate. As regards rainfall, it is high in the forest and low in cultivated regions in plains. Floods are common in hills including land slides. The sediment load is about 0 to 20 ppm in protected unpolluted forest streams. In urban and rural areas it is about 2000 to 8000 ppm. Sheet erosion and land slips contribute with high sediment load to the streams. Sediment yield of the watershed vary from 1.5 to 19.7 t/ha/year. Hence, the reservoirs constructed across the river especially in the hills are silted very badly in a short period.

Different Parts of the Basin

The entire basin for the control and management of land use and water resources can be divided into four categories namely, i) Forest land in hills; ii) Crop land in hills; iii) Dry tracts in plains including well irrigated area and iv) Irrigated land in the command area.

The first three come under catchment area and the last one is under ayacut/command area. In between, there is water spread above the dam which also requires attention since by improper land use in the catchment, the eroded soil is silting up the reservoir.

Forest land in hills

About one fourth of the basin will come under this category. The rainfall is the highest (4600 mm) and the slope is more than 50 %. The soil in the area is lateritic. Various factors influence the rate of erosion in this area such as cutting and burning of forest trees and cultivation of crops in forest land. A number of hydro-electric reservoirs have been constructed in the last 25 years and the annual silting rate per km^2 of catchment is about 944 m^3. The old species of plants are replaced by Eucalyptus in order to obtain more yield.

Crop land in hills

About one sixth of the basin area is used for growing various crops in the hills. A range of crops are grown in these lands. Crops are grown even in the slope of 50 %. Though the advise for the farmers is that they should not take up any cultivation in lands having more than 33 % slope, farmers cultivate on steep slopes. The crops grown are wheat, rice, potato, hill vegetables, tea, coffee, etc. Soil conservation measures are practiced in cultivated areas. Normally bench terraces are formed with inward slope. The excess water collected in the drains is diverted to the natural drainage courses. These works are executed by the Government and the cost is recovered in instalments from the farmers.

In some places, the farmers have their wells and pump the water for giving protective irrigation by surface or sprinkler irrigation methods. However, geological formation, disturbances, rainfall, cultivation practices, slope of the ground, uncontrolled deforestation, improper method of tillage and unwise agricultural practices have accelerated soil erosion which affects the reservoirs constructed in the hills to store water for generating electricity.

In order to reduce the soil erosion and to produce more from these agricultural lands, activities like construction of contour walls, providing walls in river embankments, forming bench terraces, introducing new high yileding varieties of crops, advising the package of practices etc. are being taken up in a planned way.

Dry tracts in plains including well irrigated areas

Nearly 50 % of the area in the basin will come under this category. The soils are red in colour, very shallow and porous in nature. The rainfall is varying from 600 to 800 mm and mostly occur in the North East monsoon period (October–December). Since the monsoon is erratic and intensity may be very high occasionally, erosion hazard is more severe in this area. The siltation in the reservoir is due to the soil erosion in these areas. Already the zones which are vulnerable to soil erosion have been identified and preventive measures are being taken up. The farmers in the area have constructed numerous open wells to extract the ground water for irrigation purposes. The strata is mostly hard rocks and as the porosity of the rocks is only about 1–3%, the storage is very limited. The crops viz., cholam, bajra, bengalgram, blackgram, horsegram, groundnut, and gingelly. In the irrigated area (well irrigation) the farmers grow cotton, millets, groundnut, chillies, banana, vegetables, sugarcane, flowers, etc.

In order to conserve and manage soil and water in these parts of the basin, the Government has taken up integrated soil and water conservation and management works on a sub-watershed basis comprising an area of 1000 to 2000 ha. The work includes construction of check dams, providing contour bunds, percolation ponds, gully control structures etc. in dry lands and water management practices like providing lined channels, underground pipe line systems, including providing control structures, better surface irrigation practices including land levelling and shaping are contemplated in the irrigated areas. Further, according to the land capability, suggestions are made to grow the trees in the barren lands and cereals in other areas. In this way, the entire sub-watershed is treated, and conservation and management works are completed in all aspects.

There are many agencies which are responbile for the development of these areas viz., Agriculture, Agricultural Engineering, Forestry, Public Works Department, Land Development Banks, Revenue Department and District Development Agencies. In order to tackle the soil erosion and water management problems, it is necessary to have a separate agency and with the full participation of the farmers the objectives could be achieved very well in these tracts.

Irrigated land in command area

About 100,000 ha are under irrigation. There are a number of small anicuts in the river apart from the Lower Bhavani Project. The main crops grown are paddy, groundnut, sugarcane, banana, cotton and millets. Soils vary from red loamy soil to loamy soil in different zones of the command area. The terrain is undulated. The paddy crop is grown during August to December which coincides with the rainy season. The efficiency of water use for both paddy and other crops are far from satisfactory. The research findings for the various crops have indicated that by adopting improved water management practices, it is possible to save about 30 to 40% of water. Further, the seepage loss is also considerable in the command area since the soil is very porous and has a undulated topography. About 25,000 open wells are constructed in the area which are used for taking 2 or 3 crops in a year since the Government is allowing water only for one crop in a year. Studies are undertaken by Tamil Nadu Agricultural University to assess the various losses in different channels and also the scheduling of irrigation for various crops. Operational research work is being done on a large scale in farmers' field to demonstrate the research findings in order to increase the productivity from unit quantity of water.

Irrigation water is controlled by the Public Works Department (Irrigation Wing) and it is responsible for the delivery of water at the outlet level including maintenance of irrigation system. The Agriculture Department is responsible for advising the farmers with reference to providing inputs using high yielding varieties to increase the productivity. The Agricultural Engineering Department is taking up On Farm Development works such as providing irrigation and drainage channels for individual fields providing control and distribution structures, land levelling etc. The Cooperative Department is helping the farmers to get loans through financing institutions. The Revenue Department is in overall charge of getting necessary help for all the activities including collection of revenue etc.

Water Management

Water management is a complex problem in view of the fact that it involves numerous farmers and various agencies in the irrigation projects. Therefore it is necessary to have a Farmers' Association involving the representative farmers and various Governmental agencies responsible for the delivery of water, giving inputs and advise for the agricultural production. The research staff of the TNAU are engaged to work out the water fertiliser production function, scheduling irrigation, suitable crop pattern taking into account the total availability of water in the basin. They are also taking up operational research projects in farmers' field selecting about 100 to 200 ha by introducing water management practices including better methods of cultivation and other practices.

The Watershed Management Board of the State is taking observations and collecting data to work out the siltation problem in the waterspread area of the reservoir. The Agricultural Engineering Department is taking up works on preventing erosion in the catchment above the reservoir in an integrated manner both in the hills and in the plains. In this way monitoring and prevention of siltation in the reservoir is beeing carried out systematically.

The following alternatives are suggested for better water management practices in Lower Bhavani Project:

a) The entire ayacut can be supplied with water at 50% of the present quantity which is supplied to half of the area in alternate years. This would enable the farmer to economically utilise the water, develop underground water resources and to adopt conjunctive water use. The excess water use leading to problems of water stagnation and salinity could be avoided and also excess drawal by the head reach farmers depriving tail end farmers. If necessary, canal modifications, structures, etc. may have to be provided. The farmers will get water for two crops every year instead of getting water only for one season in a year.

b) If it is not possible to implement the above suggestion the following alternative may be considered. The present system of water distribution to an irrigated crop from August to December to the same field could be changed as follows: Water supply to a field can be started in August. A wet crop can be raised between August and December and the same farmer can go for an irrigated dry crop like groundnut from January to April. Water shall be supplied to the same field from January to April also, but at a reduced quantity. This reduction in quantity has to be achieved by actual reduction in water supply at the sluice gate rather than by cutting of supply for a fixed period of time and then resuming supply for an equal length of time at full capacity. This will avoid the farmers' tendency of using greater quantity of water as much as 12—20 cm per irrigation even for an irrigated dry crop and the consequent problems. After the groundnut crop the farmer could be encouraged to go in for a short duration pulse or oilseed crop whose water requirement would be very much less (i.e. one or two wetting could be given).

c) During the monsoon (August—December) in view of the fact that paddy is the sole crop, ground water is available in plenty at shallow depth. At that time, electricity is also not a problem. If water should not be released continuously, rotation system can be followed. It may be four days supply and closing the canal for two or three days. Further, if water is needed for the crop, the farmer can use the ground water by pumping. That will also facilitate the conjunctive use of surface and ground water. The Government can give incentives or loans facilities wherever necessary for the construction of wells in that ayacut.

d) It is also noted that the entire cultivable command area is not brought under irrigation. Hence it is assumed that all the cultivable command area could be brought under irrigation if these measures are implemented.

e) It is estimated that about one-third of water is lost by seepage. When the acquifers once get saturated, the rate of seepage decreases to about 25 % of the total water instead of 33—35 % and the rate of runoff increases at this juncture (about 20 %). From this it is seen that there is a great possibility of utilising the seepage water as ground water and also reducing runoff if the ground water is pumped so that the acquifer is not saturated in the rainy season. This will again help to solve drainage problem and prevent salinity and alkalinity.

Conclusion

Conservation and management of water resource in the basin should be given top priority in order to increase the area of irrigation and intensity of cultivation. The management of land is also be considered to prevent erosion and siltation in the reservoirs built across the river. Research and implementation should take the entire river basin as an eco-system into account, so that people in the area may fully benefit in the coming years.

Lundqvist, J.; Lohm, U. and Falkenmark, M. (eds.):
Strategies for River Basin Management, pp. 123–130
© 1985 D. Reidel Publishing Company

Problems of Land Use and Water Resources Management in the Upper Tana Catchment in Kenya

Ongwenyi, George S., Natural Resources Development Cons., Mlima House, PO Box 5114, Nairobi, Kenya

Abstract: The paper examines the present land and water development programme within the upper Tana basin in E Kenya. It is based on a universal soil loss equation for the river basin, hillslope plot experiments and monitoring of sediment yields at the outlet of this subcatchment. It offers challenges to solutions for either management of land and water resources for the rapidly growing population of Kenya in terms of land use and multipurpose water uses.

Introduction

Kenya is basically an agricultural country and three-quarters of its 19 million population are engaged in agriculture. There have been large scale changes in agriculture in the last 75 years. In 1900 many of the presently cultivated areas were mainly haevily forested regions, but are now characterized by intensive cultivation which has led to overgrazing in the marginal areas. This has resulted in accelerated soil erosion rates in many of the agricultural and grazing areas and also high sedimentation rates. There has been overexploitation of land to the extent of cultivation. This has given rise to a number of well pronounced problems which include:

— Loss of the productivity of soil through erosion of the topsoil which contains most of the soil nutrients. This is critical because Kenya's population depends on the productivity of the soil for their daily livelihood.

— Rapid silting of reservoirs and dams which have been built by heavy capital investment. This reduces the economic life of reservoirs. In parts of Upper Tana River were reservoirs in Kamburu and Kindaruma have been constructed for purposes of generating electricity, sedimentation is much greater than was planned for. The third dam has been constructed at Gitaru and yet the same erroneous data have been used for planning against sedimentation problems in that reservoir. Heavy sedimentation rates may affect the biotic life in the reservoirs and interfere with the reservoir's ecosystem.

— Silting of river channels and other water carrying bodies thereby making them much more liable to flooding. This is presently happening in the Perkerra River and the recurring floods in the Kano plains could be attributed to the destruction of forest in parts of the Nandi Hills to give way for settlement and cultivation.

— Heavy deposition of silt at harbour sites that may have great economic effects on the country. The present sedimentary load of Sabaki River which is being deposited in Malindi has had remarkable effect not only on the development of Malindi as a tourist resort but also on the development and utilization of the Sabaki water for municipal and public water supply in parts of the Coast Province especially Mombasa.

Soil Erosion and Sedimentation

Sediment yield monitoring 1948—1965

Gauging and sediment monitoring stations are maintained by the Ministry of Water Development in Kenya. Between 1948—1965 the Ministry had established a network for monitoring sediment yields in the water areas in the country.

These catchment areas are distributed throughout a wide range of environment conditions in terms of geology, topography, climate, vegetation and land use types. The sediment yields from these catchments shown in Tab 1 for the period 1948—1965 are calculated from suspended sediment samples and stream discharge records. They show great variations of erosion between the regions of Kenya. The highest rates of soil loss are encountered in an area of very steep slopes on the eastern sides of Mount Kenya where cultivation is practised in the steep valley slopes in the upper part of the basin and cultivation and grazing are occurring in the gentler but drier hillslopes in the lower parts. On the other hand, in areas of undisturbed forest e.g. Sagana even within areas of steep slopes, soil erosion rates are extremely low. The rates of erosion increases under agricultural conditions and are much greater under grazing in semi-arid parts of the country.

Sediment production rates during the 70's

The rates of sedimentation at Tana river above Kindaruma, i.e. the present dam site at Kamburu, were estimated to be of the order of 300,000 tonnes per annum. Current investigations show, however, that the rate of sedimentation was greatly underestimated and is in fact of the order of 7,000,000 tonnes a year. This discrepancy is largely a result of using inadequate data in the previous computations. Land use has changed drastically over the last 10—15 years favouring high rates of erosion and sedimentation. Below the seven forks at Garissa the sediment load is of the order to 10,000,000 tonnes per year. This load is eventually deposited in the Indian Ocean at the Tana mouth.

Annual sediment production rates within the upper Tana catchment are to a large degree related to different types of land use. Data from period before 1978 indicates that sediment production increases steadily from the forested catchment to cultivated slopes attaining high values on the steeply cultivated slope and grazing areas (over 4000 t/km^2/year).

It has been estimated that close to 8 million tonnes of sediment are deposited at Malindi Bay at the mount of River Sabaki.

Tab 1 Soil erosion data from selected water catchment areas in Kenya

Drainage basin	Area (km²)	Total annual sediment production (tonnes)	Area rate of sediment loss (tonnes/km²/yr)	Land use type
Sagana, above Kiganjo	501	2 100	4.1	Forest, steep slopes
Sirimon, above Isiolo-Nanyuki Road	62	270	4.3	Forest, steep slopes
Sagana, above Sagawa	2 650	46 000	17	Forest/agriculture, steep slopes
Nzoia, above Broderick Falls	8 500	213 000	25	Forest/agriculture
Tana, above Kamburu Dam 4DE3	9 520	3 034 000	318.7	Agriculture/forestry
Ehania, above Thika	517	41 000	79	Agriculture/forestry
Thiba, above Machanga	1 970	156 000	80	Agriculture/forestry
Tributaries of Athi, Nairobi, Kiambu region, Nairobi	510	55 000	109	Agriculture/forestry
Kambure, Nzoia, below	1 350	171 000	126	Agriculture/forestry
Tana, between Grand Falls and Garissa	15 200	12 000 000	780	Grazing
Tana, between Kindaruma	7 700	12 000 000	1 550	Agriculture/grazing
Uaso Nyiro, above Archer's	15 300	12 000 000	780	Grazing

National assessment of soil losses and sedimentation rates

In 1969 nationwide studies of flow conditions and sediment yields in a large sample of the rivers of Kenya were commenced. The results provide a baseline against which to measure changes in soil erosion with changes in land use and climate. In 1974 a study was commenced to record the suspended sediment transport in the upper Tana river catchment above Kamburu. A proposal was also made for the development and implementation of a programme on experimental and representative basins with the following objectives:

— To analyse the hydrological consequences of further indigenous forest excision and its replacement by alternative land use.
— To examine critically the present agricultural practices in the high potential areas with regard to soil degradation and loss of sediment into streams and rivers.
— To investigate the water balance of semi-arid areas with particular reference to the estimate of actual evaporation and establishment of indices of available water.
— To evaluate the effects of soil and water conservation techniques in marginal areas in relation to their future agricultural development.

Such a programme would be located initially in three zones:

— The high potential forest boundary zone
— The medium potential rangelands and areas of marginal cultivation
— The arid and semi-arid frontiers of Kenya. These studies are still in progress.

There is need in Kenya and indeed in many developing countries for the assessment of national losses of soil and sedimentation rates in reservoirs. For this purpose there is need for quantitative studies of various types:

— Establishment of sediment yields monitoring stations in drainage basins so as to detect the intensity of erosion.
— Quantitative measurements of the spatial variation of accelerated erosion (both gully and sheetwash) within selected areas in the country.
— Analysis of the environmental conditions that control the magnitude and location of sheetwash and gully erosion including vegetation types and density, soil characteristics and grazing pressure. Specific attention has to be paid to the relative role of natural fluctuations and the influence of man.
— Analysis of the effects of accelerated erosion upon soil characteristics (e.g. supply of available nutrients, water holding capacity), the nutrient balance e.g. loss of nutrients by soil erosion, burning of vegetation and productivity of the land. In Kenya there has been a decline in the productivity, i.e. a reduction in carrying capacity of the life-support systems for livestock and people.

Problems for Hydropower Production

Sedimentation of reservoirs

Three dams have been built across the Tana river with the aim of generating electricity. These include Kindaruma, Kamburu and Gitaru. A "Super High Dam" at Masinga has been constructed above the present Kamburu reservoir. This super dam is largely intended to eliminate the current high seasonal fluctuations of the level of the water in Kamburu dam.

According to a survey in the Kindaruma reservoir by consultants for the East African Power and Lighting Company Limited, about $12,600 m^3$ of sediment has been deposited in the reservoir between 1968 and July 1977. Because Kamburu reservoir, upstream of the Kindaruma, with a trap efficiency of close to 100% was closed early in 1971, almost all the sediment in Kindaruma must have accumulated before the end of 1970. One study indicates that between July 1968 and December 1970 between 16.1 million m^3 and 31.3 million m^3 of sediment were deposited in Kindaruma. This rate of sedimentation would bring the generation of electricity at Kindaruma to a stop within a period of far less than 10 years. It will destroy the entire life of the lake, including fish, and create environmental problems such as flooding further upstream and cause many other development and social and health hazards.

The Kamburu reservoir has a total capacity of 150 million m^3 of which the useful capacity is 123 million m^3. If the present rate of sedimentation continues the originally assumed volume of 150 million m^3 will be reduced at a rate of about 4.2 to 5.1 million m^3 per year and should therefore be completely filled up within a period of 30 to 35 years. This calculation assumes the same rate of filling as in 1971–1976. If, however, the 1961–1976 weather conditions prevail, the reservoir will be filled completely within about 20 to 25 years after impoundment. This will mean a considerable reduction in storage and should bring to a halt hydropower generation at the Kamburu station.

An upper reservoir at Masinga about 40 km upstream of the present Kamburu reservoir has been constructed. Sedimentation at Kamburu has thus been reduced but the problem has been transferred to Masinga. Kamburu will however continue to receive sediment from its neighbourhood and from the Thiba arm of the catchment. By 1977, the original storage of Kamburu had been reduced by 25 to 30 million m^3.

Role of Tana River hydropower in national energy scheme

The hydro-electric production in Kenya is based on the Tana river at present. This has a total potential of 835 MW and other rivers are expected to contribute 230 MW as may be seen in Tab 2.

The next stage of development of the Tana is the Upper Reservoir Scheme which has been completed providing a further 40 MW to the national grid at the cost of K£ 50 million. This will be followed by Kiambere which is expected to be finished by 1985 at a cost of K£ 60 million and with an output of 120 MW.

	Existing	Average actual output	Planned potential	Total exploitable potential
Tana river				
Wanjii	22	20		
Kindaruma	44	20		
Kamburu	84	51		
Gitaru	145	93		
Upper reservoir			40	
Kiambere			120	
Mutonga			70	
Grand Falls			80	
Karura			40	
Adamsons Falls			50	
Koreh			80	
Usueni			60	
Total Tana	295	184	540	835
Other rivers				
Nzoia, Broderick Falls			10	
Yala, Kimundi			40	
Confluence Soudu, Soudu			60	
Arror, Kapsowar			20	
Turkwell, Turkwell Gorge			100	
Total other			230	230

Source: Development Plan 1979–1983

Tab 2 Hydroelectric potential in Kenya in MW

The Upper Reservoir Scheme has a surface area of 120 km² and is the biggest man-made lake in East Africa. It has the important function of stabilising the water supply during dry seasons and enable an additional 55 MW to 60 MW to be obtained from the downstream stations.

Need for Sediment Measurement and Evaluation

There is a need for an evaluation of the economic impact of sedimentation in all reservoirs built or planned for in the upper Tana catchment taking into account environmental implications.

Before constructing major or minor reservoirs a thorough evaluation of sedimentation problems must be assessed. Such an evaluation should not be based upon a few suspended sediment measurements made by consultants who often have little experience of field conditions in the area. It should rather be based upon a thorough and extensive investigation of hydrologic and geomorphic processes as they relate to present and probable future land use changes. Sedimentation problems cannot be adequately assessed by looking in the river channel at the site of the planned reservoir alone.

Over the last 20 years sediment transport rates in the major tributaries of the upper Tana catchment have fluctuated dramatically. During this time land use has changed, land management practices have equally changed and there have been considerable fluctuations in the weather. The resulting change in sediment yields make it difficult to predict future sedimentation rates exactly. In this paper an attempt has been made to assess the correct order of magnitude of the problem.

Considering the economic and environmental importance of these reservoirs, therefore, the network of sediment monitoring stations in the upper Tana catchment should be revised and data should be collected on a more regular and uniform basis as was the case before 1968, when sediment monitoring in the country essentially stopped. The Tana River Development Authority has accepted the need for such a programme which should be supported by all the relevant Government Departments including the National Environment Secretariat.

Different Sources of Sediment

Mathioya and the neighbouring catchments draining the Aberdares are major sources of sediment within the cultivated areas in the upper Tana catchment. Rural roads, paths and other human settlement features are important sources of sediment including the grazing areas. These are areas where concentrated soil and water conservation programmes should be vigorously promoted.

There is a need to study the distribution and nature of sediment sources within the catchments as they relate to the geomorphic and hydrologic processes. A hydrographic survey of the Kamburu reservoir should be carried out as soon as possible and should be repeated at 5 to 10 years intervals.

These processes in turn must be studied to define quantitatively their controlling factors such as rainfall, soil type, topography, vegetation cover and land use. Only when the major sediment sources and erosive processes are identified can rational decisions be made about the kinds of soil conservation measures that will be effective. The kinds of soil conservation measures needed to minimize erosion from e.g. roads are very different from those required for cultivated fields. Some attention needs to be given to the management and construction of rural roads in order to develop technology for minimizing soil loss from roads and construction sites, because it is a separate problem.

Education and Training Vital

The problem is one of designing systems of soil and water management that are effective, under subsistance and growing commercial agriculture and grazing where capital and technical expertise and public awareness are often scarce. The development of water resources in the upper Tana catchment could provide the impetus for a vigorous programme of public education and demonstration of the benefits of soil and water conservation for individual farmers and the nation at large. The significance and urgency of erosion and sedimentation problems, pose a big challenge to environmental planners in the country. This calls for the need for training and support of a large number of Kenyans in all aspects of water and soil management in related fields of earth sciences. The Ministries of Agriculture, Water Development, Environment and Natural Resources, the Office of the President as well as other private and Government agencies involved in soil and water conservation should coordinate their efforts to fight the menace of soil erosion.

Lundqvist, J.; Lohm, U. and Falkenmark, M. (eds.):
Strategies for River Basin Management, pp. 131–140
© *1985 D. Reidel Publishing Company*

2.2 Whole river as an interactive system

Upstream — Downstream Interactions as Natural Constraints to Basin-Wide Planning for China's River Huang

Mosely, P., 36, Lee Grove, Chigwell, Essex IG7 6AF, England

Abstract: This paper describes the intricate nature of the upstream-downstrem interrelations amongst the river basin problems of the River Huang. The basin is characterised by heavy silt production in the middle part — severely accelerated by far-reaching adherance to Mao's strategy "Make grain the key link" — and flood control in the lower reaches, throughout history plagued by inundations and flood breaches. The steepness of the upper reaches make them the natural centre for hydropower potential and industry. The present land use of the loess plateau and fertile plain drained by the middle reaches, is characterised by integrated soil and water conservation, withdrawal of agriculture from hilltops, and silt traps in the valley bottoms to produce new fertile land. The overriding consideration in the lower reaches is flood control and control of sedimentation to stop the river bed from rising. This creates contradictions with other land uses, as exemplified from the Beijin flood detention area, not in use for that purpose during the past 30 years.

General Conditions of River Huang Basin

Flowing 5460 km east to west through eight provinces of northern China (Fig 1), the Huang is that country's second largest river. It rates moderately amongst the large rivers of the world in terms of drainage area and, in particular, runoff, but holds first place in terms of sediment, both with respect to annual load and average concentration (Tab 1). The whole basin lies in an arid or semi-arid region and, having a continental climate, is subjected to extreme seasonal variations of temperature and precipitation, as well as uneven distribution. Annual temperature variation for example is about 30 °C and precipitation varies from about 150 mm in Inner Mongolia to 750 mm in SW Hebei. Of the total precipitation, 70 % falls in June, July and August, and less than 5 % in December, January and February. Furthermore, in the violent summer storms which frequent the middle reaches, as much as 160 mm can fall in a single day. This paper illustrates how the effects of some of the problems of this river extend over large distances, thereby linking together what may at first seem to be regions of quite dissimilar characteristics.

Upper reaches — centre for hydropower generation

The upper reaches are taken to be the 3470 km stretch above Hekouzhen in Inner Mongolia, where the river drops 3840 m and drains about 400,000 km². Of this, between Longyang gorge in Qinghai and Qingtong gorge in Ningxia, the river drops 1300 m through a series of twenty gorges which account for half the estimated 26.5 GW power generation potential of the whole river. At the same time, this would-be centre of industry is located between the inhospitable highlands of Qinghai, the poor desert scrubland of Ningxia and the loess deposits of Gansu.

Four power stations have already been completed in these gorges, with a combined installed capacity of 1.029 GW. In the Longyang gorge another project is under construction, which will have an installed capacity of 1.500 GW. Only the Qingtong gorge project has multipurpose facilities, being the reservoir and headgate for the Ningxia plain irrigation schemes.

Tab 1 Heavily silted large rivers of the world compiled from different references.

River	Country	Drainage Area $(km^2 \times 10^3)$	Runoff $(m^3 yr^{-1} \times 10^9)$	Silt Load $(t\,yr^{-1} \times 10^6)$	Average Silt Concentration $(kg\,m^{-3})$
Huang	China	752	43	1640	37.60
Colorado	USA	637	5	135	27.50
Liao	China	166	6	41	6.86
Ganges	India, Bangladesh	955	371	1451	3.92
Missouri	USA	1370	616	218	3.54
Indus	Pakistan	969	175	435	2.49
Brahmaputra	India, Bangladesh	666	384	726	1.89
Nile	Egypt, Sudan	2978	89	111	1.25
Red	Vietnam	119	123	130	1.06
Irrawaddy	Burma	430	427	299	0.70
Mississippi	USA	3220	561	312	0.56
Changjiang	China	1807	921	478	0.54
Mekong	Laos, Thailand, Kampuchea, Vietnam	795	348	170	0.49
Xijiang	China	355	253	69	0.35
Don	USSR	422	28	6	0.23
Niger	Nigeria	1112	180	40	0.22
Danube	Germany, Austria, Czechoslovakia, Hungary, Yugoslavia, Bulgaria, Romania	1165	200	28	0.14
Volga	USSR	1500	250	25	0.11
Amazon	Brasil	5770	5710	363	0.06
Congo	Zaire	3700	1400	70	0.05
Rhine	Germany, Netherlands, France, Switzerland	362	800	3	0.04

Sources: Qian & Dai 1980, Shaanxi Research Institute 1979.

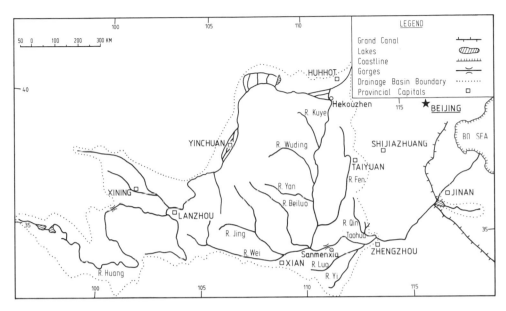

Fig 1 Major tributaries and important places

Middle reaches — loess plateau and fertile plain

The 1210 km stretch between Hekouzhen and Taohuayu, near Zehngzhou in He'nan, comprises River Huang's middle reach. This section drains the 280,000 km² loess plateau where heavy summer storms over denuded and dessicated loess terrain give rise to the silt laden flood peaks which are so troublesome to the lower reaches. Here, the silt and water flows increase from 100×10^6 t in 20×10^9 m³ to 1.6×10^9 t in 40×10^9 m³, respectively, sediment content closely following the seasonal variations in rainfall so that from July to October, 58.5 % of the annual discharge carries 83.8 % of the annual sediment load.

Over the region as a whole, an average of 3700 t km⁻² of loess is lost annually, in places over 25,000 t km⁻². This more severely eroded part comprises 247,000 km² of N Shaanxi and NW Shanxi, known as the 'gullied hillock region', where gully and sheet erosion have been so serious as to render the topography criss-crossed by gullies and cut up into an array of small round hills. The loess plateau over the years has become one of the poorest regions of China.

To the south of the loess plateau, in Shaanxi, drained by Huang's largest tributary, River Wei, is the Guanzhong plain. It is this fertile plain, 52,990 km² in area, centre of Chinese civilisation in early historical times and going back to before the denudation to the north was perpetrated, which is considered the most important agricultural part of the region. Whilst occupying only 28 % of Shaanxi's area, it contains 53 % of its arable land and produces 70 % and 90 % of its cereals and cotton respectively. Yet, owing to the poor communications occasioned by the disected terrain, the Guanzhong plain should not be supposed to support the loess plateau, which must, in the main, be self-sufficient. (Liu 1964; Academia Sinica 1958).

Lower reaches — fertile plain plagued by incessant flood breaches

The Huang, in its lower reaches, more than 700 km from Taohuayu to the mouth, drops by only 95 m. The loess bluffs of the middle reaches give way to the expansive, 250,000 km^2 and extraordinarily flat North China Plain which, itself, was deposited in the many wanderings of the Huang over as many millennia (Fig 2). For all but about 100 km in Shandong, its present course is constrained on both sides by a total of 1340 km of main flood

Fig 2 Historical courses of the Lower River Huang.
Sources: River Huang Water Conservancy Commission 1979a, Shanghai Shifan University 1974

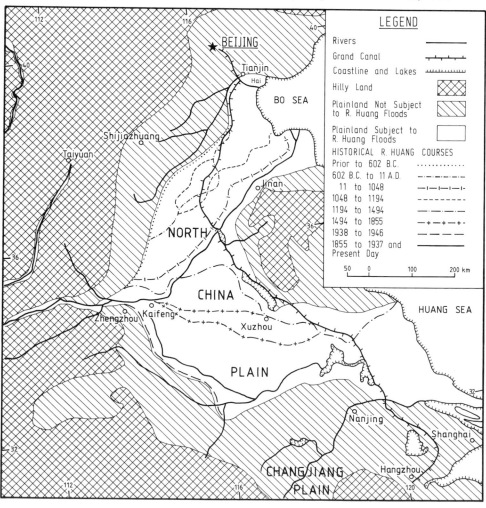

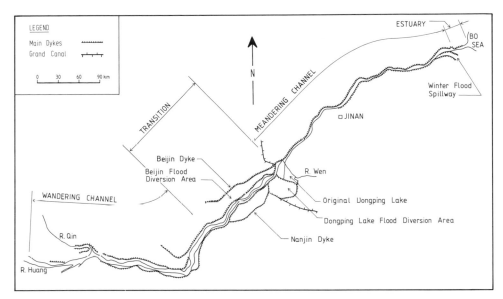

Fig 3 Some features of the Lower River Huang
Sources: River Huang Water Conservancy Commission et al. 1979b

protection dykes (Fig 3), such that 25 % of its annual sediment load is deposited within the dykes and the bed level has risen to about 4 m (10 m in places) above the surrounding land. The lower reaches are wide in their upper part, about 25 km between the dykes, and narrow lower down, about 5 km. The main channel within the dykes is about 3 km wide. The bed is at present continuing to rise at about 0.7 m yr^{-1}.

The terrain on the outside of the dykes slopes gently away at a gradient of 0.02 %. On this plain lives one tenth of the population and is to be found 15.8 % of the cultivated land of China. It is of the utmost importance therefore to prevent any repetition of the breaches which occurred in two out of three years of recorded history before the revolution and for which the river is notorious worldwide. This plain also suffers a shortage of water, compounded by a high but brackish water table.

Early efforts in basin wide approach

A basinwide approach to these problems was first advocated in 1934 and is claimed still to be the basis for the work of the River Huang Water Conservancy Commission (RHWCC), the coordinating body for all work on the river. It was formed in 1933 and under it a comprehensive data collection organisation has developed. By 1981 in the Huang basin there were 465 hydrological gauging stations and 2211 rainfall stations.

In May 1946, the RHWCC completed its proposals for flood control in the lower reaches and in the next few years solicited the opinion, in the form of other reports, of some American engineers, on the development of the whole basin. In the early fifties, the opinion of some advisors from the Soviet Union was sought and their grandiose, "Multiple

Purpose Plan for Permanently Controlling the River Huang and Exploiting Its Resources",
was officially adopted in July 1955.

It is not the intention here to expound at any length upon the reasons for, or the
consequences of, the ill-advised adoption of so disastrously inappropriate a stratagem as
the keystone for the first stage of the plan, the Sanmenxia reservoir, turned out to be
(Mosely 1982). Suffice to say that construction of a high-head reservoir 150 km upstream
of Taohuayu for the storage of water and silt and the generation of electrical power, was
so much of an error, that summer water storage was obliged to be abandoned in March
1962, after only 18 months' use, and a costly programme of reconstruction, which took
until 1974 to complete and is still considered less than ideal a solution, had to be begun.
Nevertheless this experience, or lesson, provided much useful knowledge about the fluvial
and alluvial processes of the river which, together with knowledge from more conventional
study, enabled a picture to be drawn fairly completely of the intricate nature of the interrelations amongst the problems of various portions of this most sorrowful river.

Crucial interactions between transport of water and sediments

Of crucial importance to flood control along the lower reaches are the water and sediment
discharges and sedimentary composition. The desire is for as much as possible of the
oncoming silt to be passed to the sea and to ensure that where deposition is unavoidable,
it takes place on the flood plains, between the main channel and the dykes, and not in the
main channel itself. If scouring of the main channel is also possible, so much the better.

Deposition is caused mainly by particles coarser than 0.05 mm, of which between
50% and 60% are deposited. 87.6% of fine particles, on the other hand, which enter the
lower reaches, are discharged to the sea. Under low flow conditions, deposition will occur
in the main channel. Even in winter, when the water is clear, the effect will be to move
silt from the wide He'nan stretch to the narrow Shandong stretch. Moderate and large
floodpeaks from about 4000 $m^3 s^{-1}$ to the safe limit for the lower reaches (22,000 $m^3 s^{-1}$
at Taohuayu) generally scour the main channel in the Shandong stretch and often in the
upper parts as well, whilst those which cover the floodplains (over 6000 $m^3 s^{-1}$ at Taohuayu) cause beneficial deposition thereupon (Gong & Xiong 1980, Qian 1980).

Fig 4 illustrates the sources of sediment of various sizes. It can be seen that the areas
of highest yield are also the source of the coarser grains. But, the yield of any particular
part of the loess plateau on any particular day depends upon that day's rainfall pattern.

Thus, storms over the catchments of the tributary rivers Wuding and Beiluo alone,
will give rise to sharp floodpeaks containing coarse grained silt. In the lower reaches,
because of channel storage, these floodpeaks are attenuated such that the floodplains are
not covered, and deposition from these floods account for 60% of all deposition. Con-
versely, rainfall over the fine sediment areas, usually leads to scouring of the lower reaches,
but it is slight and exerts only a restraining effect on the general trend. Simultaneous
large floods from all areas, despite their high sediment content and consequent deposition
amounting to 28.2% of the total, are actually the most beneficial to the lower reaches,
because they cover the floodplains, depositing their silt thereupon, whilst at the same
time scouring the bed of the main channel.

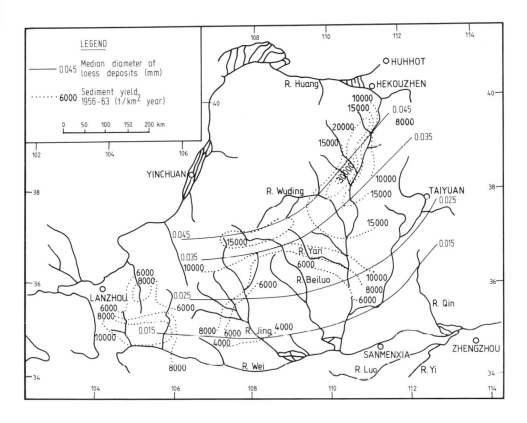

Fig 4 Major tributary and mainstream stations in the runoff and sediment gauging network for the whole River Huang basin
Source: Long & Xiiong 1981

Implications in Developmental Planning

In the gorges of the *upper reaches*, the relatively clear water has nonetheless necessitated special design features for dealing with silt on all of the structures mentioned. Yet, the principal contradiction must be the unfortunate location of this concentrated potential source of hydroelectric power in a region whose difficult communications would not otherwise be conducive to the siting of an industrial base area.

There are consequences for other parts of the basin, too, of this emphasis on industry. Lanzhou, for example, Capital of Gansu, situated in the middle of this stretch, is responsible alone for 22.5% of Huang pollution (Wang 1980). Another consideration is that, should the lower reaches request the release of clear water for irrigation, or for the creation of artificial flood peaks of clear water for bed regulation, due regard must be given to the need for fairly constant parameters (such as reservoir water level) for power generation, and to the delicate balance which obtains amongst these various reservoirs for control of their own sedimentation problems.

In the *middle reaches*, an obviously intimate relation exists between the local desire to improve production by soil and water conservation and the lower reaches' desire to be relieved of their burden of sediment. Whilst these desires at first seem united, the method of their realisation must be considered. Engineering structures on the mainstream must take into account their effects on the land upstream, even beyond the limits of the reservoir. The type of land use and the points of emphasis for soil and water conservation in the middle reaches must consider their role in flood prevention downstream, which arises from their effect on the riverbed morphology there.

What has been described with regard to channel morphology in the lower Huang, applies in similar fashion to any reservoir built therein. When first commissioned, the Sanmenxia reservoir, for example, quickly filled up with sediment, which very rapidly extended upstream, over 110km beyond the high water level in 18 months. This endangered the city of Xi'an and, by raising the water table and intensifying alkalisation of the soil, spoiled some 330 km^2 of the very fertile and agriculturally regionally important Guanzhong plain. Since the reconstructions, the sediment of the reservoir has formed itself into floodplains and main channel, rather as the lower reaches. In the latter, floodplain deposition is desirable. In the reservoir it represents a permanent loss of capacity. Channel deposition however is recoverable, so every effort is made to avoid wetting the floodplains except in winter. A very tricky balance exists between flood control upstream and downstream of the Sanmenxia reservoir, and all other of its uses must be subordinated to that requirement.

During the Cultural Revolution, a lopsided interpretation of Comrade Mao Zedong's call to "Make Grain the Key Link" and other such factors, led to agricultural production being forced upon pastural and silvicultural regions, with disastrous effects for the development of the loess plateau, and no improvement being seen towards the reduction of the sediment of the Huang. Much debate ensued, in which a backlash was felt in that it was advocated by some that agriculture be replaced by silviculture and husbandry. Yet some areas indeed are best suited to agriculture, whilst others are better suited to husbandry and silviculture (Tong 1978).

As regards the effect of these policies on the lower Huang, it has been shown that control of the high and coarse-grained sediment yield area would have the most marked beneficial effect. For this reason it is advocated that the emphasis of soil and water conservation be placed there. This is a relatively small area, 110,000 km², so the suggestion is not unreasonable. It is also an area suited to the development of forestry and husbandry but, because of the extreme topography and the need for food to be grown locally, some agriculture must be maintained, especially during the transitional period.

Many varied solutions to this problem have been found and are known collectively as 'comprehensive development'. In short, valley floors are dammed or straightened, and the land formed by subsequent sedimentation provides high and stable agricultural yields which enable the withdrawal of agriculture from the hilltops, where it is replaced by forestry or husbandry. These, in turn, protect the dams below from torrents which otherwise would accompany storms. In the initial period, when this protection is as yet undeveloped, it is best to build the dams in the upper gully tributaries, where the torrents are less severe. These schemes also provide for greater diversity in the local economy, having timber, fruit, animal and agricultural produce to offer. Thus, in the middle reaches, economic development and river control are inextricably bound by the problem of sediment.

In the *lower reaches*, the overriding consideration is flood control and, as has been shown, some requests in this respect are made of the upper and middle reaches. Even so, it will be many years before any noticeable reduction in the coarse sediment load of the lower reaches occurs. All planning here must take this major factor into account. On the other hand, the use of the river's silts, selectively chosen for grain size by careful choice of the type and part of the flood from which they are drawn, for engineering or agriculture, goes some way towards its mitigation. Power generation and, to a lesser extent, irrigation, are constrained somewhat by the requirements of flood control, which themselves are different in summer and winter.

Should a very large floodpeak occur, the designated Beijin flood detention are could be opened, but losses arising from such use have been estimated to be about Y 570 × 10⁶ (£ 200 × 10⁶). There is considerable opposition to keeping this option open. For the past thirty years this area has never had to be used and agricultural activites have continued and houses been built. Each year preparations have to be made in case it is necessary for it to be flooded, and stockpiles of materials have to be purchased.

The contradiction between flood control and other uses of land in the lower Huang extends even onto the river floodplains, where unofficial tillage takes place and the peasants build inner sets of dykes, called 'production dykes', which seriously endanger flood control work. However, despite encouragements to remove these people some 2000 km² of the floodplains are tilled by a population which still exceeds one million.

Conclusion

It can be seen that in the Huang basin, planning is made more complicated by local and distant interrelations amongst its natural attributes as well as the usual straightforward conflicts of human interest in the use of its various resources. How best to tackle such a problem is well worthy of discussion.

References

Academia Sinica: Delineation of the Natural, Agricultural, Economic, Soil and Water Conservational, Rational Land Use Districts of the Loess Plateau of the Middle River Huang. Science Press, Beijing 1958.

Gong Shiyang; Xiong Guishu: The Origin and Transport of Sediment of the River Huang. In: Chinese Society of Hydraulic Engineering (Eds.), Proceedings of the International Symposium on River Sedimentation. Guanghua Press, Beijing 1980.

Liu Dongsheng et al.: Loess of the River Huang Middle Reaches (Huanghe Zhongyou Huangtu). Science Press, Beijing 1964.

Long Yuqian; Xiong Guishu: River Huang Sedimentation Surveying (Huanghe Nisha Ceyan). Renmin Huanghe 1, 16—20 (1981)

Mosely, P.: Regulation of the River Huang — a Study of the Chinese Concept of River Basin Development. Ph.D. thesis, Edinburgh University 1982.

Qian Ning; Dai Dingzhong: The Problems of River Sedimentation and the Present Status of its Research in China. In: Chinese Society of Hydraulic Engineering (Eds.), Proceedings of the International Symposium on River Sedimentation. Guanghua Press, Beijing 1980.

Qian Ning et al.: The Source of Coarse Sediment in the Middle Reaches of the River Huang and its Effect on the Siltation of Lower Huang. In: Chinese Society of Hydraulic Engineering (Eds.), Proceedings of the International Symposium on River Sedimentation. Guanghua Press, Beijing 1980.

River Huang Water Conservancy Commission et al.: The Influence of Water and Silt from Different Regions of the Huang Basin, on Deposition and Scouring in the Lower Reaches. In: River Huang Sediment Research Coordination Group (Eds.), Selected Reports on River Huang Sediment Research. (1975, unpublished).

River Huang Sediment Research Coordinating Group: Selected Reports on River Huang Sediment Research (Huanghe Nisha Yanjiu Baogao Xuanbian). Internal Documents 1975.

River Huang Water Conservancy Commission: Ten Thousand Li Course of the River Huang (Huanghe Wanli Xing). Shanghai Education Press, Shanghai 1979a.

River Huang Water Conservancy Commission et al.: Ice Floods of the Lower Huang (Huanghe Xiayou Lingxun). Science Press, Beijing 1979b.

Shaanxi Research Institute of Hydraulic Engineering Waterways Lab. Reservoir Sedimentation (Shuiku Nisha). Water Conservancy and Electric Power Publishing House, Beijing 1979.

Shanghai Shifan University: A Concise Geography of China (Jianming Zhongguo Dili). Shanghai People's Press, Shanghai 1974.

Tong Dalin et al.: On the Construction Policy Question of the Northwest Loess Plateau. Renmin Ribao p. 2, (26. Nov. 1978)

Wang Zhimin: Appraisal and Present Situation of Water Pollution in the River Huang. Renmin Huanghe 5, 63—67 (1980)

Lundqvist, J.; Lohm, U. and Falkenmark, M. (eds.):
Strategies for River Basin Management, pp. 141–150
© *1985 D. Reidel Publishing Company*

Some Strategic Principles for Long-Term River Basin Development —
The Case of Han River

Fang Ziyun, Research Institute for Protection of the Yangtze River against Pollution,
Yangtze Valley Planning Office, Wuhan, China

Abstract: When formulating a river basin strategy economic benefits, social benefits and environmental
benefits have to be considered in combination. The selected plan for the development of a river basin
should pay reasonable attention to the benefits both for the parts concerned and for the whole area,
both for the present time and for the future, and to the consistency of productivity of nature and
human being.
   Basic strategic principles for long term river basin development as illustrated from the Han River
would include mitigation of disasters and damages, evaluation of water and land resources capabilities,
assessment of the environmental impacts of actions, protection and multiple use of aquatic environments, control and protection of land resources, and preparedness for future changes. Thus, the river
basin strategy must be formed in combining water resources planning, land use planning and environmental planning of the basin into an integral whole.

River Basin Strategy Formation

River basin is the sum of many kinds of resources. It is an ecosystem composed of fauna
and flora of the basin and their living circumstances. In this ecosystem human being
is the master. The planning of river basin development is in essence a system planning
problem. The final selection of the alternatives of the plan must be based on the integral
view of the whole basin as well as the economic view of the alternatives. That means the
economic benefits, the social benefits and the environmental benefits as a whole must be
all considered in plan selection. The plan selected must adhere to the following principles
as close as possible, namely:

a)   beneficial to the parts concerned, advantageous to the whole area and at least through
     mitigation, harmless to others;

b)   advantageous to the present, beneficial to the coming period and at least harmless in
     the future;

c)   helpful in improving the living standards of human being and harmless to the equilib-
     rium of ecosystems of the basin.

So from the point of view of river basin strategy, the resources must be multiply utilized
and the river basin must be comprehensively regulated by both engineering measures and
biological measures. As to the arrangement of each component, of the plan, the engineer-

ing project on every individual point, the cascades of the whole river and the engineering measures or the biological measures of whole river basin must be considered integrally. In addition all the upper, the mid and the lower reaches of the river must be treated in the same way. For the purpose of selecting the optimum plan we must utilize the theorem of system engineering and the mathematical method of optimization, with max. multi-purpose benefits of the whole basin and sometimes even its neighbourhood as its objective.

The essence of the optimum plan is that due attention must be paid to the equilibrium of the ecosystem within the river basin, and the movements of water and soil of the whole basin must be controlled coordinately in order to maintain their high and stable long-term productivity, at the same time the high quality of the environment of the basin should be secured. For fulfilling the plan we must reduce the disasters and damages of floods, water-logging, drought and salinity (saline); we must develop and utilize the resources scientifically and comprehensively; we must prevent soil from erosion and protect the environment against pollution so as to maintain the natural processes and the life-supporting systems of human being such as agriculture and forestry, etc.; conserve the varieties of species and keep human being to be available to secure long-term use of species and ecosystems, especially fish and other wildings.

In conclusion, the purposes for the formation of river basin strategy may be briefly stated as follows:

(1) mitigating natural disasters and damages and promoting multi-purposes benefits, comprehensive and rational utilization of resources so as to secure the stable and beautiful environment for human living and production;

(2) maintaining the harmony and consistancy of productivity of nature and human being.

A Discussion on Strategic Principles of River Basin Development

The determination of the objectives and purposes of a river basin strategy is based upon the plan of national economic, social and environmental development with full consideration of the condition and characteristics of the basin itself and the aspects of demand and possibility. The aim of the river basin strategy is mainly to set out the present and long run direction and plan of water resources development and land utilization of the basin. Its concrete principles — as illustrated from the Han river — may be briefed in the following topics:

Mitigation of disasters and damages

Most rivers of China are fed by storm rains. The disasters and damages of storm rain flood in our country are both frequent and serious and sometimes even catastrophic. Han River (Fig 1) is one of the main tributaries in the middle reach of the Yangtze River. It is the largest river emptying into the Yangtze in Hubei Province with a total length of 1542 km and a basin area of 174,000 km^2. The mean yearly precipitation of the basin ranges from

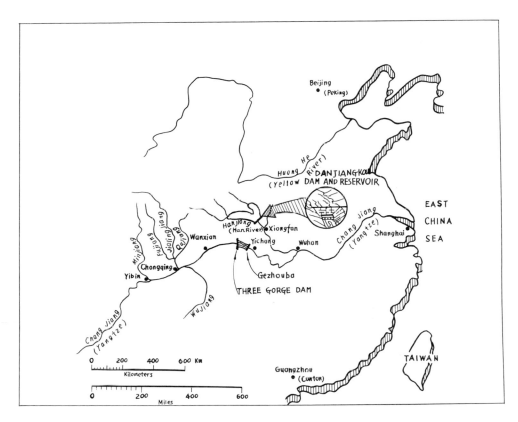

Fig 1 Map of the Yangtze river system.

700 to 1000 mm, while the atmospheric temperature ranges 15–17 °C. Within the basin there are 27.5 million of inhabitants and 2.8 million ha of cultivated lands.

Catastrophic floods occurred frequently in the Han River before the year 1949. For nine years out of ten the farmers in the middle and lower reaches of the river could not gather any harvest from the land. Poverty shaded the whole basin. But after liberation the threat of floods has been greatly diminished. Based on demand and financial possibility we have raised the standard of flood control by the following steps. After we had heightened and reinforced the dykes along the river banks, the Dujiatai flood diversion sluice was built in early 1950's so as to balance the carrying capacity of the river channel, because it was smaller below Dujiatai.

At the same time we engaged in the comprehensive planning of the whole basin with flood control as its main purpose. Hence we prepared the plan of projects for development of the whole basin and selected Danjiangkou Project as its key one and the first to be built. As for flood control, from the view of systems we chose the optimum combinations of flood control works. The scheme finally adopted was to regulate the flood discharge mainly through Danjiangkou Reservoir, divert a part of flood flow through the Dujiatai flood diversion sluice and heighten (reinforce) the dykes below Xincheng to increase the carrying capacity of the channel (Fig 2). With these measures we could guard safely against a flood of 1 % frequency, i.e. a similar flood to that of 1935.

Evaluation of water and land resources capabilities

The overall purpose of water and land resources development is to meet the needs of diverse interests to promote the quality of life. For a river basin it is essential to investigate the amount and the time and space distribution of water resources and the distribution of lands as well. According to their distributions and the conditions of the preliminary balance of water demands and supplies, we should:

Fig 2 Flood control system of the Han river basin.

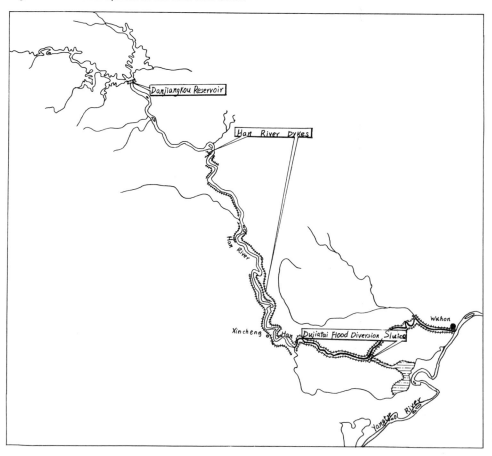

First, make a proper preliminary arrangement of land uses of the basin. Second, allocate priliminary the amount of water resources for each sector and each region, giving the highest priority to the utilization of local water resources. Third, arrange the alternative plans for development of the river with comprehensive utilization of water resources, with a good combination of developing their benefits and mitigating disasters.

For planning, attention must be paid to siting of water structures and the allocation of amount of water and the storage capacity of reservoirs for each purpose and region during development and to other utilization of water resources. The method for solving contradictions may be classified as follows:

— Rationally arranging water demand for each sector and each region. Economic analysis and allocation in accordance with the order of importance is the principle. The main purpose should have priority but the following ones must also be concerned properly. The order of importance of purposes would change as the economic development proceeds. As water power is a clean and cheap energy resource that does not consume water, the coordinate use of water and head to develop water power is a main objective for river basin strategy.

— If the water demand for various sectors and regions are contradictory and cannot be satisfied by certain plan, we must try to investigate to enlarge the magnitude of certain reservoir or a system of reservoirs and other water works with financial feasibility. It must be emphasized that comprehensive exploitation does not mean to satisfy all sectors or regions: the economic factor is of vital importance.

— Building the auxiliary reservoirs or other water works to smooth the contradictions among sectors or regions. For example, use the downstream reservoir to decrease the water level fluctuation due to sudden changes of peak flow of a power station upstreams in order to facilitate navigation.

— Investigating the rationalizations of shifting the location of outlets and the optimum plan of regional water supply. For example, whether it is beneficial to move the outlets of irrigation from the reservoir to the river reach below. After utilizing the water to produce power it could then be used for irrigation by pumping.

— Considering all the hydraulic engineering works and facilities of the whole basin as a system, raising their effects by operating and regulating integrally.

— Exploring alternative sources of water for an arid region, such as water transfer from a neighbouring basin, rational development of ground water or reuse of treated waste water, when economically reasonable.

Assessing the environmental impacts of actions

As mentioned before, we must consider the river basin as an ecosystem to enable human being and nature to develop harmonically and consistently, and to enable human being to utilize the productivity of water and soil resources rationally and permanently without degradation and degeneration of environmental quality. With the above purpose in mind we should take care of our actions.

For example, in dealing with water demand and supply we ought to evaluate the water demand not only for social activity of human being but also for the safety of nature; for land reclamation we cannot consider the harvest locally and recently without caring of loss of soil due to erosion and the sedimentation of river beds and reservoirs in the lower reaches; during exploitation of forest resources we cannot only deforest without reforestation; during catching the biological resources we cannot endeavor to obtain the high catch this year by drying the aquatic environments and emptying all the fingerlings; for drought region we should pay attention not only to irrigation but also to drainage to avoid the raising of ground water level causing decrease of products or salinity, etc. Now in many river basins the improper reclamation of lakes and deforestation are taking place on such a massive scale that their effects on water resources and the equilibrium of ecosystems are decissive and serious.

Thus integrated planning should include assessing the environmental impacts due to actions. In China, environmental impact statements have now to be prepared not merely for all large and medium projects exerting impacts to natural environment, social environment and equilibrium of ecosystems. Such assessments have been made for a number of important water resources projects such as the Gezhouba Project and the Three Gorge Project of the Yangtze River. In addition, we are now undertaking the overall environmental impact assessment and the planning of water resources protection against pollution of the Yangtze River Basin and likewise for the scheme of transferring water of the Yangtze River Basin and likewise for the scheme of transferring water of the Yangtze River system through the Danjiangkou Reservoir to the north China (Middle Route).

Protecting and exploiting multiple use of aquatic environments

Many aquatic environments have been polluted so heavily that their multi-utilities vanish. This is a common and serious problem. For protecting well the aquatic environment, both the environmental quality standard and the local water pollutant emission standard must be established and the assimilative capacity as well as the allowable effluent capacity of the aquatic environment must be determined.

Taking a river as an example, the procedure is first to select the environmental quality standard of water of the reach according to the purposes of utilization and then to evaluate the assimilative capacity of the reach for the designed hydrological and meteorological conditions; to calculate the allowable loading on the environmental quality standard of water selected and the designed conditions of the river; to sum up the assimilative capacity and the allowable loading of the outlet to dertermine the allowable effluent capacity of the reach; and to compare whether the total loading of pollutants of all sources of the reach or its allowable effluent capacity is larger. If the former is larger than the latter, the optimum allocation of allowable load of each source must be done by system engineering through certain economic criterions. Finally the local water pollutant emission standard of each source should be determined from the previous investigations.

The policy and measures for aquatic protection are complex and of comprehensive character. Its comprehensive character could be manifested into five aspects:

First, restricting waste water effluent to reduce the loading of pollutants and exploiding clear water source to increase the assimilative capacity of aquatic environment. The former is most important, especially in arid area where clear water is scarce.

Second, reducing mainly the industrial loading of pollutants during manufacturing processes; then turning the pollutants into resources by comprehensive utilization or dealing the waste water by treatment. Waste water treatment is needed not only for the protection of aquatic environment against pollution, but also for supplying the amount of water to be further used in arid area.

Third, applying the complex remedied measures such as by biological measures and engineering measures as well. Emphasizing to consider the hydraulic engineering planning and the water pollution control planning of the river basin as an interconnected system. The water pollution control plan would be strengthened sometimes by hydraulic engineering projects and as a result a lot of money could be saved.

Fourth, designing the suitable remedied or controlled measures for each river basin or each aquatic environment. Applying the system engineering to arrange all the sub-systems or the components of the plan in order to obtain the biggest benefits with the least cost.

Fifth, inhibition to transfer pollutants to others. Dealing harmonically the interrelationships between stem and tributary, upper reach and lower one, and aquatic environments each other. Solving the pollution problem thoroughly and utilizing the assimilative capacity of water rationally are very important.

Controlling and protecting land resources

Land is a valuable renewable resource, which has to be protected against degeneration, desertification and salinization. Use of land for construction should be carful and limited, for hydraulic engineering projects to decrease the inundation of cultivated lands is necessary, especially in the densely populated regions. The land use patterns of the whole river basin should be arranged reasonably and integrally as they profoundly affect the economic development, social development and the environmental quality of the basin. The interaction between land use and water quality is one of the most important ecological challenges of our time.

Many water pollution sources are land use-specific, particularly nonpoint sources. Land used for factories or agriculture causes point or non-point sources of pollution of water, respectively. Land use for urbanization, the area of impervious pavements increase rapidly so that it enlarges the volume of runoff and the magnitude of the flood peak and shortens the time of concentration of floods. Moreover urbanization makes the waste water discharge bigger. On the other hand, the field ploughing process and the changing of crops affect both the quality and the amount of water yield, too. So water quality problems are closely linked to misuse of land resources. Selective changes in land use may meet its quality objectives.

Also, in order to maintain the land to render the high and long term productivity, we must reduce all kinds of natural disasters by hydraulic engineering systems of the basin. In return, in order to develop the durable and high renewable productivity of aquatic environments we must control the land use patterns and the land resources to prevent, reduce and eliminate pollution of water. Thus to combine water resources planning, land resources planning and environmental planning of a river basin into a comprehensive one for the utmost economic, social and environmental benefits of the basin with the least cost is a strategic problem.

The objective of protecting soil resources is to establish a high quality soil ecosystem. For this, we must pay attention to: suitable use of soil, elimination of factors of low productivity, and establishing of the high quality of soil environment.

Controlling measures for soil ecosystems consist of biological, physical and chemical ones. Biological measures consist of proper use of land, suitable ploughing processes, applying organic fertilizer, planting green manure crops, etc. Physical measures consist of making surface level of the farm, building contour farms (step farms), irrigation and drainage, etc. Chemical measures consist of using chemical fertilizer, pesticides and herbicides, etc. If we only control the systems suitable to the natural and economic rule, we can certainly establish the soil ecosystem of the basin with high quality.

Preparedness for Future Changes

Finally, as mentioned before, we should not confine our view at present or near future in formation of river basin strategy, we must pay attention to the future development. Thus the future water demand and the possible change of the purposes of certain hydraulic projects must be predicated and arranged beforehand. The predication must be given to expect future conditions, 20—50 years hence, i.e., to 2000 or further. The water demands and projecting waste loadings in years of 1990, 1995 and 2000 are primarily a function of population and industrial and economic growth and land use patterns. The forecast of the above corresponding times of expected pollutions of aquatic and soil environments is necessary. Its aim is to implement the mitigating measures at the right time.

In general, the purposes of certain hydraulic engineering projects or all the ones of the basin might be changed as the economic development proceeds. We must arrange for the future changes in due time, for example, in China many existing reservoirs operated originally for irrigation purpose have been turned into water supply for industry or minicipality. Another example, during formation of the Han River Basin Strategy, the purposes of the Danjiangkou Project were determined to be flood control, power generation, irrigation, navigation and fish-breeding at the first stage of development, while in future, flood control remains to be the first role of the project, but on account of the utmost deficiency of water for irrigation and industrial uses of the north China, the purposes of irrigation and transferring water outwards of the Danjiangkou Project should be interchanged with power generation in order of ranks.

This strategic change for future is dependent on its available geographic location for water transferring to the north China. At the same time after the height of the Danjiangkou dam being raised the quantity of water transferred to the north will be increased. All the above conditions cannot be substituted by any other project. That is the superiority of the Danjiangkou Project. The decrease of power generation through the reduction of water discharged downwards can be supplemented by others. So the multi-purposes of the Danjiangkou Project would be arranged in the order of flood control, irrigation and water transferring, power generation, navigation and fish-breeding in the future.

Inconformity between water and land resources is serious in China (Fig 3). The northen part of China has more cultivated lands but less rainfall, the deficinecy of water is very serious. In order to change this condition fundamentally, in addition to the adoption of measures of comprehensive exploitation, economic utilization and quality protection of water resources, an urgent task before us is to transfer the surplus water of the Yangtze River system to the northen regions deficient in water. The Han River is the nearest to the above regions and its flow passes very close to the Yellow River. Moreover, the water can be transferred from the Danjiangkou Reservoir to the north by gravitation and surface flow. If necessary, surplus water of the Yangtze River could be transferred to the Danjiangkou Reservoir to supplement further the water resources. This is the middle route of the major water transfer scheme from the south to the north of China.

Fig 3 River systems and mean annual precipitation in China.

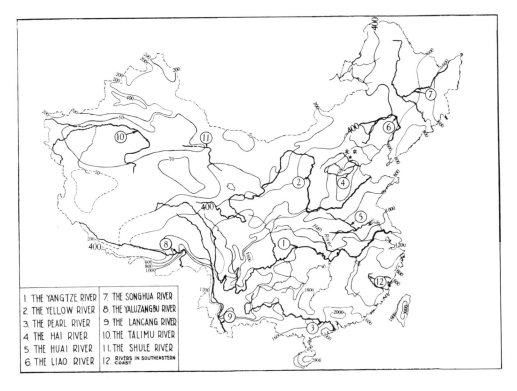

1. THE YANGTZE RIVER	7. THE SONGHUA RIVER
2. THE YELLOW RIVER	8. THE YALUZANGBU RIVER
3. THE PEARL RIVER	9. THE LANCANG RIVER
4. THE HAI RIVER	10. THE TALIMU RIVER
5. THE HUAI RIVER	11. THE SHULE RIVER
6. THE LIAO RIVER	12. RIVERS IN SOUTHEASTERN COAST

Conclusions

Strategy concerns not only the actual key problems affecting the development of the whole basin but also the long term perspective. Thus emphasis must be laid on the integral view of the whole basin and the current and future demands of economic development of the river basin. The river basin strategy must be formed by combining water resources planning, land use planning and environmental planning of the basin into an integral one with full consideration of measures to mitigate natural disasters and to facilitate comprehensive use of the resources involved, keeping the ecosystems in a state of good circulation. Water resources should be scientifically allocated to minimize the disagreement between sectors, protecting the land and soil resources, and paying attention to future demand changes.

Indices to indicate whether the strategy is correct or not are frequency of occurrence of disasters of floods, water-logging, drought, salinity or soil erosion, the degree of utilization of water and land resources and their degree of pollution. Investigation of the above problems for upper, mid and lower parts of the basin the correct direction and the main points could be found out to form the strategy for each part. For the time being we must adhere to the principle that scientific balance and allocation of water resources for various sectors and protection and promotion of the water quality have to be addressed simultaneously in forming a river basin development plan.

Lundqvist, J.; Lohm, U. and Falkenmark, M. (eds.):
Strategies for River Basin Management, pp. 151–160
© 1985 D. Reidel Publishing Company

The Integrated River Basin Development;
The Challenges to the Nile Basin Countries

Mageed, Y. A., Associated Consultants, PO Box 2960, Khartoum, Sudan

Abstract: The concept of integrated river basin development was widely recognised as early as the third decade of this century and was strongly associated with the multi purpose water development projects. This paper examines the processes of development in the Nile basin against important factors and circumstances that influenced the shaping of these processes and patterns of development. The paper then outlines the challenges to the Nile basin countries to realize the integrated river basin approach.

The patterns of development in the Nile basin were examined against the natural characteristics of the river including basin area, topographic and hydrological features and the human interferences including technical approaches, institutional and legal arrangement associated with political, and socio-economic circumstances and events that prevailed in the basin in a historical perspective.

To realize an integrated approach in this basin it is extremely important to create new attitudes between the basin states to promote cooperation and harmonization in the planning and development efforts, enhance the technological capabilities and institutional reforms necessary to cope with the intricate and complex problems of integrated river basin development. The legal framework inherited from colonial times and which created a spirit of suspicion among basin states need to be viewed through consultations and cooperation between themselves. The integrated river basin development is a long term process and must be based on conditions and realities that exist in the basin rather than copy existing models in other basins.

The paper concludes that the Nile basin has all the potential for integrated development for the transformation of the basin societies and their well-being. Such movement should be led by the intellectuals of the basin as a front line for an effective participation of the basin people and for setting up the process in the right direction to achieve its ultimate goals.

Introduction

Human interference in river basins is as old as the history of mankind. The emergence of the fluvial civilizations since times immemorial witnessed the initial forms of human interferences and control of rivers for agriculture and navigation. The approaches have developed from simple diversions to meet simple human needs to complex control works and management techniques to meet the increasing and diversified needs of human settlement, water supply, agriculture, energy, navigation, flood control and recreation.

The concept of integrated river basin development was widely recognized as early as the third decade of this century and was stongly associated with the multipurpose water development projects. In the depression years, basin-wide planning became part of a general development of natural resources to stimulate employment and economic recovery, as evidenced by the plans that were worked out in Europe and the USA. The plans were followed by the emergence of basin authorities and river boards and institutions.

During the 1960's an ongoing pressure on natural resources increased the awareness in the developed world of the threats against environmental stability. The need for conservation measures and rational management was recognized. In the developing world on the other hand, a desire grew for rapid development to bridge the economic gap between North and South and to provide the basic human needs of food, water supply and shelter.

During the decade of the seventies the world community has engaged itself in a series of UN conferences to resolve major problems facing both rich and poor nations. At all these conferences — dealing with Population, Food, Human, Settlement, Water, Desertification and Science and Technology for Development — the interrelationships and interactions between water, environment, population, food and human settlements were fully recognized.

This article will, against this background, examine the processes of development in the Nile basin together with important factors and circumstances that influenced those processes and approaches.

Present Human Interferences in the Nile Basin

Natural characteristics of the river

The River Nile is one of the major basins of the world; it extends over an area of 2.9 million km^2: from 4° south of the equator to 31° north of it, and falls from a height of 1600—2400 m to sea level (Fig 1). The total length of the river and its tributaries amounts to 37 205 km, its main lake area totals 81 550 km^2, and its swampy areas are stretched over 69 720 km^2.

The river system originates in two distinct geographical and climatological zones. The Blue Nile, Atbara and the Sobat systems originate in the Ethiopian plateau. The White Nile system originates in the Equatorial Lakes, and the Bar El Ghazal, which is a vast 'lagoon' formed by a number of streams rising at the Congo divide.

Hydrologically the river gets its water from the evaporation of water of the South Atlantic. The transfer of the moisture across more than 3000 km of Africa, which falls as rain over the highlands of Ethiopia and East Africa, is collected into thousands of streams and open lakes and swamps (Craig 1910). Out of 1300—2000 mm of rain over the two plateaus and upper reaches only 15—20% reach Aswan. The river carries 140 million tons of silt annually, mostly eroded from the land catchment in the Ethiopian highlands. This silt builds the fertile soils in the flood plain and shapes the hydraulic regime of the river down to its delta in Egypt. The water quality of the river is influenced by the different sub-catchments it crosses.

River basin development as seen in a historical perspective

Although important and significant to all the basin societies along its reaches, the particular role of the river varies. To Egypt the Nile has always been the source of life, to the Sudan

Fig 1 The Nile basin

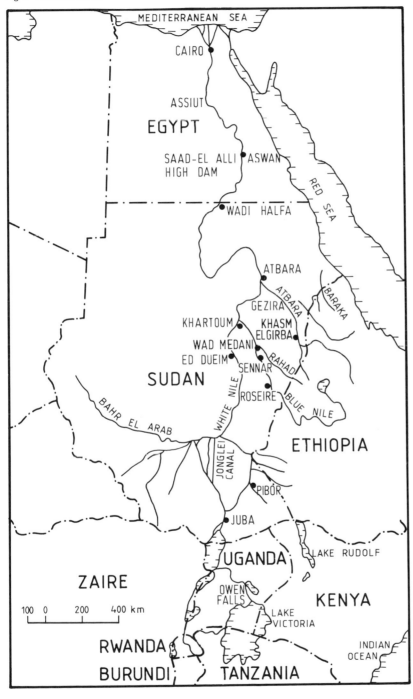

a dominating feature of the economic life of the Nilotics, especially in the semi-desert North.

"The Nile basin is perhaps the archetype of the usual historical pattern of international river basin development: early and significant development in the delta and lower basin and later — in this instance several thousand years later — development in the upper basin" (Garretson 1967). Since time immemorial, with the emergence of the fluvial civilizations, the river witnessed the initial interferences by diversion of its water for basin irrigation of the low lying fertile lands in its lower reaches in Egypt. The navigability of the river in its lower reaches also facilitated the unification of Upper and Lower Egypt. In search of information, the ancient Egyptians even pushed south to the fourth cataract and came in contact with the Meroitic civilization establishing the early forms of contact of basin societies. "Trade that flourished in ancient times between the upper and lower basins of the river proved that physical unity of the basin is stronger than political divisions" (Teclaff 1967).

Modern control works were started by Mohammed Ali Pasha at the beginning of the 19th century when the Delta Barrage was built to divert water to the fertile delta soils for production of food crops and cotton.

Three main problems continued to cause threats to the well-being of Egypt: a) the meager summer flow that falls short of the increasing demands for water, b) the occurrence of dry and c) wet years that bring famine and damage. This tempted the Egyptians to penetrate deeper into the upper reaches of the river in search of answers and suitable sites for control works. During this phase of penetration, an expanding foreign interference and influence in the basin was witnessed. It started with the exploration missions to discover the sources and mysteries of the river, followed by the exploitation expeditions that dominated the whole basin for the following decades.

The colonial period

The fall of the Turkish Empire which dominated Egypt and Sudan and other parts of the basin, marked the beginning of the stage of political influences by foreign European powers. These influences created deep rooted impacts on the socio-economic conditions of the basin societies, the patterns of development along the basin, the legal framework and the institutional systems. They indeed continue to exert influence up to our present times.

In spite of a subdivision of the basin, first between the foreign powers and lateron in the form of basin states, the hydrological unity of the basin was recognized. This was evidenced by the protocols and boundary agreements made between the British, who influenced a major portion of the basin, the Italians in Ethiopia and the Congo Free State dominated by the Belgians (the 1891 Rome protocol, the 1902 agreement for the Blue Nile and the 1906 Agreement with the Congo Free State). All these agreements prohibited, without the consent of the downstream riparians, any construction on the tributaries of the Nile obstructing the flow to or outside the Sudan.

In 1899 basin-wide plans were drawn for annual storage works. Water was to be stored from the tail of the floods to augment the summer lowflow, increase the timely water through conservation of water from the swamps at the Sudd region, flood control works and over-year storage at the Equatorial lakes and Lake Tana to guard against the occurrence of dry years. A number of commissions were established to examine those plans and the allocation of water between Egypt and the Sudan. This led to the conclusion of the 1929 Nile Water Agreement.

While this agreement recognized the need for development of irrigation in the Sudan, it stipulated that any increase of the use of Nile waters as a result of such development should not infringe on Egypt's natural and historic rights. The working arrangements based on this arrangement provided the Egyptian rights of the entire natural flow of the river during low flow season. Apart from small withdrawals from natural rivers during this period, the Sudan had to meet its requirements from water stored at the tail of the flood. So in the years that followed up to independence, the Sudan irrigation development was restricted to the Gezira project with cotton as the main cash crop in an area of about 400,000ha (one million feddans) and a crop intensity of about 50%. During the Korean War and the sharp rise in cotton prices further areas were brought under irrigation by pumping from the Blue and White Nile under flood license only.

The 1929 Agreement also required the East African countries not to constuct any works in the Equatorial Lakes without consultation with Egypt and the Sudan, and the British government to use its good offices to facilitate establishment of over-year storage in the Equatorial Lakes connected with a conservation project in the Sudd region.

In the late forties, an agreement was reached to establish the Owen Fall Dam at the outlet of Lake Victoria for hydropower generation with a concession to enable a use of the lake for over-year storage in the future. The Egyptians also presented to the Sudan government the plan for the Sudd diversion canal at Jonglei. No agreement could be reached with Ethiopia for regulation of Lake Tana.

Development after independence

With the decline of British influence within the basin, the emergence of the 1952 Egyptian revolution and the independence of the Sudan and the East African countries, a new political atmosphere prevailed and influenced important events and attitudes in the basin.

The Egyptians still face the three major problems, but in addition to those came now the increasing pressure of the Sudan after independence for more water to meet their irrigation demand. The uncertainty to resolve these problems created the concept of over-year storage at Aswan. As the impoundment upstream of the High Aswan Dam extended 150 km into the Sudan, dislocation of 50,000 inhabitants became inevitable and the two countries had to negotiate and set new legal and technical framework to the Nile question. This led to the new 1959 Nile Water Agreement (Egypt & Sudan 1959) allowing Egypt to go ahead with its plans to establish the High Aswan Dam as an over-year storage. In short it meant arresting and controlling the full discharge of the river, including the 32 billion m^3 which used to find their way to the sea during July to October as a result of the Blue Nile flood. Out of this amount, 14.5 billion m^3 became available to meet the demands of Sudan and increase the summer flow for Egypt by 7.5 billion m^3. The High Aswan Dam also provided the solution for Egypt against the fluctuations between wet and dry years. These gains are realized with a loss of 10 billion m^3 by evaporation.

A permanent Joint Technical Commission for Nile Water was established by the two countries. On behalf of the two governments this institution was to undertake the control of the river and undertake studies to increase its yield to meet future demands of the two countries. The agreement also recognized the rights of other riparian states in the river and made the arrangement of how to meet them when they arise.

Enlarged basin corporation

Following the 1959 agreement of the two downstream riparians in 1959, the British administration of the East African countries' territory brought about the questions of the water rights of those countries. It was claimed necessary to make new arrangements to supercode the 1929 agreement. Informal technical talks between the Permanent Joint Technical Commission for the Nile countries (Egypt and Sudan) and the Coordinating Nile Water Committee (Kenya, Tanzania and Uganda) followed. They led to:

— the realization of the importance of the technical cooperation between those countries
— the necessity to undertake joint studies in the catchments of Lakes Victoria, Kyoga and Albert to determine the water balance
— to identify storage works necessary to meet future demands of the riparian states.

A Technical Committee was formed by all riparian states, Ethiopia participating as an observer.

In connection with the 1959 Nile Water Agreement between Egypt and the Sudan, Ethiopia initiated studies to identify power and irrigation projects within the Blue Nile, Atbara and Sobat basins inside its boundaries. Circumstances in the area, however, continue to hinder progress in cooperation between the Blue Nile upstream riparian state and its two downstrem sisters.

Since the conclusion of the 1959 Nile Water Agreement which almost coincided with the independence of most of the basin states, the basin development in the riparian countries followed different approaches influenced by two important factors:

— The after effects of colonialism implied weak institutional, technological and economic capacities
— The challenges facing the initial national governments to build their nations rapidly and to cater for various basic needs.

The scale and magnitude of development varied from one riparian state to the other. In Egypt enormous urban and industrial centres have emerged together with horizontal expansion of the agricultural areas into the desert areas bordering the fertile soils of the Nile Delta to meet the increasing food demands of the growing population.

In the Sudan in a decade after independence, the irrigated area was increased by more than 200% through the construction of two annual storage reservoirs in the Blue

Nile and Atbara rivers. The cotton crop occupied 60% of the annually cropped area and little attention was paid to food crops. The Five and Six Year Economic and Social Plans 1971—76 and 1976—82 aimed to realize food self sufficiency and to provide the raw material for agroindustry including sugar, textile and food oils.

In the upper riparian states (i.e. East African countries and Ethiopia) no significant interference in the natural flow of river was undertaken. Vast opportunities for power production have, however, been identified in the upper reaches of the Blue Nile and equatorial Nile systems.

A schematic picture of those interferences and development of the river system is shown in Fig 2.

Some obstacles to an integrated river basin development

The Nile basin is endowed with vast natural resources: its unique climate from the tropics to the Mediterranean, its rainfall, surface run-off, lakes, swamps and aquifers, its agricultural and land potential and range lands, its forests and animal resources, its energy potential and its great navigation potential and above all its human population with their cultural heritage and old civilization. The meagre development achieved over the last 1000 decades is much below the aspirations of its inhabitants. The majority of its populations live in absolute poverty, short of food and other basic human needs.

A number of obstacles form challenges to be overcome in the future. The successive occupations by foreign powers have for a long span of time divided the basin into different influence zones. The independent nations inherited fragile economic bases, weak institutions, absence of a data base, lack of trained manpower and a deficient technological capacity. They therefore have to embark on a short term development to meet the pressing needs of their people. The legal framework formulated during colonial times, the disparaties in development and the gaps in water utilization have created an atmosphere of suspicion among the basin independent states. Today, the countries of the basin, like other African nations, are facing formidable economic problems as a result of the escalating inflation, unemployment, food deficit and rising costs of imported energy and industrial goods.

Efforts Needed

All these circumstances will continue to remain as obstacles unless concerted efforts and actions are taken by the basin countries to get the process started and to lay the foundation for dynamic socio-economic development for the optimum benefit of the basin societies.

Creation of new attitudes

New attitudes have to be created and promoted concerning co-operation, co-ordination and harmonization of planning and development effeorts. The prevailing legal framework

158

Fig 2 The Nile system. A schematic diagram of delivery and abstraction.

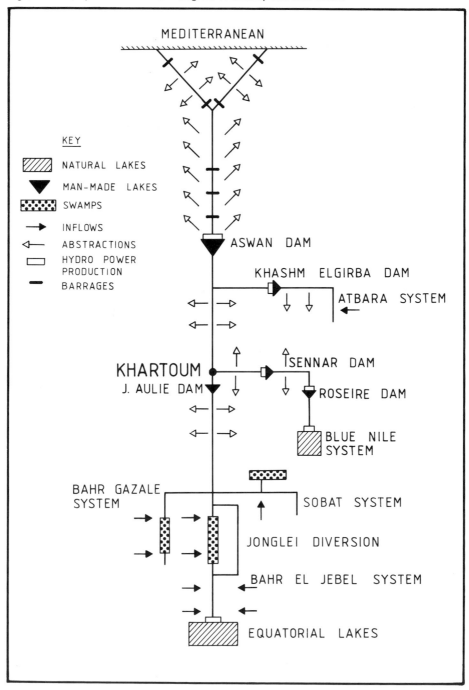

has created an atmosphere of suspicion among the basin states. It requires consultations, co-ordination and co-operation. The need for sub-basin development in the upper riparians to meet immediate and short term needs has to be recognized and encouraged to close existing gaps and disparaties in basin development. At the same time consultation and co-ordination are needed to minimize harmful effects. The search for mutual benefits is needed to encourage joint actions. The integrated river basin concept implies that such an attitude has to become deeply rooted in the minds of the basin societies.

Approach based on reality

Integrated river basin development is a long term process which must be based on conditions and realities that exist in the basin. It will not be correct to copy river basin models existing in the developed countries (such as the TVA model). The development philosophy has to be based on existing conditions of the basin societies. Unlike industrial countries who are homogenous societies, the basin societies in less developed areas lack this homogeneity. The Nile basin is faced with gross inequalities in the quantity and quality of means of production, wealth, income levels, education, social services. A dual economy exists with modern and traditional subsistance sectors, the latter encompassing the majority of the population of the basin. Therefore adaptation of integrated river basin development concepts to these conditions should ensure positive response of the societies to the process of development and enhancing their ability to adapt and improve means of production.

Enhance technological capabilities

The development capacity of nations depends on their potentiality for utilizing natural resources which in turn is determined by their ability to apply technology. Many countries and regions in the world are rich in resources but are still undeveloped due to lack of capability to apply technology to transform such resources. This is largely due to constraints imposed by social, economic, power and institutional structure both within and among nations.

While there is considerable scope for application of traditional and conventional technology for the solution of the problems of basin development, there is a great need to apply and develop new technologies for the solution of the complex and interactable problems of integrated river basin development.

The technological capacity in the basin should be enhanced. Available meagre technological capacities should be shared between the basin states through programmes of technical co-operation. The technology transfer process should be facilitated through co-ordinated international co-operation.

A number of national, regional and international scientific and research institutions within the basin states could play an effective role in this process of building technological capacities. Universities should be encouraged to show a much higher level of involvement in the process of integrated river basin development.

Conclusions

It is important to enhance and strengthen, as an urgent and basic requirement, the existing national institutions, including water engineering bodies, to enable a basin development in a wide perspective, and to ensure stronger co-ordination between sub-organ nations tackling development problems within individual basin states. At the same time basin organizations should be harmonized through adequate channels of communication. Existing basin committees or others that may emerge in the near future, could undertake consultations and co-ordination on the strengthening of national river basin organizations and the harmonization suggested.

The last few decades have witnessed great technological advancements associated with development of computers, remote sensing, mathematical modelling, system analysis and planning techniques. To benefit from these new methods calls for enhancement of technological capacities of basin institutions and manpower training.

In conclusion one could say that the basin has all the potential that is needed for a dynamic movement to combat the state of underdevelopment through the process of integrated river basin development for the transformation of the basin societies and countribute to their well-being. Such a dynamic movement should be stimulated by the intellectuals of the basin as a front line for an effective participation of the basin people for setting the process in the right direction and creating the climate for its progress towards its ultimate goals.

References

Craig: England, Abysinia, the South Atlantic, a Meteorological Triangle. Journal of the Royal Meteorological Society (1910)

Garretson, A. H.; Hyton, R. D.; Olmstead, C. J.: The Law of International Darainage Basins. Oceana Book No. 320. New York 1967. (Published for the Institute of International Law, New York University School of Law)

Official Documents

Protocol between Britain and Italy Delimiting Spheres of Influence in East Africa. April 15, 1981. Art. 3, 83 Brit. and Foreign State papers 21.

Treaties between United Kingdom, Italy and Ethiopia, Relative to the Frontiers Between the Sudan, Ethiopia and Eritrea. May 15, 1902, art. 3 Cmd No. 13070 (T.S. 16 of 1902), 23 Hertslet, Commercial Treaties 2.

Agreement signed at Brussels, May 9, 1906. art. Cmd No. 2920 (T.S. No. 4 of 1906) 24 Hertslet, Commercial Treaties 344.

Exchange of Notes, Regulating the Use of Nile Waters for Irrigation Between Great Britain and Egypt May 7, 1929, Cmd No. 3348 (T.S. No. 17 1929) 21, Martens. N.R.G. (3e ser) 97.

Agreement for Full Utilization of Nile Waters Between Egypt and Sudan, 8 Nov. 1959. Text 15, Revue Egyptian de Droit International 321—329 (1959)

Lundqvist, J.; Lohm, U. and Falkenmark, M. (eds.):
Strategies for River Basin Management, pp. 161–170
© *1985 D. Reidel Publishing Company*

2.3 Managing water quality impact in industrialized countries

Approach to Eutrophication Problems Close to an Expanding Metropolis in Australia

Isgård, E., VBB-SWECO AB, Vatten och miljöteknik, Box 5038, S-10241 Stockholm, Sweden

Abstract: The Hawkesbury river basin in SE Australia is mainly composed of forest and agricultural land. The river provides water supply for the neighbouring Sydney area and the basin — 26 % of which is national park — is very popular as recreation areas for the metropolitan population. By the year 2000, the population is expected to have doubled, and the metropolitan area to extend further into the basin, causing conflicting water interests. The author, involved in the planning as a Swedish consultant, discusses the water quality problems to be foreseen, mainly from the aspect of eutrophication, and presents the nutrient removal options available.

Description of the Hawkesbury-Nepean River Basin

The Hawkesbury Nepean river system is a multifacetted resource, which is exploited for water supply, recreation, waste transport and assimilation.

The Hawkesbury river basin is situated in New South Wals, Australia, to the north and east of the metropolitan area of Sydney (Fig 1). The river is called the Nepean in its upper reaches and the Hawkesbury in its lower reaches down to its outlet in the Broken Bay of the Pacific. It drains the Blue Mountains of the Dividing Range, extending about 150 km inland from the Pacific and a maximum extension of 300 km. The drainage area is 21,000 km².

The average annual rainfall in Sydney is about 1,200 mm, which rapidly decreases inland, at Goulburn being 678 mm.

The river is affected by tidal influence from Windsor reaching more than 50 km upstream of the outfall in the Broken Bay. The freshwater flows are highly variable with annual discharges at Penrith varying between 1 and 245 m³/s.

Only a small area of the basin has been developed to date for urban use, the remainder being forest, agricultural land and waterways, or used for water storage. The total population in the basin was as of 1981 approximately 450,000. Tentative projections indicate a future population increase by 300,000 in 1990 and by 500,000 in 2000 from the metropolitan area of Sydney, which will extend into the Hawkesbury river basin area giving rise to conflicting interests with the present users, mainly for recreation, agriculture and water supply.

Of special interest is the impact of the discharge of sewage, treated to a required level, for the future population. Responsible authority for the planning hereof is the

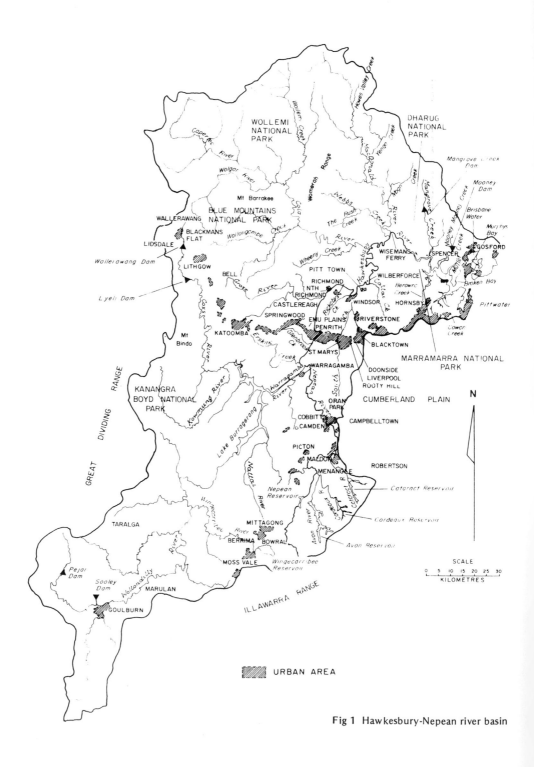

Fig 1 Hawkesbury-Nepean river basin

Metropolitan Water Sewerage and Drainage Board (MWSDB), Sydney. The Board commissioned in 1981 SWECO in cooperation with the Australian consultants SCROGGIE, Melbourne, to study and report on the Board's strategy planning activities, including a review of reports, advice on range and types of nutrient removal options available. The report was submitted in December 1981.

Simultaneously, the State Pollution Control Commission (SPCC) of New South Wales has undertaken investigations to develop a quantitive rationale for future water pollution control policies and their management within the Hawkesbury river basin. The SPCC report on water quality was submitted in September 1983. This contribution is mainly based on the SWECO-SCROGGIE Report (1981).

Competing Interests in the River Basin

Water supply

A modern metropolis utilizes great quantities of fresh water, Sydney being no exception. MWSDB supplied in 1982 $20m^3/s$ of water to its consumers. The majority hereof was taken from its waterworks in the Hawkesbury river basin, where four dams and reservoirs have been constructed for this purpose. The largest of these is Warragamba, which caters for around 75% of the present water supply capacity of MWSDB. Since there are practically no natural lakes or reservoirs in the basin the creation of dam reservoirs will enable the use of water when it is available in excess of flows necessary for keeping the natural conditions in the river.

There is an agreement between MWSDB and Water Resources Commission (WRC) that if the flow should go down below 50 Ml/d ($0.58 m^3/s$) at the Penrith Weir between Warragamba and Windsor, release for riparian use can be requested, provided that the Board does not need the water for Sydney. An evaluation of the possibility to increase this release in order to counteract the impact of future wastewater discharges has not yet been made. The future relation between nutrient load and river flow seems, however, to be so unfavourable that flushing from water supply reservoirs can only be a temporary solution to the eutrophication problem.

Recreation

The river basin is an important recreational area for the population of Sydney, including nine National Parks, which occupy 26% of the catchment area. Especially the Blue Mountains are attractive, but as usual in areas close to the divide, the impact of man on the crystal clear waters can be detrimental. In the lower reaches, especially in the zone of tidal influence, there is recreational boating and a great deal of fishing and use for shellfish harvesting.

Human impact on water quality

About 30% of the catchment area is made up of agricultural land and around 6,500 ha hereof is currently irrigated. A provisional nutrient budget shows that presently non-point sources from agriculture are the main contributors of phosphorous to the river in annual budget term. However, the major impact on the river water occurs during dry weather periods, when non-point-sources not are dominant.

The urbanization of the catchment will, of course, change the relation between non-point and point sources, so that point sources will be still more important. The existence of agriculture may be looked upon also as a positive factor, serving as possibilities for land use and disposal of effluent and sludge from treatment works.

There are a total of 56 water pollution control plants (WPCP) within the basin, of which 11 are in the Blue Mountains area (Fig 2). About 80% of the effluent is discharged downstream the tidal limit.

The MWSDB-plants provide for BOD* and suspended solids removal, although in the larger plants provision has been made for the future introduction of nutrient removal facilities. Also sand filter or oxidation ponds are provided in many plants.

License conditions according to Clean Waters Act are shown in Tab. 1. The most stringent requirements are the following:

	mg/l
BOD_5	10
Susp.solids	15
NH_3-N	5

For the treatment plants in the Blue Mountains the requirements are 20/30-standard (BOD_5 max. 20, SS max. 30).

SPCC has as a first step to improving the effluents recommended that they should be nitrified to reduce ammonia toxicity effects. Phosphorus removal, to 1 mg/l under dry-weather conditions, is the next priority at the larger plants. Nitrogen removal is considered the third priority.

As previously mentioned a future development with more than half a million people in the basin area is envisaged. If the sewage is to be discharged into the river, the impact on river quality will be substantial notwithstanding very stringent effluent standards. One important objective of the SWECO-SCROGGIE study was to find out what technical possibilites exist for water pollution control of nutrients and the experience available from the operation of such plants.

Eutrophication Problems

Limiting nutrients

Eutrophication control has to a great extent been dealing with algal growth limiting nutrients, especially phosphorus and nitrogen. A number of factors can limit the maximum

---

* BOD = biochemical oxygen demand

Fig 2 Schematic sewage loadings from urban sewage-treatment works to the Hawkesbury-Nepean system in June 1980

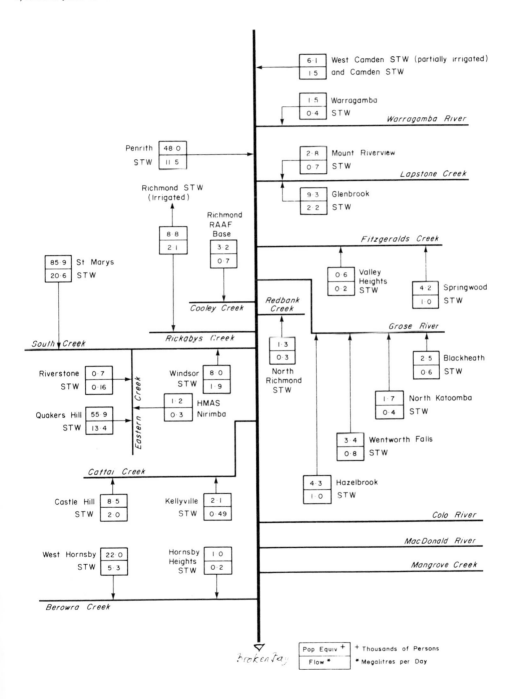

Water pollution control plant	CWA requirements mg/l		
	BOD$_5$	SS	NH$_3$-N
Metropolitan			
Camden	20	30	
Castle Hill	10	15	
Hornsby Heights*	20	30	
Kellyville	10	15	
North Richmond*	20	30	
Quakers Hill*	20	30	
Richmond	20	30	
Riverstone*	20	30	
St. Marys	10	15	5
Warragamba	20	30	
West Camden	10	15	5
West Hornsby	10	15	
Blue Mountains			
Blackheath	20	30	
Blaxland/Glenbrook	20	30	
Hazelbrook	20	30	
Mt. Riverview	20	30	
Mt. Victoria	20	20	
North Katoomba	20	30	
North Springwood	20	30	
South Katoomba	20	30	
Springwood	20	30	
Valley Heights	20	30	
Wentworth Palls	20	30	

* Conditions to be upgraded to BOD/SS/NH$_3$ 10/15/5 following commissioning of dual media filters.

Tab 1 Licene conditions according to Clean Waters Act for treatment plants in Hawkesbury-Nepean river basin

primary production in a water body; for practical purposes it is often a question of limitation by phosphorus or nitrogen. A scientifically correct approach has been formulated by Lee Jones (1980, unpublished), stating that "phosphorus would be considered limiting if the soluble orthophosphate concentration is 2 mg/m^3 or lower and for nitrogen assimilable forms (NO$_3$, NH$_4$) must be less than 20 mg/m^3". Another approach is to study the standing crop limiting nutrient. Standing crop is the "weight of organic material that can be sampled or harvested at any time from a given area". Hereby it is possible to discuss the relation between nutrients and biomass levels, without taking notice of the assimilable forms or physiological processes involved.

The ratio N:P in lake waters has proved to be a practical means of estimating the growth effect. According to Swedish experience a value of 15:1 seems to be critical (Forsberg & Ryding 1980). In the Hawkesbury river this ratio is presently lower than that, see Fig 3.

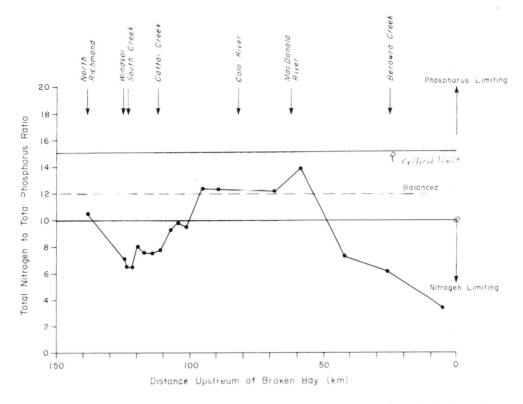

Fig 3 Ratio of mean total nitrogen to mean total phosphorus concentrations, Hawkesbury river, 1979—1981

Also the trophic state of the water body must be considered. At phosphorus concentrations about 0.1 g/m³ the role of phosphorus as a growth-limiting nutrient gradually diminishes. A concentration of 0.1 g/m³ of phosphorus may therefore be considered as a critical level, which can be seen on a diagram of the relationship between transparency and phosphorus concentration, Fig 4. Very little reduction in transparency occurs at phosphorus levels above 0.1 g/m³.

In waste receiving lakes in Sweden, nitrogen has been found to be growth limiting when N:P <10 and the chlorophyll a is more than 70 μg/l. The mean concentrations of nitrogen, phosphorus and chlorophyll a in the Hawkesbury river is shown in Fig 5. As can be seen from this figure the concentrations of chlorophyll a are presently generally below 70 μg/l.

When discussing nitrogen it is important to assess the possibilities of aquatic ecosystems to compensate a nitrogen deficiency by nitrogen fixation from the atmosphere, especially by blue-green algae.

As far as rooted aquatic vegetation is concerned, the situation is more complicated, as these plants can assimilate nutrients both via their roots and via their green parts.

168

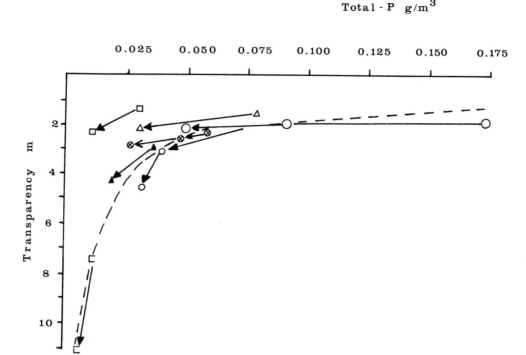

Fig 4 Total phosphorus and transparency. Summer average values. Swedish experience (Forsberg, Hawerman & Hultman 1981)

Nutrient removal options available

There is presently extensive experience in Sweden from phosphorus reduction in more than 750 plants. The results are summarized in Tab 2. A phosphorus removal to a level of 0.2–0.3 mg/l can be expected in plants with post-precipitation on two-point precipitation, followed by a filtration stage.

A single-sludge pre-denitrification system for carbon degradation and nitrification has been found to be the most economical nitrogen removal method. This system can be designed for almost complete nitrification, i.e. ammonium-nitrogen concentrations below 2 mg/l. Also a substantial denitrification will take place in the subsequent aeration tank. However, in order to reach a level of 5 mg/l of nitrogen in the effluent, it might be needed to include a separate denitrification process with methanol as an external carbon source.

The presently available treatment processes as described above would achieve an effluent of 5 mg/l N and 0.25 mg/l P, i.e. with a ratio N:P 20. From the previous discussion on limiting nutrients this means that phosphorus would be the limiting factor in the effluent as long as the dilution factor is more than 2.5, and if the phosphorus concentra-

Fig 5 Mean concentrations of total nitrogen total phosphorus, nonfiltrable residue and chlorophyll *a*, Hawkesbury-Nepean rivers, 1979–1981

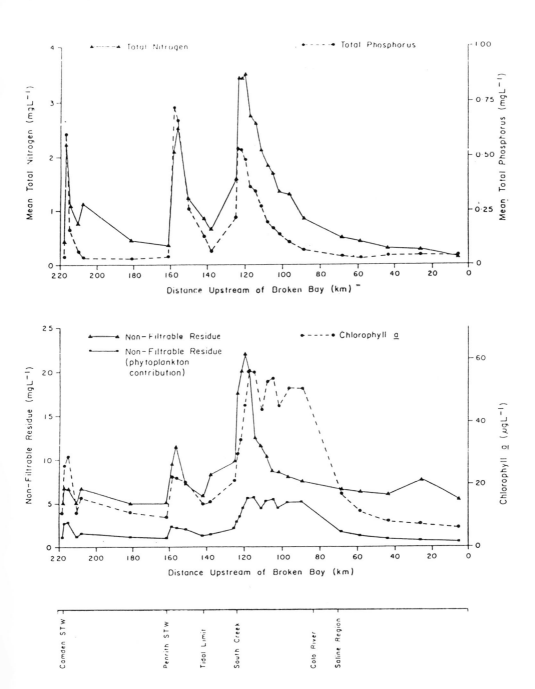

Process combination	Operational results in 1977 (National Swedish Environment Protection Board, 1979)		Expected results for plants with no significant operational problems and pH control
	BOD_7*	P_{tot}	P_{tot}, mg/l
Direct precipitation	39	0.70	–
Pre-precipitation	35	0.98	0.5–0.8
Simultaneous precipitation	28	1.48	0.5–0.8
Post precipitation	10	0.53	0.2–0.4
Post-precipitation followed by deep-bed filtration	9	0.22	0.15–0.3

* The standard incubation period for BOD analysis of 5 days has in Sweden for practical reasons been changed to 7 days.

Tab 2 Operational results from use of different chemical precipitation methods (average values of operating plants). From Forsberg et al. 1981 and Grönquist, Holmström & Reinius 1978.

tion of the receiving water is less than 0.1 mg/l. In such cases the phosphorus removal seems to be the most important factor in controlling the eutrophication.

In case of hyper-trophic conditions in the river water the nitrogen removal might also be justified not to do a bad state worse. Nitrogen removal may also be necessary to keep the absolute level of nitrogen in the water courses below hygienic limits.

Concluding Remarks

Plant scale studies are currently in hand to assess phosphorus removal by chemical precipitation with ferric chloride (pickle liquor). Pilot scale studies are also under way to find out the degree of denitrification achievable using sewage BOD as the carbon source. It seems highly unlikely that denitrification involving the addition of methanol will be implemented.

In fact, no decision has so far been made on either the degree of nutrient removal required or the methods to implement removal. Discussions are currently proceeding between MWSDB and SPCC in this regard.

References

Forsberg, C.; Ryding, S. O.: Eutrophication parameters and trophic state indices in 30 Swedish lakes. Arch-Hydrobiol. 89, 197 (1980)

Forsberg, C.; Hawerman, B.; Hultman, B.: Experience from 10 years' advanced wastewater treatment – technology and results. Water. Sci. Techn. (1981)

Grönquist, S.; Holmström, H.; Hultman, S.; Reinius, I.-G.: Experiences and process development in biological-chemical treatment of municipal wastewaters in Sweden. Prog. Wat. Techn. 10, 5/6 (1978)

State Pollution Control Commission 1983. Water Quality in the Hawkesbury-Nepean River.

SWECO-SCROGGIE: Report of Strategy of Sewerage Treatment and Effluent Disposal in the Hawkesbury-Nepean Catchment 1981.

Lundqvist, J.; Lohm, U. and Falkenmark, M. (eds.):
Strategies for River Basin Management, pp. 171–177
© 1985 D. Reidel Publishing Company

The Changing Illinois River

Stout, Glenn E., Water Resource Center, 2535 Hydrosystems Laboratory, 208 N Romine St., Urbana, IL 61801, USA

Abstract: The main stem of a river system in central USA became a stream of domestic and industrial sewage as man developed the economic base of growth for the region. Society then demanded that the stream integrity be restored. Great progress has been made since 1920. Currently, costly programs to reduce non-point pollution from urban and agricultural sources must be continued in order to restore the ecosystem for aquatic organism. Eventually, body contact sports and sport and commercial fisheries will again become a benefit to the citizens of the region.

History

Largest of the Mississippi river's tributaries in North America above the mouth of the Missouri river, the Illinois river is formed by the confluence of the Kankakee and Des Plaines rivers about midway between Chicago and LaSalle. The river flows in a westerly, southwesterly, and southerly direction for a distance of 340 km and empties into the Mississippi river at Grafton, Illinois (Fig 1).

In the early days of exploration and settlement of Illinois the rivers were the arteries of travel, communication, and commerce. It was not until the era of railroads that the people of Illinois were in a great measure emancipated from the rivers.

Little concern was shown about changes in, or the changing of, the Illinois river for the first 250 years of its use by white people. Steamboats made their way far up its reaches in the 19th century. Cities sprang up along its shores and, near the headwaters, Chicago began its growth. Events happened rapidly from the last quarter of the 19th century to the present time.

To give a simple illustration of the development in the river's basin of 75,600 km, the population of the counties which are all or in part drained by the Illinois River changed from about a half-million in 1850 to 1,629,738 in 1870. By 1964 this figure had risen to 8,537,900 of a total state population of 10,500,000.

In 1848 the Illinois-Michigan Canal was opened, and in 1900 the Chicago Sanitary District opened the Sanitary and Ship Canal as a part of a project to divert sewage and storm drainage away from Lake Michigan. In 1907 the Hennepin Canal connected the Illinois with the Mississippi.

Following the diversion of water from Lake Michigan into the Illinois in 1900, the construction of levees and the drainage of bottom lands for agricultural purposes began to change the nature of the Illinois valley. At the same time the aquatic biota ecosystem was threatened by the serious menace of the urban wastes and soil from erosion being discharged into the river. Although a small amount of pollution had occurred before

Fig 1 Illinois Waterway location map

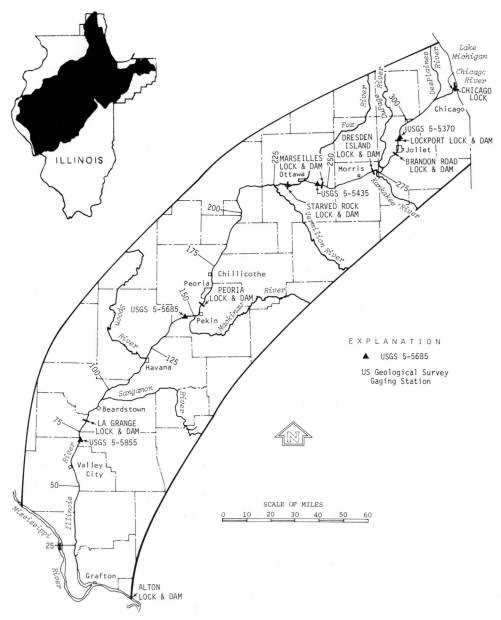

1900, the opening of the Chicago Sanitary and Ship Canal in 1900 created a problem of catastrophic proportions for the river and its backwater lakes above Peoria. At first only the extreme upper reaches of 160 km (above Marseilles) were affected. However, the zone of pollution steadily moved downstream until by 1922 the upper Illinois was essentially a dead river, devoid of important aquatic life as far south as Chillicothe.

A decline in urban and industrial pollution began with the operation of treatment plants by the Chicago Sanitary District in 1922. Pollution control was aided by the navigation dams that became operational in the upper river in 1933. These dams reduced the rate of flow, thereby resulting in bacterial decomposition of waste products within a shorter distance downstream. A gradual reduction in the urban pollution of the Illinois river has continued to the present time. Nevertheless, between 1920 and 1960, major losses of fisheries occurred for reasons not completely understood.

Another human activity has conspicuously changed the river. Before 1900, low dams were built at Marseilles, Henry, Copperas Creek, La Grange, and Kampsville. Because they were low, their greatest effect on the stream was during periods of low water. During the 1930's, higher navigation dams were built at Dresden Heights (6.6 m), Marseilles (8 m), Starved Rock (5.1 m), Peoria (3.3 m), and La Grange (3 m). Moreover, a navigation dam on the Mississippi at Alton raised water level in the Illinois river as far north as Hardin.

Barge traffic on the river is now very heavy, and consequently there is an effect on the turbidity of the water in the main stream and adjacent waters. Soil pollution has been present in Illinois river system since the recession of the last ice sheet. However, the exposure of the soil in agricultural operations has greatly increased the problem. The permanent and insidious nature of silt pollution makes it more harmful than urban pollution: although not as apparent, it is accumulative.

During the early development of Chicago, the Chicago and Calumet river system flowed easterly and discharged into Lake Michigan. As the population grew, wastes were discharged into these rivers eventually causing public health and pollution problems. Subsequently, the flow of these rivers was reversed and flushing was accomplished by diverting Lake Michigan water; this relieved many of the problems in the immediate Chicago area, but water quality conditions deteriorated downstream as a result.

Today flushing water (discretionary diversion) from Lake Michigan is drawn at three locations in Chicago. Total diversion, including that needed for public water supply, is limited by law to 90 m^3/s on an annual average basis.

The three major treatment plants discharge a dry weather average flow of 56 m^3/s.

During wet weather, combined sewers overflow and discharge throughout the Chicago waterway network.

Present Day Water Quality Problems

Yesteryear problems were much greater and more obvious than those which persist today. However, major problems do presently exist throughout long reaches of the waterway. Basically, these problems are: (1) sediment transport and sediment deposition, and (2) low dissolved oxygen (DO) concentrations during warm low flow periods.

Low DO's are reflective of active biological stabilization of organic wastes. When the DO supply is insufficient to supply the continuous biological demands, stream degradation occurs and oxygen sags occur in a stream. Low DO's still persist because: (1) the waste assimilative capacity has been reduced due to man's alterations of the natural flow regime; (2) significant organic waste loads are still discharged in the form of carbonaceous and nitrogenous compounds; and (3) bottom (benthic) sediments exert a high oxygen demand in certain locations.

The minimum DO standard for the lower river is 5.0 mg/l. This standard is routinely violated in selected reaches of the river almost every summer. However, extremely depressed DO's are no longer observed. Conditions have improved greatly in the last twenty years.

Waste Load Reductions and Water Quality Improvements

Since the mid-1800's when the Illinois-Michigan Canal was opened, overall water quality (relative to organic pollutants) has never been better than it is now.

Carbonaceous waste loads have been reduced 91% since 1922. Since 1971, a 32% reduction alone has occurred even though the load in 1971 was actually only 13% of that observed in 1922. The figures specific to the Peoria area are equally amazing — since 1925, 97% of the organic waste discharges have been eliminated (Butts 1983).

High ammonia concentrations are indicative of organic pollution, especially those associated with domestic waste. As the population increased along the waterway since 1900 a commensurate increase in ammonia occurred. Ammonia is not readily removed in the treatment processes employed up to the early 1970's. Significant increases occurred in the ammonia load up to that time, then it significantly decreased. Since 1971 over a 50% reduction has occurred. One part ammonia in water requires 4.57 parts of oxygen for stabilization. In terms of population 950,000 people equivalents are now being discharged compared to 1,950,000 about ten years ago (Butts 1983).

A tremendous reduction in ammonia levels occurred in the Peoria area of the river in the last 10 to 12 years. The apparent short term difference between the 1981–82 values and the 1982–83 flows were much higher on a sustained basis than the 1981–82 flows. These higher flows, however, do help to improve conditions, albeit only for short periods.

1982 DO usage at both the upper and lower ends of the waterway is significantly less than the 1971 usage (Butts 1983). Perhaps the most significant indication of improvement is that DO's in the upper waterway above Peoria have increased steadily from near zero conditions in 1922 to values persistently above 5 mg/l in 1982 (Butts 1983). Some localized undesirably low concentrations occasionally occur and near zero levels often occur for long time periods above Lockport but overall a tremendous improvement has been evident.

One water quality problem continues, however. Oxygen sags will still occur in the waterway even if all point sources of pollution were completely eliminated.

The cause of these oxygen sags is the relationship between sediment deposition and DO levels within the waterway. Not only do sediments reduce water volume and create physical problems, they also contribute to oxygen depletion in the form of sediment oxygen demand.

Significant increases in dissolved oxygen could still be achieved if the Chicago Calumet treatment plant were upgraded to meet the same effluents now being achieved by the other two Chicago plants. The Calumet plant is basically the last facility along the water-way that could be improved to provide detectable improvements in DO and ammonia levels downstream.

Potential Future Management and Water Quality

In addition to the upgrading of the Calumet treatment plant, two proposed projects could drastically alter the water quality of the Illinois river basin. One is the possible increased diversions of Lake Michigan water and the second is the completion of the Chicago Tunnel and Reservoir Project (TARP).

Increased Lake Michigan diversion

Proposals have been made to increase Lake Michigan diversions into the Illinois waterway to reduce shoreline erosion along the Great Lakes and to improve water quality along the Illinois Waterway.

The US Army Corps of Engineers (COE) under the Water Resources Development Act of 1976 supervised a five-year study to evaluate the beneficial and adverse effects of increased diversions.

The COE developed a computer model of the river to simulate three historic water years (representing low, average and high flows) under three increased diversion conditions:

(a) No increased diversion, reflecting the existing maximum diversions of 90 m^3/s at Lockport.
(b) 97 m^3/s of increased diversion, amounting to a total diversion of 187 m^3/s.
(c) 109 m^3/s of increased diversion, resulting in a total diversion of 280 m^3/s.

The COE identified the three major benefits of increased diversions to water quality (especially DO and ammonia in the upper reaches), commercial navigation and power generation. Economic estimates of benefits to navigation and power generation are greater at the 187 m^3/s diversion rate than at the 280 m^3/s rate. Nearly all improvement in water quality also occurs at 187 m^3/s, with only a slight further improvement at 280 m^3/s (IEC 1980).

Major adverse effects identified by the COE are due to increased flooding over exist-ing conditions and longer flooding duration in the lower reaches. Current trends of habitat destruction and loss of species diversity in natural ecosystems will be accelerated by the increased diversion, resulting in the deterioration of the natural ecosystem. Managed ecosystems such as leveed farmlands and duck clubs will suffer higher drainage costs as a result of the increased flooding. Unleveed areas may experience crop losses and flooding may ultimately preclude agriculture in such areas. These flooding impacts generally intensify at the higher diversion rate (IEC 1980).

Definition of short-term and long-term effects of this project is difficult, according to the COE. Many effects — especially the beneficial ones on water quality, navigation, and power generation — may be considered short-term in the sense that they will persist only as long as increased diversions are in effect. If the increased diversion program is successful, it could be continued indefinitely, and in this case many effects would be considered long-term.

Changes in natural ecosystem structure and distribution caused by changes in seasonal flooding regimes were considered by the COE as long-term because changes could only be reversed by natural processes over long periods of time. Flooding effects on managed ecosystems along the floodplain would also be long-term, but will probably be mitigated by more intensive water level management.

Because the COE found that beneficial impacts have little or no increase at the $280\,m^3/s$ diversion rate, while adverse impacts intensify, they believe the $187\,m^3/s$ diversion results in a great or net benefit and best achieves the stated intentions of the proposed diversions.

While not studied directly, the COE noted that one strong mitigation measure to lessen adverse effects would be seasonal diversion assignments with sharp reductions in the rates during the growing season.

Tunnel and Reservoir Project

The Metropolitan Sanitary District of Greater Chicago planned and is currently implementing the first phase of a totally new concept called the Tunnel and Reservoir Project (TARP) also known as the "deep tunnel". The plan, which has been extremely controversial, offers a possible solution to pollution problems of the Illinois Waterway and flooding problems for the Chicago area.

The key element of the plan, as originally conceived, is to intercept the storm water overflows from new and existing main sewers and channel it into vertical drains. The vertical drains would discharge into a huge underground reservoir excavated in solid rock approximately 244 m below the surface. After temporary storage underground, the storm water overflows would be pumped to a surface reservoir. By adding generating facilities to the required pumping equipment, the upper and lower reservoirs could be used for a hydroelectric pumped-storage operation. Also included in this plan would be treatment of the polluted storm water overflows before they are discharged to the waterway system at controlled rates.

If this system were expanded to include the entire Chicagoland area, it would eliminate basement and underpass flooding and prevent polluted overflows from being discharged into the rivers and channels or Lake Michigan. The proposed system could handle 100-year frequency storms.

The project was planned to be implemented in two phases. The first phase of TARP will consist of 177 km of tunnels for pollution control. The tunnels would hold 7.6 million m^3. The second phase of TARP for flood control would consist of another 32 km of tunnels and several giant reservoirs that could hold 147.6 million m^3 of water.

Controversy arose, however, that threatens the completion of phase I and that has probably completely eliminated the implementation of phase II. Various governmental agencies disputed the costs of the project with estimates ranging from $ 3.3 billion to $ 12.4 billion. Critics called the project an outrageous boondoggle, while defenders of the project believe that it is the most cost-effective solution to the pollution and flood problems of the area.

Construction of the first leg of phase one of TARP (deep tunnel) is nearing completion and will be on line by early 1985. This will include 50 km of tunnel, about 30 drop shafts, and a main stream pumping station. This first leg includes the main stream and the Calumet system. It will have the capacity to capture runoff from a 25 mm rainstorm (essentially it will have a 8 mm storage capacity for the area of the drainage basin). This first leg will reduce BOD and solids by about 80–85 % in instances of combined sewer overflow.

Funding looks dismal for completion of the rest of phase I and phase II (flood control). The US Environmental Protection Agency has indicated that $ 40–50 million would be available for a leg of the tunnel for the Des Plaines system, but no plans are being considered until other funds can be found for that project. The US Army Corps of Engineers is conducting a feasibility study for the entire project (phase I and II) to determine federal interest and benefits. That report should be completed in 1985.

Conclusion

The water quality of a stream greatly decreases when man develops a river to serve his immediate needs without management. Input into the system can readily degrade the water quality and subsequent use downstream. As a result, a major source of commercial fish was lost. Society has now demanded that the original stream integrity be restored. To avoid the degradation, sources of pollution should be avoided through corrective action at the source of the problem.

References

Butts, T.: Waste Load Reductions and Water Quality Improvement. Peoria Lake: A Question of Survival. Tri-County Planning Commission, East Peoria, Illinois 1983.
IEC: A Preliminary Draft. Environmental Impact Assessment of an Increased Lake Michigan Diversion at Chicago. International Environmental Consultants. Denver, Colorado 1980.

Lundqvist, J.; Lohm, U. and Falkenmark, M. (eds.):
Strategies for River Basin Management, pp. 179–187
© 1985 D. Reidel Publishing Company

Water-Course Monitoring in a Swedish River Basin:
Growth, Development and Future

Karlsson, G.; Löwgren, M., Dept. of Water in Environment and Society, University of Linköping, S-58183 Linköping, Sweden

Abstract: The network of sampling stations and sampling frequency in order to monitor water quality within a river basin is described. The accuracy and possibilities to detect long term trends are discussed in relation to the efficiency of the sampling programme. The development in Sweden during the 1980's implies that moves have been made from the efforts towards uniformity which characterized the 1970's to the introduction of local and regionally adapted programmes which consider the loading activities in each respective area. Therefore the recipient monitoring will be divided into basic and additional programmes.

The Post-War Development

Good quality water in sufficient amounts has long been considered as an unlimited resource in Sweden. The rapid regional structural development during the post-war years has implied new and increased demands on water availability. From originally concerning mainly hygienic and quantitative supply aspects, interest is being increasingly directed at water quality.

The turn of the century in Sweden saw like many other European countries and US a major development in the establishment of underground water and drainage systems in the towns. This development took place later in rural areas. Drainage systems were usually emptied into the nearest water-course. The peak of the load from individual, municipal and industrial discharges occurred during the late 1950's (Fig 1). The increased load resulted in an increasingly accelerated pollution and eutrophication of Swedish lakes, water-courses and coastal areas.

When human influences on water quality was highlighted in the late 1940's the need arose of a continuous monitoring system. During the first phase of the water-course monitoring the main emphasis was on bacteriological investigations to determine which surface waters could be used as water intakes for rapidly growing municipalities. During the 1950's and 1960's eutrophication of lakes and streams draw public and scientific attention to the trophic state of waters. Studies indicated that phosphorus could be regarded as the key chemical element limiting planktonic algal growth. (Pavoni 1977; Nordforsk 1980). Consequently, the monitoring activities were extended to include a number of physical-chemical variables. Simultaneously, the network of stations and the sampling

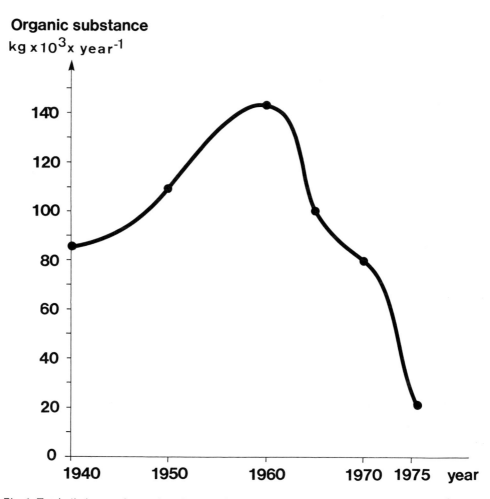

Organic substance

kg x 10^3 x year^{-1}

Fig 1 Total discharge of organic substance 1940–1975 into the upper parts of the Svartån river basin, Sweden.

frequency was strongly increased (Fig 2 and 3). The development was characterized by efforts to attain uniformity both with regard to the degree of sewage treatment and the design of the water-course monitoring programme.

As a result of the increasing interest in environmental problems during the 1960's and in order to coordinate the expanding activities, the National Board of Environment Protection was founded in 1967 and environmental protection legislation was introduced in 1969, a result of which being that companies and municipalities are legally bound to investigate the effects of their activities on the surroundings. Pollution problems also changed from being of local or regional concern of limited interest as regards aesthetics or health, to being national and sometimes international problems.

Parameter	Annual sampling frequencies (1955 → 1985)
Limnological studies	1
Benthic fauna	1
Phytoplankton	1
TCA	1
Phenoxy acids	12
Cadmium	12
Particulate phosphorus	12
Suspended solids	6/12
Water flow	12 4/12 4/12 4/12 4/12 6/12
Ammonia nitrogen	4 4 4 4 4 4 4 4 4 4
Nitrate	4 4 4 4 4 4 4 4 4 4 6/12
Orthophosphate	4 4 4 4 4 4 4 4 4 4
Alkalinity	4 4 4 4 4 4 4 4 4 4 6/12
Water clarity	4 4 4 4 4 4 4 4 4 4
pH	2 4 4 4 4 4 4 4 4 4 4 6/12
Electrical conductivity	2 4 4 4 4 4 4 4 4 4 4 6/12
Turbidity	2 4 4 4 4 4 4 4 4 4 4
Colour	2 4 4 4 4 4 4 4 4 4 4 6/12
Temperature	2 4 4 4 4 4 4/12 4/12 4/12 4/12 4/12 6/12
Total nitrogen	1 1 1 1 1 2 2 3 2 4 4 4 4 4 4/12 4/12 4/12 4/12 4/12 6/12
Total phosphorus	1 1 1 1 1 2 2 3 2 4 4 4 4 4 4/12 4/12 4/12 4/12 4/12 6/12
Oxygen saturation	2 … 1 1 1 1 1 2 2 3 2 4 4 4 4 4 4 4 4 4
Dissolved oxygen	1 2 2 4 3 2 1 2 1 2 2 … 2 4 4 4 4 4 4 4 4 4 4 6/12
Permanganate consumption	1 2 2 4 3 2 1 2 1 2 2 1 1 1 1 1 2 2 3 2 4 4 4 4 4 4/12 4/12 4/12 4/12 4/12 6/12
Biochemical oxygen demand	1 2 2 4 3 2 1 2 1 2 2 1 1 1 1 1 2 2 3 3 4 4 4 4 4 4/12 4/12 4/12 4/12
Bacteriological examin. 1)	1 2 2 2 2 2 2 2 2 2 2 1 1 1 1 2
Bacteriological examin. 2)	1 2 2 2 2 2 2 2 2 2 2 1 1 1 1 2 2 2 3 2 4 4 4 4 4 4 4 4 4 4
Bacteriological examin. 3)	1 2 2 2 2 2 2 2 2 2 2 1 1 1 1 2 2 2 3 2 4 4 4 4 4 4 4 4 4 4
Sediment studies	1 1 … 1
	1955 1960 1965 1970 1975 1980 1985

Fig 2 The changes of the monitoring programme during 1955–1985 in the Svartån river basin. The values indicate the annual sampling frequencies: 1. Thermo-stable coliforms 45°C, 2. Coliforms 37°C, 3. Total agar plate counts of bacteria, 4. The sampling frequency is increased at some of the stations from 4 to 12 times per year, 5. Proposed base programme together with an additional programme with focus on agricultural impacts.

Number of data collection stations

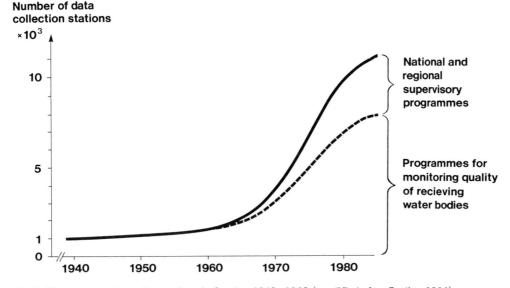

Fig 3 The number of sampling stations in Sweden 1940–1980 (modified after Ryding 1984).

During the late 1960's and the entire 1970's considerable economic inputs have been made to limit the supply of plant nutrients. The measures taken are mainly increased sewage treatment plants for private, municipal and industrial discharge water, i.e. limitation of point discharges. For example, after 1960 the discharge of organic matter into the upper parts of the river Svartån has decreased to 1/6 of the maximum load (Fig 1). Reductions of that magnitude are common all over Sweden. During 1969—1981 about 5 billion kronor (SEK) were invested in Sweden, of which the government provided SEK in state support to remove plant nutrients (Gandy 1982).

In the 1970's attention was also drawn to the need of retaining certain waters in an undisturbed state in order to function as references for environmental work and for research as well as for fishing and recreation. In this context a nationwide programme of environmental monitoring was created in Sweden (Fig 3).

Information Accuracy

The question could be raised whether the monitoring programme has been successfully designed to provide decision-makers with adequate information. An increasing amount of data has been collected from a large number of sampling stations. What do these data tell us about the environmental conditions of a water course? Do they supply information about the effects of measures taken to reduce the load of pollution? Here we present a few aspects from a research programme in progress. The examples are all taken from sampling station Li 13, located at the mouth of the River Svartån (Fig 4) where it discharges into Lake Roxen in the Motala Ström river basin.

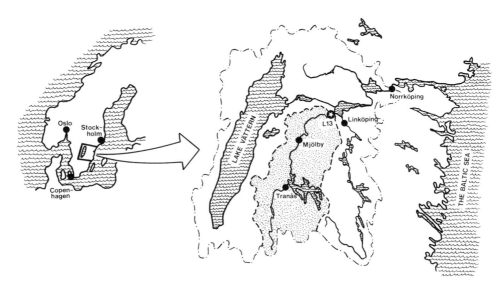

Fig 4 Map of the Motal Ström river basin. The shaded area represents the Svartån river sub-basin.

First, let us look upon the information accuracy in relation to sampling frequency. If consideration is given to the prevailing transport capacity of the water at sampling station Li 13, the amount of total nitrogen transported, as shown in Fig 5, is obtained. The calculations are based on monthly samplings of the nitrogen content and the daily water flow during the period in question. A similar pattern applies to phosphorus (to be published).

During the 1960's and 1970's the measurements and monitoring were frequently conducted during the period with high biological activity, i.e. during the summer (Fig 5). In this period the water flow in central Sweden is low (Karlsson et al. 1983) and most nutrients are bound in biologically active material (algae, plants, etc.). In addition, the activity within agriculture and forestry that may cause a diffuse load on the water-ways is at a low level during the summer. This results in that the measurements made during the biologically active period register minimum contents and minimum transports. It is obvious that transports of nitrogen and phosphorus mainly occur during the autumn and spring flood period (Fig 4), and hence the yearly transport will not be correctly estimated if one relies only on samplings from the summer period. However, the details of the pattern could vary in various parts of Sweden.

The general conclusion is that series with low sampling frequency in combination with sampling during the biologically most active period do not permit the estimation of total plant nutrient transports during the year, straight from the information gathered.

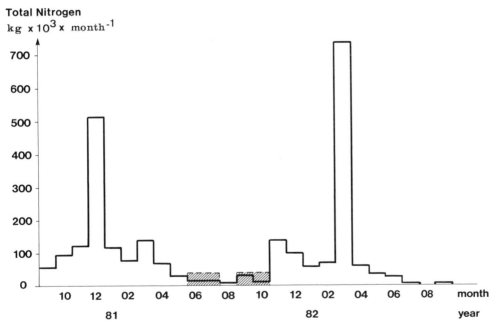

Fig 5 Monthly variation in the total amount of nitrogen transported from September 1980 to September 1982. The shaded areas represent the June-October sampling results during 1955–1980. About 70% of the total number of samples were taken during this period.

Long-Term Trends

As discussed above, individual strongly diverging analytical results in connection with very low sampling frequency, lead to considerable unreliability. The difficulty also increases owing to the seasonal variation being neither equidistant in time nor equally large in amplitude from year to year. One reason for that is the considerable variation in climatological conditions, e.g. temperature and precipitation, within and between years. Another problem is the lack of water flow information.

To a certain extent this kind of problems could be overcome. A general methodology is described by McLeod (1983) for identifying and statistically modeling trends which may be contained in a water quality time series. Techniques are worked out to deal with missing values, seasonality and serial dependence, which provides a means for testing the change over time in the relationship between constituent concentration and flow (Hirsch et al. 1982; Zetterqvist 1984).

The application of seasonal and flow adjustment methods mentioned above to the Li 13 data (Fig 6) suggests, that the level of phosphorus transports is fairly constant during 1966–1980 (to be published). During the same period large investments have been made in order to reduce the point load of phosphorus. If the preliminary analysis is correct the conclusion is that this investment made so far have not resulted in a decreased level of phosphorus transport in this river.

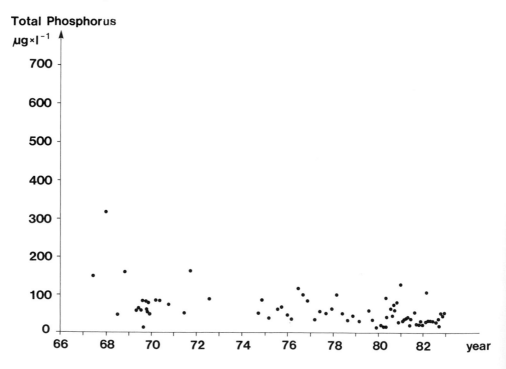

Fig 6 Contents of total phosphorus at sampling station Li 13, 1966–1983.

Environment Control in the Future

The Swedish Environment Protection Board is presently revising the general recommendation for monitoring programmes. Environmental monitoring is exposed to a general and what appears to be an unresolvable dilemma. How should a monitoring programme be designed to consider new advances of science and, simultaneously, ensure continuity and monitoring of long-term development? In the future there will be possibilities to analyze at present unknown variables and improved analytical techniques will increase the degree of accuracy in monitoring variables that have already been introduced. Priorities must also be made in the collection of data, for administrative and economic reasons.

The general opinion is that the environmental protection inputs made in Sweden during the early 1970's have had great local importance for lakes and streams (Bernes 1983). However, it is difficult to show any general improvement of water quality, at least in sampling sites that are located far away from any point source, as is the case with Li 13, mentioned earlier in this paper.

There are several reasons contributing to this. Lakes, for example, function as sediment traps, which implies that accumulation of nutrients in the sediment has taken place for many years. When a lake then unloads itself the sediment leaches whereby the lake discharges more nutrients than those supplied in the river basin (e.g. Ryding 1977). Other reasons for the lack of effect are that the load pressure within a river basin changes with time. Examples of this are found in the increase of manure and chemical fertilizers in agriculture and forestry, in aquaculture of fish with increased supply of nutrients as a result etc.

In looking for measures that can be implemented to ensure the desired improvement in water quality, the leaching of plant nutrients from the area-dependent activities has attracted great attention. The existing monitoring system which aimed at registering the qualitative changes of water in connection with point loads has been, as discussed above, found insufficient in this context.

The Swedish Environment Protection Board is presently preparing general recommendations for recipient monitoring in waters. These will replace the earlier guidelines for recipient investigations published in 1972.

The aim of the new water-course monitoring can be summarized as follows:

— to reveal large transports of elements and contributions from individual pollutant sources within a river basin,

— to describe effects in the recipient of pollutant discharges and interference with nature,

— to relate levels and developmental tendencies of pollutant discharges and other disturbances in the water environment to anticipated background and/or assessment data on environmental quality,

— to provide data for evaluation, planning and accomplishment of measures for environmental protection.

It has been proposed that the work is divided into basic and additional programmes. Whereas the basic programme will deal with a smaller number of variables with a sampling frequency of 6 times per year in a relatively sparse sampling network which is not associated with emittants, it is intended that the additional programme should be specially designed as regards variables, sampling frequency and localization of the station network (so called branch-wise additional programme).

A third type of programme, the so-called emergency programme, will be used if the water quality in an area changes for some reason. Fig 2 gives one example of a base programme and a special programme for the area-dependent activities.

In the base programme under consideration the bacteriological investigations have been omitted. Instead, priority is given to continuously recurrent biological investigations with sampling frequencies annually and every fifth year.

Concluding Remarks

Pollution control like any social policy, is a continually evolving process, altering its complexion to suit changing circumstances. Until now it has largely been a reactive activity, in the sense that it emerged as a response to unsatisfactory conditions. Thus for the most part it was predominantly an exercise in remedying damage and nuisance after it had occurred. This method of management is unlikely to deal effectively and efficiently with the emerging pollution problems. Of these, perhaps the most significant is the issue of toxic chemicals. Their numbers, quantities and latency of their effects require an anticipatory approach.

The need for the development of effective early-warning systems was agreed upon already at the 1972 United Nations Conference on the Human Environment held in Stockholm. However, we doubt that a regular monitoring programme could serve that purpose. There is a real scientific difficulty of recognizing new substances in the environment and the buildup of known chemicals. One does not know where, when, how and what to monitor to identify hazards at an early stage (O'Riordan 1981).

Still, environmental monitoring is an important tool for enforcing regulations about sewage discharges, use of fertilizers etc. The development of the Swedish monitoring programme during the 1980's implies that moves have been made from the efforts towards uniformity which characterized the 1970's to the introduction of local and regionally adapted programmes which consider the loading activities in each respective area. Compared to international conditions, the Swedish control strategy employed is approaching the more flexible system in the UK, where control at the regional and local level may enhance the implementation of monitoring programmes designed to deal with local problems as opposed to the system in the US, where statutory emission and ambient quality standards exert major influence on monitoring programmes (Gower 1980).

References

Bernes, C. (ed.): Monitor 1983. The Swedish Environment Protection Board SNV Meddelande 5/1983. 256 pp. (In Swedish).

Gandy, E.: Municipal Sewage Treatment: Evaluation of Environmental Protection Investments — Naturresurs- och Miljökommittén, unpublished report, 113 pp., 1982 (In Swedish).

Gower, A. M. (ed.): Water Quality in Catchment Ecosystems. 335 pp., John Wiley & Sons, 1980.

Hirsch, R. M.; Slack, J. R.; Smith, R. A.: Techniques of Trend Analysis for Monthly Water Quality Data. Water Resources Research 18, 107—121 (1982).

Karlsson, G.; Ryding, S.-O.; Sandén, P.: Water deficit in River Svartån? — Some Aspects on Water Demands in a Drainage Basin in South Central Sweden. Vatten 39, 376—387 (1983) (In Swedish).

McLeod, I. A.; Hipel, K. W.; Comancho, F.: Trend Assessment of Water Quality Time Series. Water Resources Bulletin 19, 537—547 (1983)

Nordforsk: Monitoring of Inland Waters, OECD Eutrophication Programme. The Nordic Project. Secretariat of Environmental Sciences Publication 2, 207 pp., Helsingfors, 1980.

O'Riordan, T.; Turner, R. K.: Progress in Resource Management and Environmental Planning, Vol. 3, 324 pp., John Wiley & Sons, 1981.

Pavoni, J. L. (ed.): Handbook of Water Quality Management Planning. 419 pp., Van Nostrand Reinhold Company, 1977.

Ryding, S.-O.: Water Monitoring Programmes — Perspectives and Prospects. Vatten 40, 216—226 (1984) (In Swedish).

Ryding, S.-O.; Forsberg, Å.: Sediments as Nutrient Sources in Shallow Polluted Lakes. In: Proceedings: Interaction between Sediments and Fresh Water. pp. 228—234. Junk publication, The Hague 1977.

Zetterqvist, L.: Trend Analysis of Water Quality Data: Test of a Non-parametric Method using Swedish Observations. Lund University. Department of Mathematical Statistics. Report 4, 33 pp. (1984) (In Swedish).

Lundqvist, J.; Lohm, U. and Falkenmark, M. (eds.):
Strategies for River Basin Management, pp. 189–199
© 1985 D. Reidel Publishing Company

3   CONFLICT MANAGEMENT; TOOLS AND PRINCIPLES

3.1 Formal and non-formal options in conflict management

Legal and Administrative Tools for River Basin Development

Cano, G. J., Arenales 2040 – 7-B, 1124 Buenos Aires, Argentina

Abstract: An analysis is made regarding the applicability or non-applicability of "the river basin approach" from the institutional point of view (institutions include law and administrative organizations).
The energy factor in connection with river basin development is considered. Functional and territorial
interjurisdictional problems are examined separately, and examples are given. Three different cases of
territorial interjurisdiction are described: a) national (non shared) river basins; b) national shared river
basins (interstate or interprovincial rivers); c) international river basins.
    The present status of international river law, and of the organization of institutions to manage
international river basins are also examined.
    The influence of water and air as factors of interdependence among all natural resources related
to each river basin preside the work. Recommendations are made of legal tools needed to reach resiliency
in the use of waters following changes in land and water use patterns.

When the River Basin Approach Works

Obviously, every river basin is defined by Nature (water divide or "divortium aquarum").
Within these units, are the waters — and in a lesser degree — the air, the natural resources
which produce interaction with and among the other natural resources, especially land,
flora and fauna. Water as well as air, is fluid and movable and it is impossible to hold it
for a long time or in great volumes. These physical aspects determine and serve as elements
and factors of primal influence in the river basin ecosystem.

The institutional problem (and legislation is an institution) arises frequently when
the geographical or territorial area of a basin does not coincide with the physical boundaries over which the political institutions have powers. However, an entire river basin
under only one political jurisdiction may not function properly, due to the inadequate
local political institutions (and not because of interjurisdictional conflict of interests).

But the most frequent fact is that of interjurisdictional conflicts. These can take
place among: a) two or more independent states, something which falls within the International Law orbit; b) two or more political subdivisions of one independent state (provinces, states, republics, or cantons). The latter case as well as that one mentioned in the
precedent paragraph are presided by the internal legislation and institutional arrangements
of the same independent state. We will deal with the three situations.

But whilst propitiating the administration of waters and connected natural resources
by river basins, we do not lose sight of the fact that a country's Government cannot become a Confederation of river basin agencies. The legislation, and the organization of the

administrative institutions are not purpose ends in themselves but only tools or political instruments. Therefore, subpolicies for each river basin must be included in the general policies of a country, starting with that which refers to water resources. A policy and a general legislation on water resources must rule in every country with the necessary modalities for adapting themselves to the particular circumstances of each river basin.

When the River Basin Approach Is Left Aside

Some uses of the water resources, or of the goods produced with them, can take place outside the river basin of origin, i.e. a transfer of water to a different river basin or the transmitting of electricity generated in a river basin for consumption in another river basin. The economic, social and political effects of the usage of said natural resources are thus expanded farther than the boundaries of its original basin. This involves also the need to expand the territorial action area of the responsible administrative organization, although only in as much as it concerns the corresponding specific subject (water or electricity provision). The same takes place when, for fluvial transportation effects, the navigable channels of two different river basins are interconnected (this is the case of the St. Lawrence Seaway System between the USA and Canada).

Some people object to the fact that international law does not bind to adopt the river basin concept (e.g. Bourne 1969; and UNDP 1975). In my opinion the above mentioned cases are the exceptions which only confirm the rule, and which do not prevent to adopt the basin as an action unit without prejudice of expanding outside it, the responsibilities of the administrator organization, in the fields that might be necessary.

The transfer of water from one basin to another creates particular juridical problems. Generally it requires consent from the people of the areas who might be deprived of present or potential usage of the water and compensations may consequently be called for. However, in the case of the Lanoux lake (Martin-Retortillo et al. 1975) France's unilateral right to transfer water from an international river by recompensating the water withdrawn with the future addition of similar volumes was recognized.

As recalled three years ago (Cano 1982) there is in each river basin a hydrological sub-cycle integrating the global cycle. This implies a need of including underground waters and also atmospheric ones, both in the administration of the water resources of the river basin and also in the pertinent legislation.

Surface and underground river basins may sometimes not coincide. We have therefore patronaged (Cano 1980) a legal solution in which the sites of recharge are regulated as well as of those of extraction. In these exceptional cases, the river basin concept must be displaced due to a physical fact.

Atmospheric waters are of special concern in very barren regions such as Israel. Legislation does there consider every year's precipitation and the place of occurrence in order to assign the total amount of water (of any source or origin, including recycled waters) which each farmer can dispose of in said year. Under such circumstances, the river basin concept is not considered primal either.

Acid rains which fall in the country of origin, shall be presided by that country's legislation. But if, as it happens frequently, the acid rain falls over another country, the International Law starts to intervene. It is not always water which is the vehicle of the noxious agent: in the first and famous case of the Trail Smelter (Widstrand 1980) between the USA and Canada, the carrying agent was the air simply. The idea of the river basin played no part either in this case. But there are other examples where it matters: in Cubatao, Brazil, the extreme concentration of noxious fumes and steams due to the establishment of a large industrial park, within one river basin, could be solved by the national or local legislation and administration refered to it.

The Region Approach

In the cases considered above the river basin approach has been substituted by that of region. Water may cease playing a role as an interdependent natural factor, but it can be used as a tool to induce interdependency both physically and economically and even politically (such the case of the St. Lawrence Seaway). See Teclaff's (1967) intelligent considerations in this respect.

Some countries which adopted the river basin authorities system, where there are many small and neighbouring river basins of similar physiographic and sociological characteristics, choose to consolidate in a single agency the management of said various small basins, due to administrative "scale economy" reasons. France is one of them.

Energy Considerations Independent of the River Basin Approach

Energy is the natural resource determinant of the very existence of human life. It is linked to the existence of the biotic natural resources — and to the feeding chain — and to the inert resources. Energy is linked to entropy and the explainable human desire to prolong life as much as possible and that of the planet we inhabit, has to do with the use of energetic sources which produce low entropy. This can certainly be obtained within the boundaries of a river basin, but it is not subject to said limits, and therefore the role of the energetic factor can be and must be contemplated at nationwide scale and even at planet wide scale.

The hydroelectric generation is one of the energetic sources of lowest entropy. It would be a good policy to induce its use instead of burning the nuclear or fossil fuels, particularly if the alternative sources are within a same river basin. But also if they are outside at accessible economical distance. The transbasin water diversions for use of existing slopes in neighbouring river basins have this purpose.

National (non-shared) River Basins

A river basin can be managed in various ways: a) under an ad hoc regime, b) under a normal or common administrative regime (ministerial or departmental) existing in the country. The inconveniences of the latter system are graphically expressed in the figure at the left in Fig 1.

It is taken from a document which I wrote 25 years ago for the United Nations (UN 1959), but is still relevant. Yet this regime admits the simple coordination – at the river basin level – of the action of the different responsible Ministries (Departments). This coordination is still lacking in the majority of river basins.

In these cases, conflicts arise because of functional jurisdiction conflicts rather than territorial ones; for instance: among water and agricultural resources Departments (such as it recently happened in Mexico) or of energy, or of transportation, or of provision of fresh water. River Basin Committees in Argentina coordinate only programming functions but not operative ones.

The management idea for river basins was strongly impulsed through issuance of a document from the United Nations printed in 1958 (UN 1958) and reprinted in 1969, which has become a classic. But the ways in which it can be materialized are many. The above mentioned document (UN 1959) amplifies the ideas outlined and the different possible solutions. Of course, the TVA case (Tennessee Valley Authority USA) was instrumental for divulging the idea although it was largely debated, not because of its institutional aspects but due to financial ones. Anyway, the TVA was not the first agency of this kind since the Hydrographic Confederations, created in Spain (Cano 1978) five years before (in 1928) responded to the river basin administration concept.

The river basin agency idea has been opposed or denaturalized by three factors:

a)   The strong resistance of the bureaucratic "spirit de corps" to admit cuttings or alterations of their powers. Every form of disconcentration or decentralization of powers is resisted by all World's bureaucracies. In spite of two favourable reports, the President

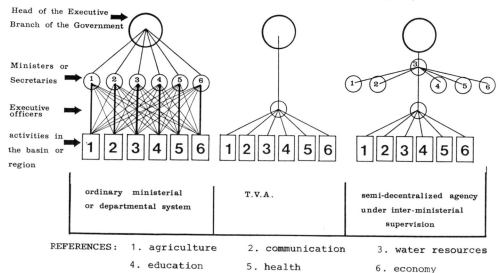

Fig 1 Comparison of the various types of organization for the administration of river basin regions. Source: UN Doc. E/CN.12/503 (1959)

of the United States has been unable to turn the Interior Department into one of Natural Resources.

b)   The tendency to create agencies responsible for hydraulic works instead of agencies responsible for the whole basin. Sometimes, the economic or technical magnitude of said works is a determinant factor. A typical case is that of the "De la Plata river basin" in South America. Although we are now referring to the subject strictly within national boundaries and the example given by me is international, it is worthwhile to point out this because of its complexity. Eleven agencies have powers within the boundaries of a given basin without any coordination. The following table shows the complex organizational structure.

	River	Participant countries	Organization
1	Del Plata Basin	Argentina, Bolivia, Brazil, Paraguay, Uruguay	CIC, Intergovernmental Committee Coordinator of the la Plata Basin Countries
2	Paraná	Brazil, Paraguay	Itaipú Binacional (Work: Itaipú)
3	Paraná	Argentina, Paraguay	COMIP, Argentine-Paraguayan Mixed Commission for the Paraná River (Work: Corpus)
4	Paraná	Argentina, Paraguay	EBY, Yacyretá Bynational Entity (Work: Yacyretá)
5	Uruguay	Argentina, Brazil	Water and Electrical Power (Argentina) and Elettrobras (Brazil) Coordinating Commission
6	Uruguay	Argentina, Uruguay	CTM, Salto Grande Argentine-Uruguayan Technic Mixed Commission (Work: Salto Grande)
7	Uruguay	Argentina, Uruguay	CARU, Administrative Commission of the Uruguay River
8	de la Plata	Argentina, Uruguay	Administrative Commission of the River Plate
9	de la Plata	Argentina	Sea Front of the River Plate Mixed Technical Commission
10	Paraná	Argentina	Water and Electric Power State Enterprise, Management of the Paraná Medio Works
10	Paraná	Argentina	Interprovincial Coordinating Committee for the Paraná Medio Works
11	Paraná	Brazil	Elettrobras, "holding" Enterprise from the Brazilian Federal Government

When the basic frame legislation for a whole country allows the adoption (through complementary administrative regulations) of different solutions for problems related to different river basins, it is obvious that the degree of resilience — essential for facing the need of such changes — may increase.

The same can be affirmed not only regarding the legislation but also regarding water administration. In 1968 Iran changed its legal institutional system based on the country's religious feelings, adopting a Water Code (Iran 1968). This reform ordered that the new system be gradually introduced also in the administrative aspects, basin per basin. The reform was frustrated for political reasons strange to the subject under discussion here, but nevertheless, it shows a good way.

In both cases, the use of the river basin approach far from opposing this, provides the necessary resilience to meet with success.

The following institutional tools are available to obtain resilience within the frame of a river basin.

a)   Law frame planning; i. e. planning conditioned by legal requisites of binding execution such as the study of the environmental impact; or the anticipated fulfilment of certain given hydrological, technical, social and economical studies.

b)   Land use zoning, which allocates the use of land in each zone and to this the necessary waters for said uses, and which foresees previously regulated compensations when the authorized use of land or water is changed whether favourably or against a pre-existent situation. A good example is that one offered by the Netherlands. But the problems caused by water do not only have to do with its scarcity but also with its excess, and the land use zoning is one of the institutional tools more fit to mitigate floods on floodable plains. The problem just mentioned is typical of the river basins.

c)   Anticipated foresights of the effects which would derive from a change in legislation and of the eventual compensations to be payed when effective — rather than hypothetical — damages occur (Moisset & López 1980).

d)   Re-allocation of water rights for which the original allocation of rights was made for pre-determined periods of time so allowing reallocation at expiration time without any compensation what so ever.

e)   Pollution control based on quality models, both geographical as well as chronological, which allows restrictions up to certain limits. At a given time and place these should not cause environmental damage, but which nevertheless, could provoke it elsewhere or in the future. ("Permissible Damage" theory). This involves the necessity of allocating a given use to each river basin section (Castagnino 1983).

Two of the most developed countries — France and the United Kingdom — have recently adopted the river basin approach in order to organize their water resources administrative institutions such as had been done already by Spain since 1928. Although Italy has not adopted the river basin concept, it has, however, "regionalized" the administration of its water resources based on its main basins.

National (shared) River Basins

In certain federal countries — not all — the problem of adopting a river basin approach originates when river basins are shared by more than one of their internal political subdivisions (states, provinces, cantons, republics). This is the case in Argentina, Australia, Canada, FR Germany, India, Switzerland, and the USA). In other federal countries (Brazil, Mexico, the USSR, Venezuela etc.) the central government has the necessary power to solve such conflicts, or the rivers are simply under the full national authority and, therefore, territorially based interjurisdictional conflicts do not exist.

Two alternative solutions have been used among the countries comprised in the first group, namely:

a) The celebration of agreements, treaties or compacts between the political entities involved (Witmer 1968). In the case of the USA said "compacts" require consent by the National Congress, something which involves the Federal Government in the negotiations as an interested party.

b) Judiciary decision of the controversies when such agreements are lacking, decision which in Argentina, FR Germany, Switzerland and the USA concern to the country's Supreme Judiciary Courts. In India, where the Supreme Court lacks such attribution, they resorted to the creation of River Disputes Boards (India 1956,b), which are to prevent conflicts but which must be consented by the states involved and where dissentments must be solved by an arbitrator named by the President of the Supreme Court. Simultaneously the Interstate Water Dispute Act (India 1956,a) establishes a procedure for solving already formalized disputes, which excludes from its regime the case of the rivers where River Disputes Boards exist. When a party denies to appear before a judge the cases are also decided by a unipersonal Court, named by the President of the Supreme Court.

In any case, in these disputes the International Law principles are applied by analogy. So it was decided by the USA Supreme Court (1907) as well as by Germany's Supreme Court (1927). Experience in interstate river basin agencies is plentiful, both in the USA and in Argentina, India and Australia, and from it a doctrine can be obtained for international scope, see Cano (1960).

International River Basins

This ist the *vexata question*. Progress on this subject is slower than desirable (Widstrand 1980).

In the legislation field, sufficient progress has been made in what Alexander Kiss calls the "soft law": legal rules recommended by international organizations which are not juridically binding but which have, nevertheless, such moral weight that they are being gradually imposed. As Kiss says: "the only thing they (the Governments) cannot do is to ignore them".

Among these are in the first place those which come from international Governmental organizations. In this respect we might mention Stockholm's Declaration (UN 1972) and the Mar de Plata Plan of Action (UN 1977). The latter recommends Governments that, for international basins, "joint Commissions be established among countries when it correspond with the agreement of the interested parties, in order to cooperate in those aspects that have to do with the compilation, normalization and exchange of data, of the shared waters, pollution, prevention and control etc." (Recommendation 86 b).

The UN International Law Commission — pressed by the General Assembly — started at last to work on the subject of the international rivers legislation for uses other than navigation, basing its activity on the river basin concept. The first Commission's Rapporteur, Ambassador Schewbel (USA), produced two reports. Present Rapporteur, Mr. Jens Evensen, produced a third one (UN 1983) which abandon the idea of "basin" substituting it for "river systems", and includes the subject of the River Basins Commissions. But the time of materialization of same into a Convention of binding rules seems still to be quite far away yet. There are other valuable studies on both legal and institutional aspects (Garretson et al. 1967; Chapman 1963; Utton 1977; Hayton 1975).

The international Law Association — a Non Governmental Organization — took eight years to adopt in 1966 the Helsinki Rules, which are a body of legal principles based on the river basin area as legislation and management application unit. They reject the irrestrict sovereignity principle and are based on the equitable apportionment of the waters and benefits of the international river basins. Some Governments unilaterally proclaimed its adoption (Argentina), or incorporated its rulings to their national legislations (Colombia), or apply them in their internal practice (South Africa: "Kriel Report" 1978). In Asia and Africa, the regional political organizations have recommended their adoption. These rules were added or complemented in successive Conferences: New York (1978) Flood Control rules; New Delhi (1974) maintenance of navigable rivers; Belgrade (1980) interrelations of water resources with other natural resources. Some authors have refused to admit any value to the conventions and general declarations of principles in this subject and prefer to deal with these, case by case. I have defended those.

In the international ambit there are two basic rules which have won universal acceptance (although there are a few countries still which refuse to accept them entirely):

a)   The exchange of data which should be reciprocal, ample and on good faith, among all the co-riparian states or (as defined by the Helsinki Rules with more juridical accuracy) the co-basin states. This is a fundamental factor for preventing conflicts and even damages. The flood alarm system achieved in 1983 by the joint action of Argentina, Brazil and Paraguay is an eloquent example in this respect.

b)   That of consultation prior to the construction of water works which might cause harm to other co-basin states. Consultations which some states wish to bring to a "consent" level, and which other states are trying to temper in order that they may not result in the delay in the purposes of a State to construct works in its territory. In the American ambit the 1933 Montevideo Declaration recommends consultation and consent as well as rules to achieve same.

The subject of the transbasin diversion in the international ambit has been already dealt with above. And as to the international river basin administrative organizations, there has been considerable progress. On one part, the International Law Association, in its Madrid Conference in 1976, recommended a set of rules — written by Dante Caponera — for the organization of the administrative institutions.

In fulfilment of a resolution of the Mar del Plata Conference, the United Nations called for an Interregional Seminar in Dakar (1981) which gathered not only Governments but also a selected group of about 25 agencies of the type hereby described, in order to interchange expertise. The author of this document was one of the rapporteurs in said meeting and refers to the analysis carried out there of the different institutional solutions adopted in this field. From that meeting arose also the semiannual publication of a Newsletter (International Rivers and Lakes) by the United Nations Secretariat, which keeps up to date field news. A clear conclusion came forth: that international cooperation is indispensable, but also that in this matter, the Golden Rule is that there does not exist any Golden Rule. It is necessary to adopt an ad hoc solution for each river basin.

References

Bourne, C. G.: The development of international water resources: the drainage basin approach. Canadian Bar Review, p. 62 (March 1969)

Cano, G. J.: Los tratados y convenios entre divisiones politcas de paises federales como fuentes del derecho fluvial internacional (Treaties and compacts among political subdivisions of federal countries as sources of international rivers laws). La Ley 98 s.d. 755 (April 1960)

Cano, G. J.: The river basins as optimal units for water resources planning and administration: Users participation. In: Ministerio de Obras Publicas y Urbanismo, Spain (ed.), Conferencia internacional sobre las organizaciones para la gestión autónoma del uso del agua, Madrid 1978.

Cano, G. J.: Frontier underground waters. Water International 5, 7–9 (1980)

Cano, G. J.: Laws of Nature and Water Laws. In: Lohm, U. (ed.). Water legislation and realities of Natural Law. Department of Water in Environment and Society, University of Linköping. Report B1, 11–17, 1982.

Castagnino, W.: Criterios para el control de polucion del agua a nivel regional. In: CIDAA (ed.), El principio contaminador pagador. Aspectos juridicos de su adopcion en America, Buenos Aires 1983.

Chapman, J. D. (ed.): The international river basin. Vancouver 1963.

Garretson, A. H.; Hayton, R. H.; Olmstead, C. J.: The Law of international drainage basins. Oceana Publications, New York 1967.

Germany Supreme Court: Württemberg vs Prussia and Baden, June 18, 1927.

Hayton, R. D. (ed.): Management of international water resources: institutional and legal aspects. UN Doc Sales Nr. E.75.II.A.2, New York 1975.

India: The interstate Water Disputes. Act 33, 1956a.

India: The Rivers Board. Act 49, 1956b.

Iran: Law of Waters and its Nationalization 1968.

Martin-Retortillo, S. et al.: Aspectos juridicos del trasvase del Ebro. In: Caja de Ahorros de la Inmaculada (ed.), Zaragoza 1975.

Moisset de Espanés, L.; López, J.: Derecho de Aguas. Régimen transitorio y normas de conflicto (Water Law-Transition regime and rules for conflict. Universidad de Cordoba (Argentina), Cordoba 1980.

Teclaff, L.: The river basin in history and law. Martinus Nijhoff, The Hague 1967.

UN: Integrated river basin development. Doc. E/3066, Sales no. 1958 II.B.3 & E.70.II.A4.

UN: Systems of administrative organization for the integrated development of river basins. Doc. E/CN 12/503, 1959.

UN: Report of the UN Conference on the human environment, (Stockholm). Doc. E/conf. 48/14, 1972.

UN: Report of the UN Water conference (Mar del Plata). Doc. Sales n⁰ S.77.II.A.12, Doc. E/conf. 70/29, 1977.

UN: First report on the law of the uses of international water courses for purposes other than navigation. International Law Commission. Doc. A/CN.4/367, 1983.

UNDP: Proceedings of the Interregional Seminar on Water Resources Administration. DP/UN INT-70-371, New Delhi 1975.

USA Supreme Court: Kansas vs. Colorado, 206. US 46, 1907.

Utton, A. E.: Some suggestions for the management of international river basins. Doc. Confagua/C14/16. Mar del Plata 1977.

Widstrand, C. (ed.): Water conflicts and research priorities. Pergamon Press, Oxford 1980.

Witmer, T. R.: Documents on the use and control of the waters of interstate and international streams, compacts, treaties and adjudications. US Government Printing Office, House Department 319, Washington 2nd ed., 1968.

Lundqvist, J.; Lohm, U. and Falkenmark, M. (eds.):
Strategies for River Basin Management, pp. 201–208
© *1985 D. Reidel Publishing Company*

Legal, Administrative and Economic Tools for Conflict Resolution —
Perspectives on river basin management in industrialized countries

Williams, Alan, Department of Economics, University of York, Heslington, York YD1 5DD, England

Abstract: Alternative mechanisms for the resolution of conflict between competing uses and users of river basins need to be judged by their respective effects on efficiency, equity, and macro-economic policy. Efficiency should go beyond financial considerations, and assess true real resource costs and all benefits whether or not reflected in revenues. Equity concerns compensation for loss of legitimate expectations as well as equal treatment of equals and distributive justice. Macro-economic policy concerns effects on inflation, unemployment, balance of payments, etc. Legal and administrative measures involve a prior assignment of rights, duties, powers, etc., and then some expectations about the nature of the transactions that will occur between the various parties in the light of that prior assignment. Market mechanisms can be analysed in the same way. Various possible structures and types of interaction are then explored, from which it is concluded that no system is likely to work well unless it ensures that all decisionmakers are put in a position where they have to weigh all the costs and benefits generated by their actions, and not just those that accrue to them themselves.

Activities to Co-ordinate and Judgement of Success

The volume's remit covers "land and water conservation and management", which, with respect to land, I take to be restricted to those aspects where water use is important, otherwise we will be taken too far afield. That restriction still leaves us with a formidable array of activities to co-ordinate, namely:

— Collection, storage, treatment and distribution of "clean" water for a variety of users (domestic, industrial, commercial and agricultural) for a variety of uses (drinking, cleaning, food preparation, hygiene, cooking, industrial processes, irrigation, etc.) with a variety of quality requirements.
— Collection, treatment, reuse and disposal of "dirty" water from all of the above sources and uses.
— Land and highway drainage, land reclamation, flood protection, sea defences.
— Navigation on inland waterways.
— Recreational use of water space and surrounding land.
— Fisheries.
— Hydro-power.

It requires no great feat of the imagination to envisage the manifold conflicts of interest that can arise between these different uses and users, and which will therefore need to be resolved in some way or other in co-ordinating river basin management.

Efficiency and equity

By what overall criteria are alternative mechanisms for such conflict resolution to be judged? It is usual to consider two kinds of criteria, those relating to efficiency, and those relating to equity. But in the present context I think it may be necessary to consider also a third kind, namely, those relating to macroeconomic management of the economy (i.e. policies concerning inflation, and/or unemployment, and/or the balance of payments). A whole river basin will often be so large an economic entity that its jurisdictional structure, and the policies of the bodies co-ordinating it, will need to be judged in the light of all the government's objectives, and not just those concerned with microeconomic efficiency and equity.

These criteria are difficult to operationalise, even in the context of a unitary nation-state. Economic efficiency requires the comparison of the value of benefits with the value of resources used, these valuations being on a society-wide basis. Budgeting systems and accounting conventions typically throw up a very distorted view of resource costs. Capital investment and current costs are e.g. usually handled differently, accounting conventions over valuation are not consistent with the relevant economic principles, historical debt structures distort the economic data. Moreover, revenues are often accepted unthinkingly as adequate measures of benefit.

Alternatively, the judgment of benefits is treated purely as an exercise in professional judgement, or a matter for political persuasion, which carries with it the danger that the mechanisms for conflict resolution will be judged by the satisfaction they generate for the people working the system rather than by their effect on the welfare of the community at large. For the latter to be established, a necessary, but not sufficient, condition is that when inspecting the outcome of the conflict resolution mechanism used, at the margin the benefits should equal the costs. To make the efficiency condition complete, however, total benefits must also exceed total costs, all judged across all affected parties.

The concept of equity (or fairness, or justice) is equally difficult to operationalise, partly because it has so many different strands.

One element in it is that of compensation for loss of legitimate expectations, as in cases where land is compulsorily taken from somebody, or someone's livelihood is jeopardised by some change in the way the system works, (e.g. the diversion of water for some new use puts an existing user out of business). It is a nice judgement which of these instances is a loss of "legitimate" expectation and which is an ordinary commercial risk.

Another element in equity is the equal treatment of equals (so-called "horizontal" equity). But equal in what respect? Does horizontal equity rule out discrimination between users because of where they are? Or who they are? Or how much water they want (or actually use)? Or when? Or of what quality? If the only form of discrimination ruled out by horizontal equity is that which is strictly personal (i.e. not based on any measurable characteristics of the user, but only on the user by name) then it is a pretty vacuous concept. If it is to have real substance, it must be given greater specificity, by stating the dimensions along which the "equalness" of "equity" may be established and therefore what kinds of discrimination are regarded as equitable. It then follows that anything which is omitted from that list of relevant dimensions is a kind of discrimination which is to be held to be inequitable.

A third element in equity is that which concerns the unequal treatment of unequals (so-called "vertical" equity). This is usually taken to mean that the rich should pay more than the poor for the same good or service. In the water industry this discussion usually gets transformed into the related issue as to whether providing for some particular water use should be seen as a commercial undertaking or as a social service. Finally, it may be noted that some people consider that equity mainly requires that people should themselves meet the costs they impose on the system (or upon others), in which case equity and efficiency criteria generate precisely the same requirements at the margin. It will be noted, however, that this equity concept is diametrically opposed to the preceding one for it rules out systematic redistribution from one group to another, though it still permits one group to have priority over another provided the full costs of so doing are met by the priority group.

Macroeconomic management

I will not elaborate further the criteria stemming from macroeconomic management considerations. However, it is to be noted that, since any co-ordinated river basin management strategy is bound to be in the hands of a governmental or quasi-governmental body of some kind, then it is almost certain to be liable to political direction (or 'persuasion') to ensure that its policies at any particular time are at least compatible with (if not actively promoting) the general stabilisation policy of the government.

It will be obvious that just as there may be conflicts of interest between uses and users in respect to the various activities so may there be conflicts between the various objectives. In that situation it is useful to have some common test of the extent to which 'tradeoffs' between objectives are being accepted, as preferable to allowing any one objective to dominate. This can be done by calculating the 'efficiency' cost of pursuing the other objectives, that is the extent to which the pursuit of (say) equity increases the resource costs of the solution over and above what these costs would have been if no such equity objective existed. This has the advantage of giving some numerical value to each objective, so that it is possible to judge the strength of people's preferences about non-efficiency objectives.

A further point which needs stressing is that all mechanisms for conflict resolution have 'operating costs'. These costs may be associated with the gathering of information, arranging and conducting meetings of the interested parties, promulgating decisions, exhorting people to abide by such decisions, monitoring behaviour, enforcing agreements, rules, laws, etc. punishing offenders, and so on. These 'transaction' costs clearly use up resources and therefore represent efficiency costs, and the way they fall on the different parties may also generate a sense of inequity, so these transaction costs need to be judged against the same criteria as are being applied to the outcomes they generate.

A somewhat different problem I need to consider here is that of divided political jurisdictions within a river basin. In this context there may be no political body which sees its role as attempting to enforce consistent efficiency, equity, or stabilisation policy requirements upon conflict resolution mechanisms which span the whole community. This lack of political integration will obviously arise where several nation-states share a river basin, but it also arises when, within a nation-state, decentralisation of power has fragmented geographical responsibility for any of the functions listed in the beginning of this paper. Moreover, alongside all these functions, and intimately associated with them, is control over land use and building developments (housing, roads, railways, commercial and industrial development, public services such as schools and hospitals, defence establishments, etc.) which typically involves a further proliferation of geographical jurisdictions. In these circumstances I doubt whether there is (or ever will be?) any sizeable river basin anywhere which really manifests effective control by a single political jurisdiction.

So my perspective on the problem is:

— Water has many uses and users.
— Conflict of interest is inevitable.
— Mechanisms for conflict resolution can be judged by any or all of the following criteria:
  1. Efficiency
  2. Equity
  3. Compatibility with macroeconomic policy
— Each of these criteria needs careful elucidation before being applied in any particular situation.
— Comprehensive effective political integration of responsibility for all aspects of river basin strategy is not feasible.

Mechanisms for Conflict Resolution

Since the economic system is itself a mechanism for conflict resolution, the legal and administrative tools should be judged against the economic tools (i.e. pricing rules and investment criteria) to see in what circumstances, and in what senses, they might generate 'better' incentives to optimal river basin management. There are two distinct stages in the establishment of coordinating mechanisms, the prior assignment of rights, duties, powers, responsibilities, etc., and the nature of the interactions that then occur between the parties once these 'rights' have been assigned. Let us consider each stage separately to begin with, though they are clearly interdependent.

One possibility is to rely on enabling legislation to permit private individuals or organisations to compete in the supply of the various services listed in the beginning of this paper, subject to such regulations concerning safety or quality as the government thinks fit. This is what happens with food production and distribution in most economies, where the health and safety hazards are similar to water, so, though uncommon in the water industry in industrialised countries, should not be unthinkable. Individuals or organisations may simply supply their own needs, or they might offer services to others, so that each service might have its own array of private suppliers competing with each other. Alternatively each supplier might be awarded an exclusive territorial franchise, which could be taken away and offered to someone else if performance is poor in any respect.

A second possibility is to assign responsibility for each service to a separate public body, which has no direct taxing powers (but might nevertheless be wholly or partially financed out of tax revenues) and which is left with a great deal of managerial discretion within a broadly drawn constitution. Such a body would typically be given the responsibility of providing some particular service in the manner best designed to serve the welfare of its customers, but subject to an externally determined budget constraint and/or charging policy.

A third broad possibility is to set up a separate body for each service but in this instance to give it taxing power so that it does not have to rely entirely on the revenues obtained from selling its services to its customers (or from subsidies from the government). The principle of no taxation without representation is then usually adduced to ensure that such bodies have at least some elected people on the management board. This, in turn, may diminish the rigour of the constitutional constraints imposed on such bodies, relying instead on the democratic process to protect everyone's interests.

A fourth possibility is that the government runs each service as part of its own activities, without separate user charges or taxation, rather as it might run police services, defence, education, etc., but with each service being run by a separate government department. A fifth possibility is that one or more of the services is run by an international or supranational agency on behalf of several interested nation states.

A variant of each of these five possibilities is that many, if not all, of the services are brought together under unified control, which means that some of the conflicts now arise within organisations rather than between them. In this case the bodies concerned will be called "intra-organisational entities".

When we further consider that private individuals may also be parties to disputes (with each other and with any of these organisations) this leaves us with a formidable array of types of user, each with access to a different range of mechanisms in order to seek a resolution that favours their particular interest. Our list of types of user thus looks like this:

— Private individuals
— Private firms
— Public bodies without taxation powers
— Local authorities with taxation powers
— National governments
— International and supranational agencies
— Intra-organisational entities

In general, I would expect any one river basin to manifest a heterogeneous pattern of assigned powers, such that some services are run one way, some another, and some by a mixture.

Mechanisms for conflict resolution

Turning from the constitutional issue of who has what rights, powers, duties, etc. to the nature of the interactions which might occur between them when they are in conflict over water use, it is useful to distinguish nine different mechanisms which may operate (though more than one may be brought to bear in any particular dispute). The nine mechanisms are:

1   Negotiation
2   Seeking an alternative supplier/user
3   Changing the tax system to improve incentive/disincentive effects
4   Changing the grant/subsidy system with the same objectives
5   Litigation through courts or tribunals
6   Applying informal non-violent coercion (e.g. political power)
7   Seeking arbitration by superior authority
8   Seeking public support for a change in regulations, constitutional powers, or the legal framework more generally
9   Use of violence or threats of violence.

Where the process begins depends on the powers of the aggrieved party relative to the other party. Where it finishes up depends also upon who the parties are but also upon the nature of what is at stake and the extent to which their respective interests are affected.

If the conflict of interest is one which regularly and predictably occurs (e.g. seasonal shifts in priorities as between different uses or users of water) then it is likely that a good administrative or technical procedure will have been worked out which will be applied routinely without any great tension. As circumstances change, growing resentment, or the occasional grossly undesirable outcome, may lead to review of the procedures. This may lead initially to attempts to renegotiate the rules, a process which will normally be conducted quietly and in a businesslike way between the professionals. If things get rough, and one party has access to more powerful adjustment tools than the other (e.g. can tax and give grants) it may use these to manipulate things in its favour. Alternatively, it may develop into a power struggle and escalate to stages 5, 6, 7, or 8, which may resolve matters, or lead ultimately to 9, depending partly on the nature of the institutions concerned.

Of the above mechanisms, it is clear that 3 and 4 have a strong economic element, but so also do 1, 2 and 5. This can be seen more clearly if we subdivide economic elements into two types, those which operate through some kind of pricing mechanism, and those which operate expost as forms of compensation or punishment on an ad hoc basis. Recourse to financial comensation or punitive fines is most likely to occur where there has been a breach of some other type of mechanism.

A market transaction is then being used essentially as part of an enforcement process rather than as a mechanism for conflict resolution in its own right. But if the enforcement of, say, anti-pollution regulations is weak, and fines for those who are caught are low, then polluters may regard the occasional fine as a "price" they are prepared to pay. They could then use a method of waste disposal which, though "against the rules", remains the cheapest as far as they are concerned. The other subclass of market transactions uses charges and subsidies more directly as a means of resolving conflicts of interest, by attempting to ensure that these charges and subsidies will direct scarce resources to go where their value is greatest, which usually means to the user with the greatest willingness and ability to pay. If such a mechanism is operating efficiently, it should also generate revenue out of which the "losses" may be fully compensated. Unlike the use of regulations (backed up by occasional fines) a fixed tax/subsidy scheme has greater certainty as regards amount, and can be taken into account prospectively, hence has a more reliable incentive or disincentive effect.

Returning then to the question posed at the beginning of this chapter, namely, "what mechanisms are there for conflict resolution?" the answer is "an almost infinite variety", but this "almost infinite variety" can nevertheless be analysed (or at least described) with respect to the various elements described earlier. Understanding and analysis of modes of conflict resolution with a river basin therefore require not only knowledge of what the dispute is about, but also about the nature of the parties to the dispute, and their chosen type of interaction. Even with this simplified picture of the world we have a rich set of categories with which to think about what goes on in any particular river basin.

Choosing a Strategy for Conflict Resolution within a River Basin

For a river basin to serve the interests of those whose welfare depends in any way upon it, ideally all those making decisions which affect it should take into account correctly the impact of those decisions on all the parties. Put in economic terms, they should weigh all the costs and benefits which flow from their actions, and not just those which affect the interests of the decision making body itself. In general, this is an impossible task, because there are always effects external to the immediate parties which it may be very difficult, or costly, to discover and value, and which, even if known, the decision maker has no incentive to pay regard to (indeed, they would not be "external" effects if he had such an incentive!). Since it is also not feasible to have one single body conducting a continuous comprehensive review of all aspects of river basin management, a practical way forward in an inevitably decentralised system is sketched out in the following paragraphs.

If an activity manifests no significant economies of scale, so is unlikely to become monopolised, and if it generates no significant benefits other than those enjoyed by the people who buy its output, and imposes no significant costs that it does not meet itself, and if the willingness and ability to pay of the individuals served is an adequate test of the value of that output, then such an activity can safely be left to private firms, and any governmental interest in macroeconomic management can manifest itself through fiscal and monetary policy in the usual way.

If these rather stringent conditions are not met, then a variety of other possibilities arises. External costs and benefits may be "internalised" by appropriate tax or subsidy mechanisms. Similarly, a firm may be paid by the government to provide service to people who would not, or could not, pay for it themselves, if it is believed that they should have it. In the presence of monopoly power due to greater efficiency, the government might regulate pricing policies or structures, or, in the extreme, take the monopoly into the public sector. Legal or administrative rules might be used instead of tax/subsidy incentives to regulate behaviour, and then the issue turns on the costs of operating alternative mechanisms and their effectiveness in controlling behaviour.

One important dimension of this effectiveness is the sensitivity of the behaviour-modifying device to changes in the underlying circumstances (e.g. reliance on a rule which says that pollution must never exceed a particular level is less sensitive to changes in the costs of pollution than a pollution charge which varies with treatment costs and the amount of pollution). It seems likely that legal and administrative rules will work well with professionals or with people all within the same organisation, since even complex procedures will be understandable and enforceable in that context. Financial incentives or penalties may work better where the individuals or bodies concerned lack the expertise or the motivation to understand or observe complex regulations, but will conform to more simple rules if failure to do so has obviously adverse financial consequences.

It is important to appraise rival mechanisms in the context of uncertainty. A market mechanism has resilience when prices are free to move quickly and accurately in response to changing demand or supply conditions, thus conveying key information to the respective parties, and eliciting appropriate responses from the various 'actors' in the system. It is a system which can work on very little information. It may sometimes be too sensitive, and too rapid, or exaggerated, responses to changing conditions may then generate additional uncertainty, and speculative behaviour which may perversely destabilise the system or delay the establishment of a viable long term equilibrium.

Can legal and administrative rules do better? They may generate greater certainty in some respects if they are very precise and rigorously enforced, and if they require time consuming and ponderous political or managerial procedures before they can be changed. But how then does the system equilibriate in the meantime? By bribery and corruption (which is a kind of free market)? By hoarding or rationing? By favouritism? Persistent disequilibria are likely to lead to general inefficiency, because people begin to plan defensively, to over-insure against disruption, to get cynical and disillusioned, and generally, to become defeated by the inequity, as well as the inefficiency, of a system which lacks rational adaptive capacity.

Returning to my opening theme, all mechanisms are to be subjected to the same criteria, namely how much do they cost to operate, and how well do they enable the system to perform according to broad efficiency, equity, and macroeconomic policy objectives. It is by that test that the reader should judge the systems discussed elsewhere in this volume.

Lundqvist, J.; Lohm, U. and Falkenmark, M. (eds.):
Strategies for River Basin Management, pp. 209–218
© 1985 D. Reidel Publishing Company

Environmental Conflicts: Structure and Management

Bateld, K., Dept. of Water in Environment and Society, Linköping University, S-58183
Linköping, Sweden

Abstract: Biological, material and psychological claims on natural resources imply human interventions
in the natural systems. The expanding demands may create complex and severe conflicts.

This article deals with competition of water resources within political and social systems. Incompatibility of interests, goals, principles or norms of the parties involved are discussed. A major part of
environmental dispute management occurs within legal institutions. It is argued, however, that environmental protection measures as well as conflicts in that connection through unconventional methods,
may be highly rational and effecient.

Claims, Demands on Water and Power to Get Access to It

Nobody owns the water. People may only have the right to use it according to certain
provisions. In most countries water issues are regulated by a large number of laws and
regulations, and administratively they are handled by several departments and agencies.

Considering the flow of water within the limits of river basins, there are two main
difficulties. One arises from the fact that administrative and natural boundaries do generally
not coincide. Administrative and legal bases of division are usually not based on physical
properties of a landscape. Another difficulty is related to the mutual influence between
land and water utilization, a complexity which is rarely considered in terms of coordinated
legislation or planning.

The water within a river basin is held in common by the nations/states/communities
etc. involved. In most cases there is a common interest in finding a solution to the different
problems. The overall issue is the tension between the specific interests of the individual
users — the necessity of collective measures for protection from the indivisible effects of
individual exploitation.

Water can be diverted and retarded, but it is misleading to say that it is consumed. It
is important to note, however, that the qualitative properties of water undergo change
when it is used as, for instance, a solvent for the transportation of substances. Land can
be put to a variety of uses and thereby causing changes in the quality of the passing water.

Water is consequently a connecting link between various activities and land use
patterns (agriculture, industry, recreational facilities etc.) and between society and the

c) A tendency to add to the river basin agencies other additional responsibilities, such as agricultural or industrial, and even urban development. This weakens — and even complicates — the water responsibilities. The la Plata River Basin case (see above) is a case in point and the institutional mechanisms established for the international waters do not work adequately because of it. As an example of the opposite we could mention the Guayana's Venezuelan Corporation (Corporación Venezolana de la Guayana). This is a "holding" agency covering the whole of one geographical region (the Venezuela Guayana) but where the sectorial responsibilities (navigation on the Orinoco river, hydroelectric generation, urban development, iron ore mining) are vested in autonomous branches.

Population Growth and Technological Progress

Since the second half of the 20th century both the rapid population growth as well as the technological progress are responsible for the frequent and constant changes in land use patterns. The allocation of water resources have consequently also changed.

One the one part, the growth of cities — and consequent water requirements for domestic and municipal use — generally takes place at the expense of waters previously allocated to agricultural uses. Likewise it happens with the new industrial uses. Both new urban as well as industrial uses, synergically increase the water pollution diminishing availability of water for agricultural purposes.

The increase in hydroelectrical generation may preclude use in agriculture — because of temporality of use — creating difficulties in navigation. An example of the latter is a treaty signed in 1979 between Argentina, Brazil and Paraguay in order to make compatible the operation of the Itaipú Dam (destined only to the generation of electrical power) with navigation over the Paraná River and other dams of multiple uses under construction downstream. Thermic or nuclear electrical power generation requires water for cooling, competing with their use in agriculture, usually lead to incompatibilities. In the Danubian Austrian span, the need to open locks frequently (with consequent water discharges) in order to take care of an ever increasing fluvial traffic, diminishes hydroelectrical possibilities.

Recreation uses of high social value (and also economic in connection with the tourist industry) make other uses problematic or impossible, especially if the pollution takes place upstream.

With regard to all the factors which I have pointed out, there are two legal institutions which could become bottlenecks, namely:

a) The rigid and permanent allocation of water to specific uses, thus producing vested rights which must be respected, or compensated in the event of a deprivation.

b) The standing rigid priorities system between conflictive uses or users. Conflicts which would originate among the various different ways of usage or among different users, or even in connection with other natural resources (for instance: hydroelectrical generation vs. thermoelectrical generation; i.e. water vs. petroleum).

natural environment. Man demands water to satisfy various biological, material and psychological needs such as clean drinking water, irrigation and possibilities to recreation. Within and between the different activities there are various and at times irreconcilable claims on water, with conflicts arising as a result. Conflicts may arise in connection with practically any intervention in the natural system. The various interested parties differ in their viewpoints and consequently employ a variety of arguments and lines of action in making claims on the water resource.

> "Conflicts over water tend to mobilize a very high degree of energy and to be very intense in character, probably because actors and group members are deeply involved with one another and highly dependant on the water supply." (Widstrand, 1980)

One aspect is the relationships between upstream and downstream uses of water. A non-consumptive upstream use may involve a degradation in the quality of water, and the downstream user may not be able to use that water. Alternatively he has to invest to be able to use it. Likewise, activities downstream might lead to harmful results, such as flooding, upstream. Another potential conflict area is when water demand and use in the urban areas may cause conflicts with the needs of water for irrigation in the rural area. As the global average for need of water for irrigation is about 80 % of the total water use, conflict situations might arise between the agricultural and urban sectors as well as within agriculture. Irrigation is a costly development which involves substantial technical problems. Therefore, heavy subsidies or regulations are often called for to secure water for irrigation purposes.

With increasing demands for water for various purposes, conflicts between different states sharing the same river basin or aquifer and conflicts between different nations are increasing and may intensify further in the future. Access to, and the right to use or not to use the water is a political question. It is a question of power.

Each conflict includes a number of actors whose values and interests are incompatible. The actors may operate at the local, regional, national or international level, on arenas of various dignity. Which actors who have access to which arenas might be either a formal (i.e. through legislation) or an informal (through well established practice) question.

Power can be seen as an actor's ability to pursue his interests and control the occurrences or ability to prevent others to pursue their interests and control the occurences. The relative power of the actors involved is decisive for the outcome of a conflict. This power is not just canalized through instruments of economic pressure, but also through propaganda, information control, skills and knowledge, and sanctions of varying strength.

The actors may be private individuals, private firms, public bodies, local authorities, national governments, or international entities.

A conflict over a water resource might comprise actors on the same or on different levels. Different actors may act on their own or establish different kind of coalitions. However, the interests of a party in a dispute are not always easily identified. Sometimes they are hidden in complex issues and positions which might be previously stated.

Generally conflicts are more pronounced and difficult to handle in cases where there is a sudden and unpredictable change in the demand for water. For such situations there are normally no established routines for how to solve them.

The Structure of Conflicts

A classical definition of a conflict is given by Kenneth Boulding:

> "Conflict occurs when two or more people believe they have mutually incompatible goals."
> (Boulding 1963)

Conflicts can be seen as compound processes, and to describe them empirically means a generalization of certain characteristics by analysing their dynamics (Tägil 1977). A fundamental assertion is that conflicts are not necessarily static or uniform over time.

Expanding the concept of water conflicts to "environmental conflicts" it can be stated that environmental and resource conflicts are both multiple and complex, around overlapping sets of problems (Wehr 1979). Often there is a conflict over visions of the future, something which is often difficult to pinpoint (Widstrand 1980).

Among the characteristics of natural resources conflicts are very strong ideological bases among the involved, and the judgements and evaluations of technical expertise and general apprehension of various interest groups may differ dramatically.

Wehr claims that people who are engaged in environmental peacemaking, have personal conceptual frameworks guiding their perceptions, strategies for intervention, and evaluation of results. Some would be quite transdisciplinary, others more disciplinbound. The field of conflict management lacks a standard body of knowledge for training peacemakers. Fragmentation characterizes conflict management theory,

> "... and only the smallest part of that which does exist has been applied in environmental dispute settlement."
> (Wehr 1979)

Moreover, primacy of jurisdiction or interpretation of law may not be established, long range effects may be uncertain and decisions may be irreversible (Moore 1983).

Despite all this, it is important to have in mind that conflicts might be productive; producing new standards, and new patterns of relationships. "Conflicts may be necessary in the pursuit of justice" (ibid).

As mentioned earlier, there are within sectors such as agriculture, industry, fishing, energy, nature conservation and recreation different and sometimes irreconcilable claims on water, with different types of conflicts arising as a result:

a) *Quality conflicts:* far-reaching claims on water quality as opposed to recipient claims. When one activity makes water unfit for use in, or causes damage to, another activity.

b) *Preservation conflicts:* claims for intervention in the water system as opposed to claims on undisturbed nature.

c) *Quantity conflicts:* when one use makes difficult/prevents another use, i.e. urban water use makes difficult/prevents another use, i.e. urban water supply as opposed to irrigation.

d) *Mixed quality/quantity conflicts* (SOU 1980).

These conflicts often become more complex through the balancing of the political and other kinds of goals against one another. Water conflicts do often involve political, economic, social interests, such as employment, in opposition to interests which are difficult to measure, such as scenic beauty. Even if water, or more broadly environmental, issues have strong political consequences, they do not attract political priority unless there is a feeling of crises. In many industrialized countries where growth and expansion are the driving principles, environment has never been a primary issue, it is always a kind of back-door issue. Concern for development and environment are like a seesaw, they never run parallel with each other.

States and nations do often have different political strength, in terms of power, and the levels of economic development may also differ. This leads to very complicated conflicts. This is true all over the world, within river basins in Asia as well as in Africa or the United States.

Environmental Conflict Management

Talking either of resolutions of water conflicts or more broadly on natural resources conflicts it must be stated that there are an almost infinite variety of mechanisms for conflict management. It is important to bear in mind that there is a difference between manifest conflicts where tools are needed to *solve* them, or about potential conflicts where you need instruments to *prevent* or *avoid* them.

Differences between actors at different levels in legislation, in political, economic and social prerequisites, in the administrative structure, traditions, valuations are, of course, of great relevance. Methods and instruments for conflict resolutions vary; from uncomplicated agreements, to very formal, heavy international treaties; from settled diplomacy to unconventional negotiations. The results vary likewise; from successful cooperation to unsolved, continous conflict.

The major part of environmental and natural resource dispute management occurs within legal institutions. That means in the courts themselves or through some adversary outside negotiation. It is to be assumed that the intention with water legislation is to pursue an allocation that guarantees that usage leeds to a maximum of benefit for individuals as well as for the entire society.

Because water is a scarce natural resource, rules and regulations are necessary instruments to allocate the supply. How much water is used, where and for what purposes are political questions. It is the classical question of distribution expressed by Harold Laswell: Who Gets What, When, How? Some people think that the best way to allocate water would be through a free maket without any governmental interference. Others realize that this might insufficiently imply an amount for basic needs, and that governments must be held responsible for policy planning. Both groups work within the political system, with their information and propaganda.

Wehr (1979) discusses conflict management through environmental policy planning. It is an approach stating that meaningful participation of citizens in the various stages of planning is the most successful way of resolving and sometimes even avoiding policy conflicts. Planning occurs in three stages: policy analysis, policy determination and policy implementation.

The Citizen-Based Planning includes 1. Alternative Invention, which means that planning agencies and environmental organizations are encouraging the formation of citizen constituencies to assist in defining the problem and to generate new alternative responses to it, 2. Value Mapping, a method to determine value preferences, constructed by citizens and planners and where each scenario has a different set of water requirements and goals based on desired growth, and a set of factors identified as important for the community, 3. Future Images and 4. Judgement Analysis Panels, the use of citizen and expert judgement panels. This method combines value and technical judgements to select a citizen-based policy.

For conflict management elsewhere than in courts, the following measures could be tried (the division is based upon Wehr 1979):

Conflict Avoidance

For instance compensatory agreements among the opponents that eliminate the opposition of the disadvantaged party. It is increasingly used by energy companies. In Japan, however, this buying-out approach is losing ground as potential sellers have started to increase their demanded price beyond what the market will bear.

Another method is buying-in — often used in the United States by federal water planning agencies to avoid conflicts. By this mode, projects are formed so attractive to so many that opposition is eliminated incrementally as potential opponents "climb on board".

Conflict Prevention

Contrary to conflict avoidance, conflict prevention can offer some protection of the natural environment or the less powerful individuals and groups who may suffer from the adverse social impacts of development projects. One approach is internalization of project costs, which means that the planners build into the project both the normal production costs of a commodity, and its social and environmental costs. Further need a development in how to measure and evaluate negative externalities such as air and water pollution is admitted.

Third Party Intervention

There are several forms for third party intervention. In principle the third party should be unconnected to the conflicting parties, or someone with more than a mediator's interest in the dispute but who is accepted by all parties. A third party may act in two ways:

a)   Negotiation and Mediation

A mediator is the one who identifies parties in the dispute, who clarifies central issues, selects participants for negotiating sessions and helps to assure implementation of reached agreements. According to Wehr a mediator in an environmental conflict must be at hand at a stage of development where parties are organized and capable to form clear negotiating goals and strategies. This might be the case when the parties are at an impasse they wish to break and therefore turn to mediation as a last resort. Implementation, Wehr asserts, is the weakest part of the environmental mediation process.

b) Conciliation

Conciliation is used to avoid an impasse, and differs from mediation in its lack of a formal negotiation process. Moreover, its medium is informal discussion rather than formal negotiation.

Impact Analysis and Mitigation

The federal Environmental Impact Analysis (EIA) is used in the United States since a couple of years. An EIA includes comprehensive descriptions of the natural and social environments to be impacted by a proposed action, as well as its likely positive and negative effects.

> "Leaving aside the question of how well this procedure actually protects the environment, we should note that it sometimes works to resolve conflicts. It can help to prepare a dispute for negotiation by clarifying the issues, settling disagreements over facts, and suggesting mitigating measures that might satisfy opponents. On the other hand, an EIA may just as easily stimulate conflict since it must vigorously question the proponent's contention that the need for the project exists and that proposed action is the soundest way to meet it."   (Wehr 1979)

Irrespective of the means used as methods or instruments for conflict management, the most interesting aspect is the struggle for power that can be observed behind said methods and instruments in order to avoid, prevent, mitigate or solve the water conflicts.

Accord Associates — An Example

As an example on third party intervention the organization Accord Associates (former ROMCOE), Center for Environmental Problem Solving, situated in Boulder, Colorado, is quite illustrative.

The center is a private, non-profit institution that specializes in environmental conflict management. Their services include:

> "... long-term and short-term mediation; conflict management, training programs for corporations, government agencies, communities, professional associations, and citizen-advocacy groups; workshops to help small communities prepare for population growth and other changes; and consultation to organizations that want to deal effectively with conflict."   (Lansford 1983)

One of their projects is about Colorado's capital Denver, and its problems with water supply. Denver's enormous population growth has produced two related demands; for more water and for more place to store it. Therefore the Metropolitan Water Roundtable was formed in the fall of 1981 for the purpose of exploring possible concensus resolution on Denver metropolitan water supply needs. The Roundtable is chaired by the State Governor and staffed by Accord Associates. The Roundtable does not have power to make decisions on what projects or water conservation measures should or should not be undertaken.

> "Its function is to examine issues and fashion possible means for conflict resolution, through concensus of affected interests, and then to make recommendations to those empowered to make decisions."   (Hobbs 1982)

The participants have developed a list of interests and have divided themselves into task groups to identify possible methods of meeting Denver's water needs. Membership is held by members of the Denver Water Board, mayors, county commissioners, farmers — all of them from the east and west slopes. Furthermore a representative of the Colorado River Water Conservation District and representatives of the Homebuilders Association, the League of Women Voters, the Environmental Defense Fund, the National Wildlife Federation and Denver neighbourhood groups (ibid).

Reports in the following areas have been written with assistance from a number of professional experts in the water field who have volunteered their time in different committees:

— Projected Water Demand for the Year 2010
— Water Development Projects
— Water Use Efficiency and Recycling
— Ground Water Integration (ibid).

It leads too far to describe in detail how the Roundtable has worked things out. It is however important to stress that it is an unconventional organization which has replaced the tranditional atmosphere of mistrust, animosity and conflict. It seems as if the Accord Associates has created a model where mutually agreeable decisions can be reached.

Legal and Administrative Institutions and Tools

It is impossible to discuss all institutional arrangements for management of land and water resources since they differ widely from one nation to another. Also the natural conditions vary from basin to basin. Special judicial problems can occur when water is transferred from one basin to another, or when the surface and underground basins do not coincide.

Mostly, land use planning and management is conducted by other agencies than those which have responsibility for water resources planning and management. Questions of the principle of integration have been debated over many decades. One thing is clear though; the decision process concerning land and water involves interaction of many entities and rules. Sometimes there are problems with lack of adequate motivating mechanisms among and within institutional arrangements. Even in situations where administrative structures seem to be ideal, rules and regulations seem perfect, things do not work as expected. Or, excellent legislation

"... can fail if the system established for its implementation is inadequate. The fault lies with the manner in which the administrative machinery is organized." (Cano 1977)

This inefficient implementation may be due to various factors; lack of will, lack of internal coordination, shortage of staff or insufficiently trained staff, lack of proper equipment, lack of awareness of the objectives of the legislation.

All over the world it seems to be problematical to achieve desired economic and social development goals because of the strong slugishness of the bureaucracy. In the United States it has been shown that proposed bills on environmental issues are today passing many more committees and sub-committees than 15 years ago. The process is therefore much more time-consuming and complicated. Even if it might be good that many people handle important issues, the process is experienced as unnercessarily unwieldly. Moreover, many of the elected people within the political system have so many assignments that in reality the staff members within the administration are running the business. This is especially true within policy questions on back-door issues such as the environment, which is generally not met with too much attention in mass-media.

Water issues are complicated both technically as well as from the viewpoint of information. Between the relatively few who have the information and the many thousands or hundreds of thousands who are to put it into practice, there is accordingly a complicated network of legal, political and administrative structures. But the problems are not only concerned with issues of facts but equally, or to a greater degree, with how information and directions can be disseminated and implemented.

The trend in recent legislation is to stress the need for multiple, coordinated planning (Cano 1977), and it seems rational as one should consider water resources problems as interdisciplinary problems, requiring cooperation of experts and scientists in different fields. But do we have enough legal experts and administrators possessing the right kind of education and training and with ability to meet the trend towards a more interdisciplinary character of water law? And what about the preparedness among the politicians?

The experts on river basin management are usually engineers or lawyers. There is a need to give the politicians other bases than technical or legislative facts for their decisions on how to formulate the public water policy. People with other more long-term apprehensions than political resources in terms of votes, jobs or influence have only in exceptional cases access to the decision-making arenas.

The legal and administrative tools acting as incentives or disincentives for water resources management can assume different forms; rules on property and water uses, regulations (fines and other penalties in case of violation), water withdrawal charges, pollution charges, user fees or subsidies. Unfortunately are those creating quality problems very rarely benefited by stopping or even correcting their activity. The strategy, what kind of tools will be used to manage water issues within a river basin, shared or not, is first of all a question of political structure and policy.

Problems within national (shared) river basins are to be solved through different kind of agreements, treaties or compacts between the political entities involved. Interstate river basin agencies are in use all over the world, e.g. in the United States and Argentina as well as in India and Australia. But everywhere, most writing on water ignores the political battles behind different kinds of treaties, even though the exertion of political power might be the most important determinant of when, how, and in what form treaties are made. Arizona's acting in connection to the Colorado River Compact gives an example.

Arizona — An Example

First it must be stated that the major purposes of the Colorado River Compact was to provide for the equitable division and apportionment of the use of the waters of the Colorado river system, and to avoid conflicts within the area.

The compact (still in force) was signed in 1922. Ratification by the legislatures of six of the seven involved states followed within a ten years. Arizona refused to ratify until 1944. The representative from Arizona objected to the agreement for two reasons. First, he desired a definitive allotment of water to each state. Secondly he objected to including water from the Gila river as a part of a total Colorado river water.

Houghton (1951) tries to explain the reasons for Arizona's point of view regarding the development of the Colorado river. First of all he explains that in the early 1920's, Arizona's economy was incompletely developed and largely dependant upon mining. The economy was lacking in financial capacity for large-scale power and water development. There existed a general confidence that no major dam construction would start until the compact was ratified by all seven states. So the refusal to ratify the compact would mean a safeguard of Arizona's rights in the river, rights she was not ready for yet. Secondly, it seems obvious that self-centered minority group interests were behind the so-called "State of Arizona" objections to the Colorado River Compact. The principal objecting and opposing groups were central arizona farmers, private power utilities, and mining companies.

The mining interests also opposed the building of Hoover Dam, and was a part of a general policy of utility companies to oppose all so-called "socialistic" extension of public ownership of utility services. They also feared that a government-owned plant would compete seriously with local power sales in Arizona.

Houghton also gives one explanation, among others, to the change in Arizona's policy which led to ratifying the Colorado River Compact in 1944:

> "... by the time of the final collapse of efforts to block River developments through court action in the middle of the 1930's, the stronghold of the opposing groups upon public opinion in Arizona was obviously weakening."   (ibid)

Ever since, the state's history is filled with court actions. The Arizona controversies, such as the bitter campaigns against California, are closely associated with the 1922 Compact. In the future there will not be less politics in Arizona or elsewhere in the Colorado river basin as there is not enough water to meet all the needs of competing interests.

Concluding Words

There is no need to once again point out the impossibility to give detailed guidelines on water conflict management, valid to all river basins. But it is important to be emphatic about the need for, not only traditional, but also fearless, daring or even totally unconventional solutions in the struggle for a better use of our water resources.

It is likewise important to linguistically reform the material, such as the scientific findings, on how to conserve the water resources so that the decision-makers and the public will be able to grasp the issues.

References

Boulding, K. E.: Conflict and Defense. A General Theory. Harper & Row, New York 1963.

Cano, G. J.: Water Law and Legislation: How to use them to obtain optimum results from Water Resources. Proceedings from UN Conference 1977.

Hobbs, G.: The Metropolitan Water Roundtable: Building Consensus. Paper prepared for Colorado Bar Association Annual Convention, Water Law Section, 1982.

Houghton, N. D.: Problems of the Colorado River as Reflected in Arizona Politics. Western Political Quarterly 4, 634–643 (1951)

Lansford, H.: The Metropolitan Water Roundtable. An Interim Case Study in Environmental Conflict Management. Accord Associates, Boulder, Colorado 1983.

Laswell, H.: Politics: Who Gets What, When, How? Cleveland 1958.

Moore, C. W.: Negotiating, Bargaining and Conflict Management. Romcoe, Center for Environmental Problem Solving, Boulder, Colorado 1983.

SOU: 1980 Vattenplanering. Betänkande av vattenplaneringsutredningen. Stockholm 39, 1980.

Tägil, S. (ed.): Studying Boundary Conflicts. A Theoretical Framework. Studies in International History, Lund 1977.

Wehr, P.: Environmental Peacemaking: Problem, Theory and Method. Research in Social Movements. Conflicts and Change Vol. 2 JAI press INc. 1979.

Widstrand, C. (ed.): Water Conflicts and Research Priorities. Water Development Supply and Management, 8, 2. Pergamon Press, London 1980.

Lundqvist, J.; Lohm, U. and Falkenmark, M. (eds.):
Strategies for River Basin Management, pp. 219–227
© 1985 D. Reidel Publishing Company

3.2 Importance of time perspective

Problems of Water Reallocation in the Colorado — America's most fully Utilized River

Skogerboe, Gaylord V., Utah State University, International Irrigation Center, Logan, UT 84322, USA

Abstract: The basin development of the Colorado river started 2000 years ago and accelerated rapidly in the late 1800's. In 1922, the river flow was appropriated between upper and lower basin, based on estimated virgin flow at the dividing point at Lee Ferry. Today the river waters are highly controlled with major reservoir facilities. The paper treats the constraints of increasing downstream salinity and vast reallocation problems in the basin caused by energy development and growing domestic water needs of competing urban regions. With completion of the Central Arizona Project, the lower basin will be fully utilizing their entitlement, and further development may enforce a conversion from irrigation to domestic water supplies. In the upper basin there are still serious questions about their full entitlement, due to varying estimates of the Lee Ferry flow, and thus about the need for future water transfers from the irrigation to water supply of energy complexes. Many questions also remain unsolved regarding Indian water rights.

Colorado River Basin Development in a Historical Perspective

Early History

There is archeological evidence that some 2,000 years ago irrigation canals were built and maintained by the ancient Hohokam tribe in the Salt River valley of the lower Colorado river basin (LCRB) near present-day Phoenix, Arizona, USA (Fig 1). The Hohokams probably began settlement of the valley as early as 300 BC and abandoned it about 1400 AD. Exactly how many people were here is conjectural, but certainly several tens of thousands must have lived and worked in this economy. However, by the time Spanish explorers entered the region in the middle 16th century, this prior civilization was gone and only small settlements of Indians remained, but the economic base was still agriculture. For example, the Navajos, who are situated in the lower basin, only arrived during the last 500 years.

The next irrigators of the Colorado river basin were Jesuits who established themselves at the old missions of Cuevavi and San Xavier in Arizona in 1732. In the period of 1768–1822, considerable irrigation was practiced along the Santa Cruz river near the missions and the Spanish presidios of Tubac and Tucson.

Fig 1 Colorado river basin.

By the treaty concluding the Mexican War in 1849, and by the Gadsden Purchase of 1853, the United States acquired the territories of New Mexico, Arizona, and California. Discovery of gold in California in 1849 brought hordes of adventurers westward. They crossed the Colorado river near Yuma, Arizona, and at Needles, California. In 1857, Lieutenant J.C Ives traveled 60 km up the river by boat to the Black Canyon, present site of Hoover Dam. He reported the region to be valueless.

In 1869, Major John Wesley Powell explored 800 km of the Colorado river system from Green River, Wyoming, to the mouth of the Virgin River within the present area of Lake Mead. Powell's studies and recommendations were the first and for many years the most significant in shaping policy and legislation for adapting the arid lands of the West to agriculture.

Breckenridge, Colorado, on the basin's eastern rim, was settled in 1859 by miners and prospectors pushing over the mountains from older mining districts on the eastern slope of the Continental Divide. Within the next decade, other mining camps were established nearby. Unsuccessful miners turned to farming and supplied agricultural products to the mining communities. Settlements grew downward from the mountains to the valleys, the advance being slowed somewhat by conflicts with the Indians who occupied the territory. Grand Junction, Colorado, now the largest community in the upper basin, was not settled until 1882. The greater part of the Uinta Basin in northeastern Utah was established as an Indian reservation in 1861, and lands unoccupied by Indians were not open to settlement until 1905.

Indian water rights

During the last half of the 19th century, the Indian conflicts had been settled in the basin and Indians were placed on reservations. Through the years, there has been considerable litigation regarding the rights of Indians under the various treaties that established these reservations.

A very famous court case in 1908 was Winters versus United States, in which the court decreed:

The Indians had command of the lands and the waters, command of all their beneficial use, whether kept for hunting, and grazing roving herds of stock, or turned to agriculture and the arts of civilization.

The Winters case, which is often referred to as the Winters Doctrine or Reservation Doctrine, is the basic law of entitlement for Indian peoples. An important consequence is that Indian water rights, to all intents and purposes, were made paramount over non-Indian water rights by this doctrine, because nearly all Indian reservations were created prior to the diversion of waters from the rivers of the western USA by non-Indian interests.

Conceivably, the Indian reservations located in the Colorado river basin could lay claim to a majority, if not all, of the waters in the basin. In fact, this has not been the case. However, there is litigation presently underway to provide more water for some reservations than is presently being used. More importantly, there is a real legal basis for Indians to claim much more water than they are presently utilizing.

Industrial development

The early history of the basin has its roots in the mining industry. As has already been mentioned, the discovery of gold and other precious metals led to an influx of prospectors and miners and the establishment of numerous early settlements. Mining activity and commercial requirements of the booming populations associated with the industry led to the construction of a great network of railroads, mostly narrow gauge to cope with the mountain conditions. These in turn produced a demand for wood for railroad ties and bridge timbers and for fuel. Coal replaced wood as a domestic and industrial fuel source and led to the coal mining industries of Colorado, Utah, and Wyoming.

In the late 1800's and early 1900's the growing populations, both within the region and in the adjacent metropolitan areas, provided an expanding coal market for heating and industrial uses. For a time coal production was of major economic importance. After World War II the substitution of gas for coal as a fuel and the adoption of diesel power on the railroads caused a major decline in coal mining. The decline in coal production was precipitous and many mines, even whole camps and towns, were closed and abandoned. The trend has been reversed in recent years as soaring demands for electric power have led to the development of strip-mining techniques and the construction of mine-mouth power-plants. These have resulted in increased coal production but with only little recovery of coal mining employment.

Mining of molybdenum in western Colorado was started during World War I. Production grew rapidly and now about half of the free world's production is obtained from the area.

Uranium-vanadium deposits have been mined sporadically since about the turn of the century. Exploration and mining boomed during and following World War II with the development of atomic fission and the demands for atomic energy. Radioactive mineral deposits in the region are among the greatest known in the world today.

Production of oil and gas in the region dates from the early 1900's. Petroleum booms came with the discovery of the Rangely field in western Colorado in the 1940's and the Greater Aneth field in southeastern Utah in the late 1950's.

Timber harvesting began with the early settlers who produced lumber for home and business construction, rail ties, mine props, fuel wood, and poles. The accessibility and abundance of this forest product were major factors in the completion of the transcontinental railroad and its subsequent expansion to the early settlements of the region. In recent years, with new methods of utilization and processing, uses for the local timber resources have been greatly expanded and timber has become of major importance to the local economy.

Impetus to hydroelectric power generation was given by the mineral industry. The first hydroelectric development was at Aspen, Colorado, in 1885. In 1891, the Ames Plant, located in the upper portion of the Dolores drainage in Colorado, was among the first hydroelectric plants to transmit alternating current at high voltage. As the region became settled and the need for electricity grew, several small hydro-electric plants were built. It was not until the 1950's that steam-electric power production had significant growth.

During World War II, a very diversified and expanding economy began to evolve in the lower basin. Hoover Dam's power plant provided a source of plentiful electricity for industry. The yearlong climate was favorable to military training activities and the advent of air conditioning tempered the harshness of the hot summer months; Las Vegas, Nevada, with its flamboyant entertainment industry; Lake Mead, Lake Havasu, and Lake Mohave on the Colorado river; and the lakes formed behind the dams on the Salt, Verde, and Gila rivers in central Arizona invited recreationists, hunters, and fishermen. The affluent economy that developed during the war started a tourist influx to the arid Southwest to escape the cold winter months occurring elsewhere. Light industry followed to capture a labor pool and to utilize the yearlong working environment.

Agriculture and forestry produce less than 10 % of the total economic output from the Colorado river basin. Manufacturing provides less than 10 % of the economic output in the upper Colorado river basin, but more than 20 % in the lower Colorado river basin. In contrast, mining is very important in the upper Colorado river basin (one-third of economic output), but only provides 10% of the economic output in the lower Colorado river basin. Noncommodity-producing industries, which include outdoor recreation and tourism, provide more than half of the gross economic output in the basin.

Tremendous growth in noncommodity-producing industries is expected in the decades ahead throughout the basin. Mining activities are expected to show rapid growth in the upper Colorado river basin, while manufacturing activities are expected to show significant growth in the years ahead in the lower Colorado river basin.

Upstream-Downstream Competition

Colorado river compacts

In the early 1900's, a series of actions were instituted which led to interstate compacts and international treaties, State and congressional legislation, and Supreme Court decisions, which today in aggregate constitutes the "Law of the River".

The first action comprising the Law of the River began in 1922 with approval of the Colorado River Compact by representatives of the Colorado river basin states. The compact appropriates the waters of the Colorado river system between the upper and lower basins but did not divide the water among the states. The Boulder Canyon Project Act of 1928 approved the compact and authorized the construction of Hoover Dam and the All-American Canal system.

The Mexican Treaty of 1944 obligated the United States to deliver 1.85 billion m³ of Colorado river water annually to the Republic of Mexico. The Upper Colorado River Compact of 1948 divided the upper basin Colorado river compact apportionments of the Colorado river for beneficial consumptive use among the upper basin states. This, in turn, led to the Colorado River Storage Project Act of 1956 which established an Upper Basin Development Fund and authorized the initial phase of the comprehensive upper basin plan of development.

Today, the waters of the Colorado river are highly controlled. Major reservoir facilities have a combined storage of 80 billion m³, which is more than four times the mean annual flow of less than 18.5 billion m³.

Relations to Mexico

Before the Colorado river was controlled by Hoover Dam, its annual fluctuations from snow melt torrents to meager late summer flows limited Mexico's ability to use Colorado river water to a maximum of about 0.925 billion m^3 per year. Following completion of Hoover Dam in 1935 with regulated releases and reduction of floods, Mexico began to greatly expand its usage of Colorado river water, reaching a reported use of 2.22 billion m^3 in 1943. A treaty was negotiated wherein the Republic of Mexico was guaranteed an annual delivery of 1.85 billion m^3.

Between 1945 and 1961, there were no major problems with respect to the river, as the salinity of the water delivered to Mexico at the international boundary was generally within 100 milligrams per liter of the water at Imperial Dam, the last major diversion for users in the United States. In 1947, the Wellton-Mohawk Project in Arizona was authorized by Congress, with construction being completed in 1952 by the US Bureau of Reclamation. The project lands soon became water-logged and a pumped drainage program was implemented to lower the water table under project lands. The wells, pumps, and drainage water was discharged into the Colorado river below the last United States diversion, but above the Mexican diversion at Morelos Dam. It included a substantial proportion of highly saline ground water that had been concentrated through reuse during the previous half-century. Initially, it had a salinity of around 6,000 mg/l. As a sharp reduction in river flows to Mexico occurred around the same time, the combined impact of the Wellton-Mohawk drainage water and reduction of dilution water was a sharp increase in the salinity of the water delivered to Mexico, from an average of around 800 mg/l in 1960 to plus 1,500 mg/l in 1962. Mexico raised strenuous objections to receiving the drainage waters.

This international problem precipitated action on the part of the US Government, which resulted in the passage by Congress on June 24, 1974 of the Colorado River Basin Salinity Control Act, Public Law 93-320. This act not only provides measures for alleviating the salinity problem in Mexicali Valley, but also contains numerous measures for salinity control throughout the basin in order to allow for full development of the water entitlements to each state.

Reallocation Problems

Salinity control

The water supplies of the Colorado river are the most fully utilized of any major river basin in the USA in this century. The salinity concentrations in the lower main stem have been steadily increasing. Although the salinity problems in the Republic of Mexico (Mexicali Valley) brought international attention, salinity control measures would still be required without the international problem. In fact, the political attention focused upon solving the problems of international relations actually facilitated salinity control measures that will allow for full development of each state in the upper basin to their entitlement, while protecting the utility of the water supplies in the lower basin. However, future water development will have to be done under the constraints of salinity control.

The most cost-effective salinity control measures are on-farm improvements (improved irrigation methods, flow measurement, irrigation scheduling, etc.) on irrigated croplands. Such programs are presently underway in the Wellton-Mohawk Irrigation and Drainage District near Yuma, Arizona, the Grand Valley in western Colorado, and the Uinta Basin in northeastern Utah. Following on-farm improvements, the lining of laterals (water-courses) or conversion to pipelines is the most cost-effective salinity control measure, which is followed by canal lining. In some cases, desalination of irrigation return flows or natural point sources (e.g., mineralized springs) is cost-effective.

In the future, water development projects (whether for municipal, irrigation or energy development purposes) will have to mitigate any increases in salinity concentrations in the Colorado river, no matter how small. A non-degradation policy should be flexible so that salinity detriments by a particular project can be mitigated off-site (e.g., by lining a length of canal in a nearby irrigation system). Mitigating salinity detriments will add to the cost of future water development costs, but not prohibitively.

Energy development

The Colorado river basin contains a vast supply of energy resources. In particular, potential energy sources consist of oil shale, coal, uranium, oil, gas, hydropower, and geothermal resources. Eventual development of these resources depends upon the economics of processing each particular resource deposit. In addition, resource extraction and conversion must be compatible with environmental restraints. The actual resources ultimately developed will depend upon a complex interchange of available energy resources, water resources, economics, environmental safeguards, political intervention, and private resource ownership.

The resources to be considered for large scale commercial development with significant impact upon the water resources of the Colorado river basin are oil shale, coal and uranium. Known reserves of both oil and natural gas will be tapped and new technological developments for increasing reservoir yields will permit recovery from deeper wells. All in all, the oil and gas fields are not expected to be a major contributor to the total energy output of the basin.

The water resources of the Colorado river basin are already heavily utilized so continued development of any remaining water may well demonstrate the problems of over-allocating a vital resource. Future energy developments within the upper basin will require large amounts of water (perhaps 1 billion m^3 by the year 2000) in addition to significantly altering the chemical quality of any remaining downstream water supplies. Several legal compacts and treaties, which regulate the flow of the Coloradeo river, stipulate water deliveries which the upper basin must make to the lower basin as well as deliveries which must be made to the Republic of Mexico. A growing awareness for environmental protection will also play a significant role in determining future resource developments.

In the overall economic analysis of synthetic fuel developments, the cost of water supplies and water related facilities represents a small part of the total energy cost. However, the demand for water at increased energy production levels creates a marginal value of water to the energy industry in the order of $ 40,000 per million liters. This value is three orders of magnitude greater than the value of irrigation water in the upper Colorado river basin indicating that additional energy related water supplies might be purchased from agricultural water supplies.

Growing domestic water needs in competing urban regions

The present population in the upper Colorado river basin is more than 4,000,000 but the population is projected at only 7,000,000 by the year 2020. In contrast, the population in the lower Colorado river basin is presently about 3,000,000 which is projected to reach 7,000,000 by the year 2000. The present population in Mexicali Valley is more than 500,000 and although projections are not available, this population will likely increase to more than 1,000,000 by the year 2000.

A significant portion of the Colorado river waters are exported to southern California via the Colorado river aqueduct to serve various municipalities including Los Angeles and San Diego. The population of this service area is more than 9,000,000. However, if the population of southern California does increase, this will not impact the Colorado river because their water entitlement is legally fixed. Consequently, southern California is faced with having to import additional water supplies from nothern California or elsewhere.

The population projections disclose that meeting future domestic water requirements will be particularly crucial in the lower Colorado river basin, particularly the metropolitan areas of Phoenix and Tucson in Arizona and Mexicali in the Republic of Mexico. The Phoenix and Tucson areas are already seriously mining the surrounding ground water reservoirs. Unfortunately, the construction of the multi-billion dollar Central Arizona Project, which will be completed in a few years, will only alleviate a portion of the present water supply problems.

Conclusions

The large reservoir capacity in the system (more than 80 billion m^3) in comparison with the mean annual flow provides long-term storage that allows the collection of flood flows and minimizes the impacts of drought years. The long-term estimates of mean annual virgin flow at Lee Ferry, Arizona (the dividing point on the Colorado river that separates the upper basin and lower basin) varies from 16.1—18.1 billion m^3, with 17 billion m^3 frequently used in planning studies.

With the completion of the Central Arizona Project during the 1980's the lower basin will be fully utilizing their entitlement to the Colorado river. The Republic of Mexico is already utilizing their full entitlement of 1.85 billion m^3. In contrast, the upper basin is presently consuming 4.9 billion m^3, but the question is their full entitlement, with estimates varying from 5.1—9.2 billion m^3; however, there is not sufficient water supplies to exceed roughly 7.154 billion m^3. There are still serious questions to be answered regarding the degree of water development that will be allowed in the upper basin.

The population projections for the lower basin show that domestic water demands will increase by 1.23 billion m^3 by the year 2020. Considering that ground water depletions exceed recharge by 31.1 billion m^3 annually, and that the lower basin entitlement is already nearly fully utilized, the only viable alternative appears to be the conversion of irrigation water supplies to meet domestic demands in the future.

The greatest demands for new water supplies in the upper basin will be the result of energy development. Since the water supplies in the upper basin are not fully utilized, there is considerable capacity to accomodate new energy developments. In many cases, water transfers will be made from irrigated croplands to energy complexes. However, the upper basin states have in the past been exporting significant portions of their entitlements to meet growing water demands in the metropolitan areas of Salt Lake City, Denver and Albuquerque. These areas are still growing very rapidly, so there will be great demands upon the system to meet increasing water demands.

Finally, many of the questions regarding Indian water rights are yet unresolved. Even very reasonable demands would create serious water allocation problems throughout the basin and adjoining basins.

Lundqvist, J.; Lohm, U. and Falkenmark, M. (eds.):
Strategies for River Basin Management, pp. 229–244
© *1985 D. Reidel Publishing Company*

Conflict between Flood and Drought Preparedness in the Colorado River Basin

Dracup, John. A., Rhodes, Steven L. and Ely, Daniel, UCLA, Civil Engineering Department, School of Engineering and Appl. Science, Los Angeles, CA 90024, USA

Abstract: The Colorado river dominates water resource development in the seven states of the southwestern US. It also is one of the most carefully managed river systems in the world. The basin has been arbitrarily divided by the Colorado River Compact into the Upper Colorado Basin and the Lower Colorado Basin for purposes of interstate administration. A wide range of climate occurs throughout the basin because of differences in altitude, latitude, and topographic features. Considered here are current and projected scenarios of management use. Examined are the combined legal, institutional, hydrologic and water demand constraints, and how they are managed under circumstances of increasingly competitive uses. As a case study, the current analysis examines the 1983 spring and summer flooding in the Lower Basin and whether the US Bureau of Reclamation adequately managed the heavy spring runoff from the Upper Basin. This analysis stresses that the reasons for the flooding go beyond the climatic events of the year and the Bureau of Reclamation's response to them. Rather, the flooding was the result of the convergence of three factors: (1) the 17-year period of filling Lake Powell behind Glen Canyon Dam ended and the system of water storage reservoirs on the river basin considered full; (2) during the filling period, physical encroachment into the Lower Basin flood plain accelerated; (3) the streamflow forecast procedures were not suitable for the extreme climate variability of the Colorado river basin, as exhibited in 1983.

Introduction

The Colorado river dominates water resource development in the seven states of the southwestern US. It also is one of the most carefully managed river systems in the world. Its multi-purpose uses as water-supply, hydroelectric power, and water-based recreation compete with its supposed management priority of flood control. With the exception of the deserts of the Great Basin, this 243,000 square mile (631,800 km^2) basin has the greatest water deficiency (average precipitation less potential evapotranspiration) of any basin in the coterminous US. Yet, more water is exported from the Colorado river basin than from any other basin in the US. (This paper is based on some original research which appeared in the Bulletin of the American Meteorological Society, Rhodes, Ely & Dracup 1984.)

The basin has been arbitrarily divided by the Colorado River Compact into the Upper Colorado Basin and Lower Colorado Basin for purposes of interstate administration. The upper basin drainage includes those areas of Arizona, Colorado, New Mexico, Wyoming, and Utah that drain into the Colorado river above Lee Ferry, Arizona. It is bounded on the east and north by mountains forming the Continental Divide, and on the south it

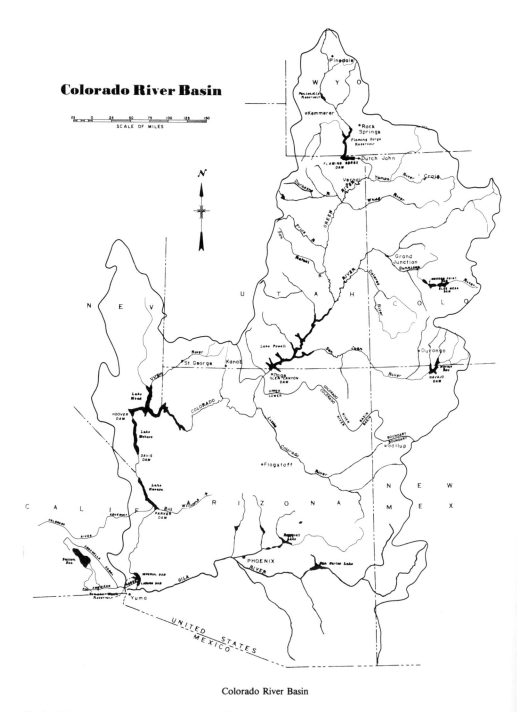

Colorado River Basin

Fig 1 Colorado river basin upstream the inflow to Mexico.

opens to the lower Colorado region. The lower basin drainage includes most of Arizona, parts of southeastern Nevada, southeastern Utah, southeastern California, and western New Mexico.

The focus of this paper is on the current water usage priorities and control practices utilized by the Colorado river operating agency, the Bureau of Reclamation, in the management of the river. In particular, the actual priorities of water resource management, that is, flood control, water supply, and hydropower generation, are examined and compared with stated policies and procedures. The management of the 1983 spring runoff is utilized as a case study.

Compact Regulation and Equitable Water Apportionment between Upper and Lower Basin

Water availability estimates

A wide range of climate occurs because of differences in altitude, latitude, and topographic features. In the north, summers are short and warm, winters are long and cold. In the southern part, the summers are longer and the winters are moderate at low altitude, but colder temperatures occur in the mountains.

About 83% of the water that flows in the Colorado river basin comes from the upper basin. The average annual precipitation throughout the entire upper basin is about 16 inches (40.6cm), which amounts to 93,440,000 acre-feet per year ($115 \times 10^9 \, m^3$). Approximately 15% of the precipitation runs off, as most is lost to evapotranspiration in the upper basin.

One of the most famous and controversial hydrologic records in the United States is that of the virgin flow of the Colorado river at Lee Ferry, Arizona. Lee Ferry is defined as a point on the Colorado river, one mile below the mouth of the Paria river. Estimates of virgin flow have been made of the upper basin since 1896; however, run-off has been measured and recorded only since the first gauging station was established at Lee Ferry during the summer of 1921. The importance of this flow is accentuated by the Colorado River Compact, which anticipates that the upper basin can deliver 75 million acre feet (maf) ($92.5 \times 10^9 \, m^3$) at Lee Ferry each 10 years. Estimates of the long-term annual average flow vary from 11.8 to 16.8 maf (14.5 to $20.7 \times 10^9 \, m^3$) depending on the time period selected (Colorado River Board of California 1969). Recent tree-ring analysis dating back to 1512 has indicated the long-term mean to be approximately 13.5 maf ($16.6 \times 10^9 \, m3$) (Stockton 1977).

The current estimates of available surface-water supply within the upper basin are less than those at the time the Colorado River Compact was negotiated. This is because of the abnormally wet period that occurred during the early part of this century. The range of annual flow at Lee Ferry has varied from a low of 5.6 maf ($6.9 \times 10^9 \, m^3$) in 1934 to a high of 24.0 maf ($29.6 \times 10^9 \, m^3$) in 1917. Some argue that the average flow of 13.1 maf ($16.1 \times 10^9 \, m^3$) per year that has occurred since 1931 is closer to the long-term mean (Jacoby 1975).

The law of the river

The laws governing the Colorado river have been presented in detail by Meyers (1966) and Hundley (1975, 1983). Only a brief summary of the major treaties, laws and compacts will be presented here.

The allocation of the Colorado river is based on the concept of beneficial consumptive use. The allocation system operates at four levels: international, interregional, interstate, and intrastate (Weatherford and Jacoby 1975).

The international allocation was accomplished by the Mexican Water Treaty of 1944. Mexico was guaranteed an annual amount of 1.5 million acre-feet (maf) (1.8×10^9 m^3) except in times of extreme shortage. However, this treaty contained no provision for water quality. Thus, joint agreements in 1965 and 1973 called for a temporary agricultural drainage water bypass and eventually a desalting plant to improve the quality of water crossing the border.

The interregional allocation was achieved when Congress approved the Colorado River Compact, which became effective in June 1929. Sectional rivalry has caused the states included in the drainage basin to agree to an equal apportionment of the Colorado river waters among the states of the Upper Basin (composed of the states of Colorado, New Mexico, Utah, and Wyoming and a portion of Arizona) and the states of the Lower Basin (composed of the states of Arizona, California, and Nevada) (Colorado River Compact 1922, Article III (b) and Article III (d)).

Traditionally, the fertile lowland valleys in the Lower Basin states (Arizona, California, and Nevada), have developed economically more rapidly than have the mountain headwater "areas of origin" in the Upper Basin states (Colorado, New Mexico, Utah, and Wyoming). The Upper Basin states insisted that an equitable apportionment of the river be made to them prior to the expenditure of large sums of federal money, which might result in a modification of equities adverse to the Upper Basin states. This is in essence what was achieved in the 1922 Colorado River Compact.

The intent of this landmark document was to give each basin the perpetual right to the "exclusive beneficial use of 7,500,000 acre-feet (9.2×10^9 m^3) of water per annum ..." However, the Lower Basin was assured that depletion in the Upper Basin would allow at least 75 maf (92.5×10^9 m^3) flow to the Lower Basin at Lee Ferry in each successive ten-year period. Thus, the Lower Basin received a guaranteed ten-year, not annual, minimum flow, and the Upper Basin assumed the burden of any deficiency caused by a hydrologic dry cycle. However, there are differing viewpoints concerning this allocation.

Current Colorado river management techniques

As a result of the "Law of the River", each basin currently possesses many storage facilities including a large linchpin reservoir: Lake Powell for the Upper Basin and Lake Mead for the Lower Basin (Tab 1). The reservoir system now stores about four times the annual flow of the river. This volume of water in storage reflects the determination

Reservoir	Dam	Stream	Capacity			
			Gross		Usable*	
			maf	$10^9\,m^3$	maf	$10^9\,m^3$
Upstream of Lee Ferry, Arizona (Upper Basin)						
Fontenelle	Fontenelle	Green River	0.35	0.43	0.34	0.42
Blue Mesa	Blue Mesa	Gunnison River	0.94	1.16	0.83	1.02
Morrow Point	Morrow Point	Gunnison River	0.12	0.15	0.12	0.15
Flaming Gorge	Flaming Gorge	Green River	3.79	4.66	3.75	4.61
Navajo	Navajo	San Juan River	1.71	2.10	1.70	2.09
Lake Powell	Glen Canyon	Colorado River	27.00	33.21	25.00	30.75
Total in Upper Basin			33.91	41.71	31.74	39.04
Downstream of Lee Ferry, Arizona (Lower Basin)						
Lake Mead	Hoover	Colorado River	28.54	35.10	26.16	32.18
Lake Mohave	Davis	Colorado River	1.82	2.24	1.81	2.23
Lake Havasu	Parker	Colorado River	0.65	0.80	0.62	0.76
Total in Lower Basin			31.01	38.14	28.59	35.17
Total in Upper and Lower Basins			64.92	79.85	60.33	74.21

* Capacity above dead storage.
Note: 1 maf = $1.23 \times 10^9\,m^3$

Tab 1 Major reservoirs in Colorado river basin

of the Basin states to conserve as much water as possible, thereby strengthening their claim to it and at the same time providing a margin of safety in the event that a run of dry years occurs. The Upper Basin states prefer that releases to the Lower Basin be the absolute minimum required by law. Furthermore, the current operational requirement states that the river's manager, the US Bureau of Reclamation, maintain-s Lakes Mead and Powell at or near equal volumes of water in storage at the end of each water year (30 September). Each basin is thus assured "equal ownership" of the river.

Flood control protection is provided to residents, farms, and business below Hoover Dam. Flood control operations rest on two central elements:

1. scheduled dedicated water storage space made available to catch the spring runoff in Lake Mead;

2. a forecast of how much water will enter Lake Powell from 1 April through 31 July, produced by the National Weather Service Colorado Basin River Forecast Center in Salt Lake City, Utah.

The objective of the flood control procedure currently in effect is to create enough storage space, through reservoir releases from August to January, to catch the predicted April—July runoff. The plan, in action since 1968 and slightly modified in 1982, utilizes a monthly streamflow forecast generated for the period January—May that predicts the spring inflow to Lake Powell. Adjustments in the storage forecasted space for the inflow can then be made to keep downstream releases below damaging levels and at the same time conserve as much water as possible.

The US National Weather Service's (NWS) Colorado Basin River Forecast Center uses monthly estimates of error to arrive at forecasts of the maximum probable and minimum probable April—July runoff. Values are added to and subtracted from the most probable runoff forecast to obtain the maximum and minimum probable runoff forecasts, respectively. The added and subtracted values decline from January through July because later forecasts improve in accuracy. Tab 2 displays the estimates of error that were used in Water Year 1983.

To meet these goals, the Bureau of Reclamation increases flood control space in Lake Mead on 1 August of each year in order to have 5.35 maf (6.6×10^9 m^3) available on 1 January (Tab 3). According to the flood control plan for the Colorado river, Lake Mead is the only major basin reservoir with an explicit flood control space schedule. Prior to the construction of Glen Canyon Dam and prior to the US Army Corps of Engineers 1982 report, the standard flood control procedure was to have 5.8 maf (7.1 $\times$

Tab 2 The estimates of error to obtain the maximum probable or minimum probable April-through-July runoff forecast (used in water year 1983)

Date of Forecast	Lake Powell Inflow				Lake Mead Inflow			
	Maximum Probable Forecast		Minimum Probable Forecast		Maximum Probable Forecast		Minimum Probable Forecast	
	maf	10^9 m^3	maf	10^9 m^3	maf	10^9 m^3	maf	10^9 m^3
January 1	+ 3.970	+ 4.883	− 3.010	− 3.702	+ 4.415	+ 5.430	− 3.255	− 4.004
February 1	+ 3.130	+ 3.850	− 2.400	− 2.952	+ 3.510	+ 4.317	− 2.620	− 3.223
March 1	+ 2.310	+ 2.841	− 1.815	− 2.232	+ 2.645	+ 3.253	− 2.020	− 2.485
April 1	+ 1.600	+ 1.968	− 1.300	− 1.599	+ 1.885	+ 2.319	− 1.490	− 1.833
May 1	+ 1.040	+ 1.279	− 0.900	− 1.107	+ 1.280	+ 1.574	− 1.075	− 1.322
June 1	+ 0.557	+ 0.685	− 0.557	− 0.685	+ 0.770	+ 0.947	− 0.720	− 0.886

Source: National Weather Service, Salt Lake City, 1983.
Note: 1 maf = 1.23 $\times$ 10^9 m^3

Tab 3 Lake Mead flood control available space schedule

Date	Available Flood control storage space	
	(acre-feet)	(billion cubic meter)
1 August	1,500,000	1.845
1 September	2,270,000	2.792
1 October	3,040,000	3.739
1 November	2,675,000	3.290
1 December	3,963,000	4.874
1 January	5,350,000	6.581

Source: US Army Corps of Engineers (1982).
Note: 1 maf = 1.23 $\times$ 10^9 m^3

$10^9\,m^3$) of storage available on 1 January, as recommended by Debler (1930). This storage requirement was increased each month until a maximum requirement of 9.5 maf $(11.7 \times 10^9\,m^3)$ was reached on 1 April. These procedures were formalized by the Corps of Engineers in 1955 and continued in effect until 1968 (Holburt 1983b).

The Bureau's scheduled outflow release rates through the dams, and storage space availability based on the NWS inflow forecasts, have worked well in recent decades in minimizing water lost through unneeded anticipatory releases and through flooding below Hoover Dam. This has helped to maximize hydroelectric generation, water conservation storage, and flood control. However, the conditions prevalent in the Colorado Basin in the 1980's have become radically different from those of the 1960's and 1970's. The changes in those conditions are significant contributors to the flooding which occurred in the spring and summer of 1983.

Colorado River Flooding during 1983

River management

On 1 January 1983, there were 6.6 maf $(8.1 \times 10^9\,m^3)$ of storage space available in Lake Mead and upstream, more than the required 1 January target of 5.35 maf $(6.6 \times 10^9\,m^3)$. Yet even with the surplus storage space available, the reservoir system was overwhelmed by the magnitude of the spring inflow to Lake Powell. Because of late precipitation and cool weather throughout the Upper Basin, snowpack continued to increase during April and May. Fig 2 shows the rapid and unusual changes in the forecast inflow to Lake Powell from January through June 1983.

Tab 4 presents the projected inflows to Lake Powell for the April—July period, the required Hoover Dam release rates according to the flood control operating procedures, and the actual Hoover Dam release rates decided upon by Bureau officials in the Lower Colorado Basin. Fig 3 illustrates the relationship between Lake Powell inflows, Lake Powell outflows, and Hoover Dam releases from April through July 1983. Because of the massive influx of water into Lake Powell, Bureau officials downstream had to increase outflows from Glen Canyon Dam. This, in turn, obliged Lower Basin officials to raise outflows from Hoover Dam. The releases at Hoover Dam, which historically had been held to approximately 25,000 cubic feet per second (cfs) $(708\,m^3)$, were elevated to over 40,000 cfs $(1,132\,m^3)$ in July. This is a critical point, for 40,000 cfs $(1,132\,m^3)$ was the targeted maximum outflow rate from Lake Mead under the 1968 revised flood control procedures. Indeed, Bureau of Reclamation operators successfully limited Hoover release rates to 40,000 cfs $(1,132\,m^3)$ except for the month of July, during which Hoover releases averaged 41,854 cfs $(1,184\,m^3)$. However, flooding downstream of Hoover Dam begins when the flow exceeds 19,900 cfs $(538\,m^3)$.

The rapid sequence of meteorological events occurring late in the spring, coupled with the problem of attempting to move massive amounts of water through Lakes Powell and Mead in a short period of time, resulted in streamflows greater than those experienced during the previous two decades by Lower Basin residents and businesses.

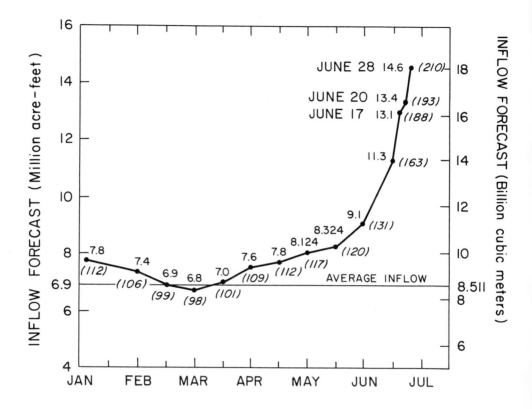

Fig 2 Forecasted inflow to Lake Powell from January through June 1983. Source: Dozier and Brown 1983.

The unusual and unexpected flooding along the Colorado River during 1983 was the result of three converging factors: the sudden required operation of a full river system, an encroachment into the downstream flood plain, and climate variability in the basin. Each of these are discussed below.

Cause 1: The Full River System

The filling of Lake Powell behind Glen Canyon Dam began in 1963 and was completed in 1980. During that seventeen-year period, there were virtually no required flood control operations on the Colorado river. Runoff in excess of downstream water supply and hydropower generation was easily stored. However, in 1980 Lake Powell became full and the 1 January storage requirements for flood control at Lake Mead were reduced from 5.8 to 5.35 maf (7.1 to 6.6 × 10^9 m^3), an eight percent decrease.

These changes required that the river now be operated in a careful, prudent manner; there was little room for forecast error. The forecasted inflow to Lake Powell not only had to be accurate but carefully monitored on a real-time basis. Monitoring inflows

Forecast		Date	April—July Runoff Estimate		Percent Average	Minimum Required Hoover Release		Actual Hoover Release	
			(maf)	(10⁹m³)		(cfs)	(m³/s)	(cfs)	(m³/s)
January 1983	Preliminary	12/30/82	7.8	9.6	112	19,000	538	19,130	541
	Final	01/06/83	7.8	9.6	112	19,000	538		
February 1983	Preliminary	02/02/83	7.4	9.1	106	None		6,590	186
	Final	02/04/83	7.1	8.7	102	None			
	Mid-Month	02/15/83	6.9	8.5	99	None			
March 1983	Preliminary	03/02/83	6.8	8.4	98	None		10,270	291
	Final	03/04/83	6.7	8.2	96				
	Mid-Month	03/16/83	7.0	8.6	101	2,200	62		
April 1983	Preliminary	04/01/83	7.6	9.3	109				
	Final	04/06/83	7.9	9.7	114	12,900	365	17,810	504
	Mid Month	04/15/83	7.8	9.6	112	10,800	306		
May 1983	Preliminary	05/03/83	8.124	9.933	117			19,800	560
	Final	05/06/83	8.124	9.933	117	14,600	413		
	Mid-Month	05/17/83	8.324	10.239	120	16,200	458		
June 1983	Preliminary	06/02/83	8.853	10.889	127			31,700	897
	Final	06/07/83	9.103	11.197	131	19,000	538		
	Mid-Month	06/17/83	11.3	13.9	163	47,300	1,339		
	Special	06/17/83	13.1	16.1	188	52,900	1,497		
	Special	06/20/83	13.4	16.5	193				
	Special	06/28/83	14.6	18.0	210				
Actual runoff:	April 1983		1.124	1.383					
	May 1983		3.225	3.967					
	June 1983		6.725	8.272					

Note: 1 maf = 1.23 × 10⁹ m³   1 cfs = 0.0283 m³/s

Tab 4 1983 runoff forecasts — Colorado river above Lake Powell

allows corrective management responses if conditions permit. However, in 1983, Lake Powell inflows rose so rapidly that there was no time for mitigating responses. For example, on 24 May 1983, the unregulated inflow to Lake Powell was approximately 37,000 cfs (1,047 m³). Eight days later, on 1 June, the unregulated inflow was 102,000 cfs (2,887 m³) (US Bureau of Reclamation 1983).

Even the availability of real-time data may not have been sufficient to manage a wet year such as 1983, since it takes substantial time to move water through dams with structurally-limited release rates. The 1983 April—July inflow into Lake Powell was more than 14 maf (17.3 × 10⁹ m³). Approximately 140 days would be required to discharge that quantity at a rate of 50,000 cfs (1,415 m³).

Cause 2: Physical Encroachment into the Flood Plain

Physical encroachment into the Lower Basin flood plain is a function of the defined flood plain boundaries and of the relative stability in the annual streamflows. Thus encroachment into the Lower Basin flood plain, which would not have been possible in the absence of the upstream storages, was encouraged by a combination of technolo-

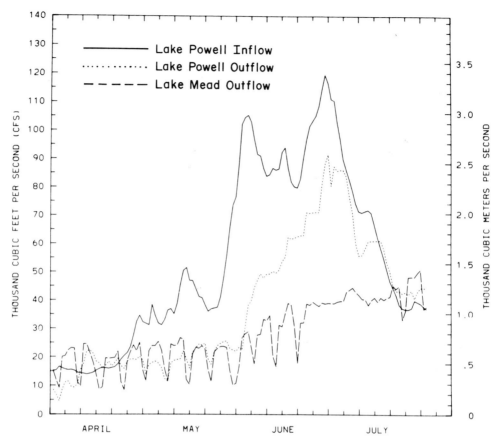

Fig 3 Relationship between Lake Powell inflows, Lake Powell outflows and Hoover Dam releases from April through July 1983.

gical fixes and lax zoning practices in counties bordering on the Colorado river (Arizona Republic 1983a, b).

The two major dams on the Colorado river have performed as planned in controlling the variability of streamflow rates (Fig 4). Glen Canyon and Hoover Dams have consequently provided substantial protection to the Lower Colorado flood plain in terms of their ability to reduce the river's meanderings and the flooding associated with high spring streamflows. Even in 1983, releases at Hoover Dam did not significantly exceed 50,000 cfs (1,415 m³).

The history of development in the flood plain roughly began with the construction of earthen levees in the area around Yuma, Arizona. The levee system was constructed to protect agricultural land (fertile flood plain soil) from the annual rush of spring snowmelt (US Army Corps of Engineers 1982). With the completion of Hoover Dam and the subsequent decrease in the spring streamflow's variability, more flood plain acreage

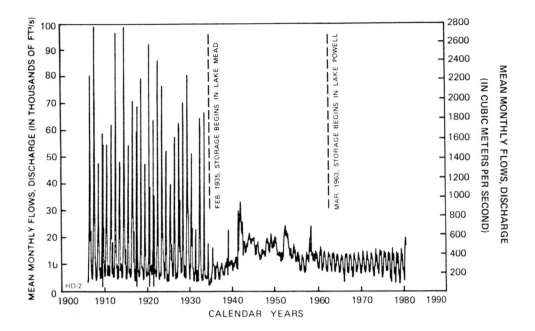

Fig 4 Mean monthly flows downstream Hoover Dam in Colorado river.

became available for development. This resulted in the construction of residential and commercial structures in these areas. As the 1982 review of flood control operating procedures notes (US Army Corps of Engineers 1982):

> Few, if any, structures were located in the 40,000 cfs (1,132 m³) flood plain in the lower Colorado river at the time of the closure of Hoover Dam (1935) and for some years thereafter. For many years the flood control operation plan for Hoover Dam has incorporated a "target maximum" flood control release of 40,000 cfs (1,132 m³).

With the completion of construction of Glen Canyon Dam in 1962, streamflow variability was sharply narrowed. This coincided with the period of the greatest physical encroachment into the flood plain, including construction in and development along the streamflow profile of less than 28,000 cfs (792 m³) (US Army Corps of Engineers 1982).

The period of time that the Colorado reservoir system was filling constituted a period during which true exposure to climatic impacts, i.e. precipitation variability, did not exist. It was not representative of a new climatic regime in the basin, but only of anthropogenic interference with the flow of the river. The encroachment into the flood plain was possible because water was in storage upstream, and also because the period of filling Lake Powell was drawn out for almost two decades. Two decades are more than sufficient to affect societal perceptions of climate stability.

Cause 3: Climate Variability

Perceptions of climate stability in the Colorado river basin are not borne out by historical data. In fact, the third factor that contributed to the 1983 Lower Basin flooding was the variability of the climate in the arid American Southwest. As Fig 5 demonstrates, the variability of the river's streamflow is substantial. As recently as 1977, the western United States was hit with a severe drought which resulted in a significant drop (to 5.6 maf) $(6.9 \times 10^9 \, m^3)$ in the estimated virgin flow of the Colorado. (Upper Colorado River Commission, 1982)

It is interesting to note that the 1977 drought also played a key role in the decision among the Colorado river basin states to defer any action regarding revision of the river management scheme (Broadbent 1983). The concern in the Colorado river basin was

Fig 5 Estimated virgin flow in Colorado river.

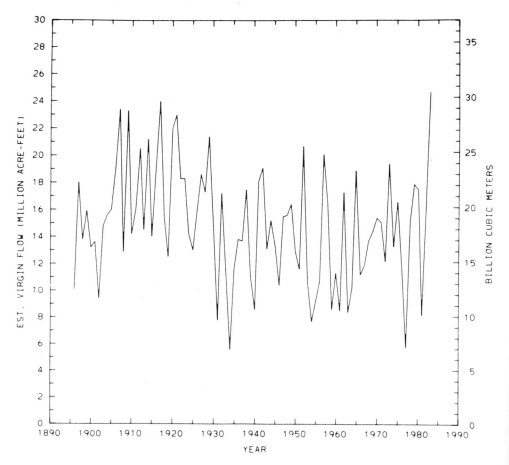

again reinforced in terms of the major adverse climatic impact anticipated, i.e. drought is of concern, while a wet year is not.

What is of greatest importance in Fig 5 is the fact that the river's virgin flow has not only been below the historical average of approximately 14 maf (17.3×10^9 m^3), but has frequently greatly exceeded that average as well.

Indeed, because of the effect of a full system and a variable climate, a Bureau of Reclamation official had noted in 1979 that "the present operation strategy ... involves an 85 % risk that damaging floodflows will occur between 1980 and 1984" (Freeny 1981). Furthermore, because of climate variability, it has been estimated in a preliminary study by the US General Accounting Office that if the river continues to be managed by keeping reservoirs full or nearly full, controlled flooding similar to that which occurred in 1983 can be expected to be repeated once "every 10 to 15 years" (Arizona Republic 1983c).

Thus, one can argue that it is not the streamflow variability in itself that caused the 1983 spring flooding and the associated damages. Flood plain encroachment and a full reservoir system, in conjunction with streamflow variability, converged to create appropriate conditions for the events of 1983.

River Management Priorities

Few advocates for flood preparedness

After the large storage facilities along the Colorado river were completed, there developed the potential for conflicts among those requiring flood control protection, which is only achieved by reserving reservoir storage space, and those desiring water supply and hydropower (Freeny 1981). Other conflicts in storage requirements occur among the recreational users, hydroelectric power purchasers, and municipal and agricultural users. All of these users have priorities for reservoir release rates and timing which are optimal for their own needs.

A cost-benefit analysis run by the US Corps of Engineers, using nine different operating strategies, produced the optimal benefits using the current operating strategy. Therefore, the river management scheme currently in effect has remained virtually unchanged since 1968 (US Army Corps of Engineers 1982). Yet the original flood control operating plans of 1935 and 1968 were intended, as a recent Bureau of Reclamation report states "to control flooding to the greatest extent possible" (US Army Corps of Engineers 1982). Furthermore, the Boulder Canyon Project Act (1928), which authorized the construction of Hoover Dam, established that the river be managed:

> First, for river regulation, improvement of navigation, and flood control; second, for irrigation and domestic uses and satisfaction of present perfected rights ...; and third, for power.

Since the total available storage on the river is about four times the average annual flow, a flood of any magnitude could be controlled if sufficient storage space were placed in reserve. However, the only advocates for keeping a large storage space reserved for flood control are the individuals living in the flood plain below Hoover Dam. All other interested parties throughout the Upper and Lower Basins, including recreational users, strongly advocate keeping the reservoirs as full as possible or within a narrow range of fluctuation. Therefore, the relatively few entities in the Lower Basin's flood plain may have become the "policy orphans" (Daneke 1979) of the river's operations.

Risk exposure has shifted

One might argue that storing water has become the dominant agenda in the management of the Colorado river. This can be viewed as the result of efforts to satisfy the interests of beneficiaries other than those interested in flood control in the Lower Basin. Obviously, the risks to beneficiaries of the water storage objective were reduced as the reservoir system became full. It is evident that as the reservoir system has filled, risk exposure has shifted from those interested in maximal water storage (consumptive users, hydroelectric generation contracters, recreational interests) to those concerned with flood control. As one discussion (Freeny 1981) notes:

> Regulation of the lower Colorado system requires that a delicate balance be maintained between three diametrically opposed communities of users. One of the problems faced by system operators arises from the attitudes of the users toward risk: all three groups demand zero risk. The flood control group wants zero risk of damage. Power users want zero risk of generation losses. Water users want zero risk of water shortage 30 years in the future. Somehow, through technology or political science, some way should be devised to change these zeros into more realistic figures so that values can be created.

The interplay of the three factors which contributed to the flooding in 1983 in the Lower Colorado Basin could recur in the future. Because climate variability is beyond the control of the river's managers, attention must be focused on flood plain encroachment and on the question of how much storage space should be dedicated to flood control in Lake Mead and upstream.

Strong support for drought preparedness

The issue of dedicated flood control storage space thus becomes a critical point in the management of the Colorado. Accepting the Bureau of Reclamation's assumption that the Central Arizona Project will help to alleviate the flooding problem by providing an additional water diversion point, a temporary change in the dedicated flood control space could help to protect property in the Lower Basin (Broadbent 1983; Freeny 1981). However, since 1983's precipitation and runoff were so abnormal compared with recent years, proposals for increasing storage space will not necessarily be received warmly (Broadbent 1983).

The 1977 response to drought indicated that there is substantial support for maintaining full reservoirs upstream, and that flood control for Lower Basin residents and other interests is not given high priority by the other beneficiaries of the river. Water resources in the American Southwest are managed for dry years, not for extremely wet years such as 1983. This is precisely the justification for the reservoir system to be maintained at a nearly full level. The dominant preference to keep the reservoir system full may continue to expose Lower Basin interests to flood damage during very wet years. This preference is reflected by Holburt (1983a):

> Requiring release of water in excess of beneficial use increases the probability of reduced future use because the water that would have otherwise been in storage will not be there. This fact must be considered in reviewing flood control criteria.

Conclusion

Unanticipated flooding of the Colorado river in the lower Colorado river basin caused substantial damage to homes and businesses in the spring and summer of 1983. The abnormal meteorological events, i.e. greater than normal precipitation, contributed to the flooding but were not solely responsible for it. Other factors which contributed to the flooding are: 1) the practice of maintaining a system of full reservoirs, in response to societal functions, such as consumptive users, hydroelectric generators, and recreational interests, which prefer maximal water storage; and 2) physical encroachment into the flood plain, made possible by the dams along the river. Although these factors could be managed physically, thereby averting the flood risks seen in 1983, there are many contrary interests which may cause mitigating or preventive steps from being taken in the future. Although priorities for Colorado river management are mandated by US law, such management has historically been the product of political and economic constraints created by the river's many beneficiaries.

References

Arizona Republic: The Flood Facts. p. 42 (22 June 1983a)
Arizona Republic: Residents Who Built Near River Should Have Expected Trouble, Watt Says. p. A2 (1 July 1983b)
Arizona Republic: $ 80 Million Flood Was 'Unavoidable'. p. B1 (3 September 1983c)
Boulder Canyon Project Act: 45 Stat. 1057, 1928.
Broadbent, R. (Commissioner, US Bureau of Reclamation): Prepared statement for the US House of Representatives. Committee of Interior and Insular Affairs, Yuma, Arizona (7 September 1983)
Colorado River Board of California: California's Stake in the Colorado River. Los Angeles, CA 1969.
Colorado River Compact: 70 Congressional Record 324, 325, 1928; and H. Doc. 605, 67 Cong., 4 Sess. pp. 8—12, 1923.
Daneke, G. A.: Solar Futures: A Perspective on Energy Planning. In: Steinman, M. (ed.), Energy and Environmental Issues: The Making and Implementation of Public Policy. Lexington Books, Lexington 1979.
Debler, L. B.: Hydrology of the Boulder Canyon Reservoir with Reference Especially to the Height of the Dam to be Adopted. US Bureau of Reclamation 1930.
Freeny, G. B.: Managing Conflicts on the Lower Colorado River System. In: Proceedings of the National Workshop on Reservoir Systems Operations. University of Colorado at Boulder, 13—17 August 1979, American Society of Civil Engineers, New York 1981.

Holburt, M. (Chief Engineer, Colorado River Board of California): Prepared statement for the US House of Representatives Committee on Interior and Insular Affairs. Needles, California, 8 September 1983a.

Holburt, M. (Chief Engineer, Colorado River Board of California): Personal Communication, 1983b.

Hundley, N., Jr.: The West Against Itself: The Colorado River in Historical Perspective. Prepared for the Colorado River Working Symposium: Management Options for the Future. Santa Fe, New Mexico, 23—26 May 1983.

Hundley, N., Jr.: Water and the West. University of California Press, Los Angeles 1975.

Jacoby, G. C., Jr.: Lake Powell Effect on the Colorado River Basin Water Supply and Environment. Lake Powell Research Project Interim Rep., Institute of Geophysics and Planetary Physics, UCLA 1975.

Meyers, C. J.: The Colorado River. Stanford Law Rev. (19 January 1966)

Rhodes, S. L.; Ely, D.; Dracup, J. A.: Climate and the Colorado River: The Limits of Management. Bulletin of the American Meteorological Society 65 (July 1984)

Stockton, C. W.: Interpretations of Past Climatic Variability from Paleoenvironmental Indicators. In: Wallis, J. (ed.), Climate, Climatic Change, and Water Supply. National Academy Press, Washington, DC 1977.

Upper Colorado River Commission: Thirty-Fourth Annual Report. Salt Lake City, Utah (30 September 1982)

US Army Corps of Engineers: Colorado River Basin — Hoover Dam. Review of Flood Control Regulations. Final Report, 1982.

US Bureau of Reclamation: Status of Reservoirs — Colorado River Storage Project, 1983.

Vandivere, W. B.; Voster, P.: Hydrology Analysis of the Colorado River Floods 1983. GeoJournal 9.4 (1984)

Weatherford, G. D.; Jacoby, G. C.: Impact of energy development on the law of the Colorado River. National Resources J. 171—213, U. of New Mexico, School of Law, 1975.

Lundqvist, J.; Lohm, U. and Falkenmark, M. (eds.):
Strategies for River Basin Management, pp. 245–254
© 1985 D. Reidel Publishing Company

Water Resources Conflicts in Integrated River Basin Development
The Case of Kävlinge River

Castensson, R., Dept. of Water in Environment and Society, University of Linköping, S-581 83 Linköping, Sweden

Abstract: In this paper the river basin concept and the development strategies of the Kävlinge river basin are analyzed. The analysis indicates that the chosen development strategies are heavily dependent on socio-economic processes as changes in production conditions, technological renewal, legal priority, political and economic superiority. The strongest development factor is the urban growth and consequently the rising demand on water. Isolated single-purpose water development schemes have expanded to basinwide and multi-purpose water management problems. Specific local water demand conflicts have expanded to an inter-basin level. Lack of adaptation in water management rules have caused unforeseen ecological damages and suboptimal allocation of water resources.

Introduction

The River Basin concept has a wide application and a long history. The interpretation and the use of the river basin as an integrative concept varies from single-purpose, small scale development schemes e.g. of hydropower to large multi-purpose development schemes (White 1969; Wengert 1980). The use of the river basin concept seems to need reinterpretation and specification. Much criticism has been presented to the idealistic use of the river basin concept as a "natural law" for water management organization (McKinley 1964). On the other hand increasing water demands and water development calls for integrative river basin development strategies.

The purpose of this paper is to give a condensed geographical and hydrological background of conditions and factors of special importance for the Kävlinge river basin development. The paper is organized in three sections. The first is a brief description of the hydrological and geographical characteristics. The second is an overview of the development history with some future orientations. The third is concluding remarks.

Some Characteristics

The Kävlinge river basin represents a basin with a total water abstraction for municipal and industrial uses of more than 10% of the water availability (Hjorth et al. 1979). Its central location in a densely populated and water short area in the southern part of Sweden makes the Kävlinge river basin extremely strategic for industrial and municipal water

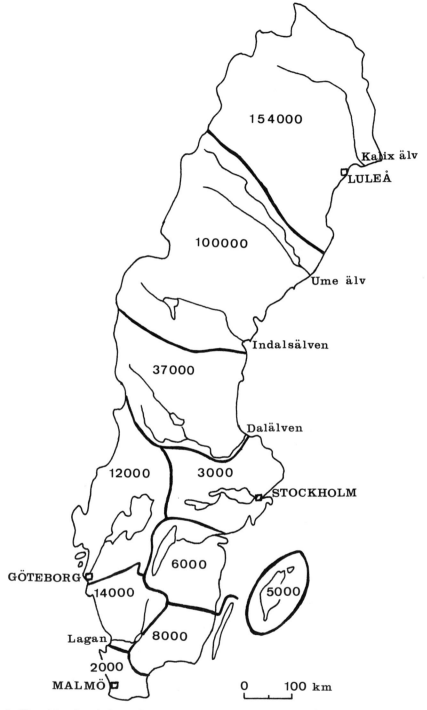

Fig 1 The regional variations of the available water resources in Sweden expressed as m^3 per capita and year (from Falkenmark 1979).

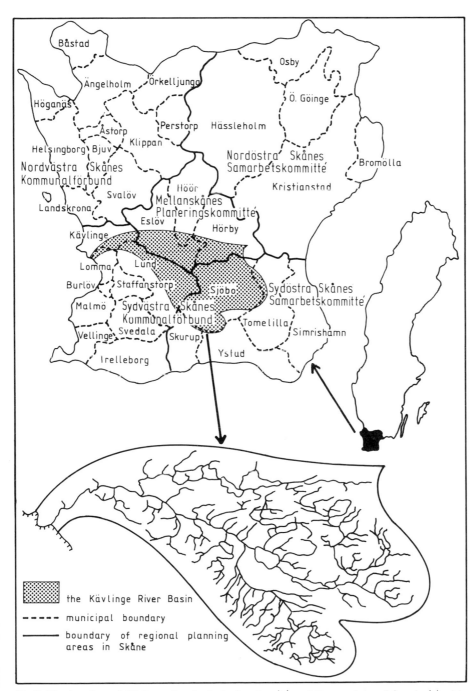

Fig 2 The location of Kävlinge river basin in Sweden (b) and the province of Scania (a) with municipalities and regional planning areas.

supply. As can be seen from Fig 1, showing the relative amount of water available per capita in Sweden, the southern part of the country represents a minimum area with only 2000 m^3/inh. and year available from the hydrological cycle.

The geographical location of the Kävlinge river basin and its administrative context is shown in Fig 2. The total area is 1217 km^2. The altitude varies from 160 m for the source lakes to sea level. The lake percentage is quite low for Swedish conditions, only 2.2%. The precipitation varies from 550 mm/year in the western lowlands to 750 mm/year in the eastern highlands. The main natural reservoir is the Lake Vomb, located in the middle of the basin. It is one of the most fish productive lakes in Europe.

There are three main tributaries. The River Björka drains the eastern slopes of the highlands (260 km^2). The land use in this area is mainly forestry and cattle intensive agriculture. The River Klingaälv drains the southern sandy flatlands (196 km^2). This area is under the International Convention of Wetlands and most of the basin is an environment conservation area implying heavy legal and administrative regulations on the land use with priority to pasture (Castensson et al. 1984). The third main tributary is the River Braa (157 km^2), which drains the northern parts, and is an area of intensive agriculture with some smaller municipalities.

The run off conditions at different locations in Kävlinge river basin are characterized in Tab 1.

Four Development Phases

Phase I: before 1936 (Land reclamation)

Historically the Kävlinge river basin is an agricultural area. During the medieval time the monks at the monastry of Öved used the river for irrigation of the flatlands. The rapid changes in agricultural technologies during the 19th century had strong implications on the ecological balances, the strongest impacts emanate from land reclamation. These activities were mainly directed to the drainage of wetlands in the upper parts of the basin. The resulting losses of natural water reservoirs have caused long term effects. Wolf

Tab 1 Characteristics of discharge, Kävlinge river basin (m^3/s)

Q	Upstream Lake Vomb	Downstream Lake Vomb	At Sea
Extreme HQ	45	36	122
Mean HQ	20	18	55
Mean Q	3.0	4.1	11.0
Q 50% duration	1.9	2.7	6.7
Q 75% duration	0.9	1.4	4.0
Mean LQ	0.3	0.6	1.8

Source: Hjorth et al. 1979

(1956) claims that losses of natural water storages from drainage activites have caused heavy erosion damages with soil losses of 750.000 kg/year or 100 ha arable land/year (ibid, p. 38).

The use of water during this development phase was single purpose. The hydro-energy was used for small industries like saw-mills, flour-mills, tanneries and dye-works. In accordance with the growing industrialization, the claims for water increased. Examples on big water users during this phase within the basin are sugar-factories, textile mills, and dairies.

The intensive land reclamation activities and its soil erosion effects led during the early 1930's to frequent and heavy floods. The harvests on the lowlands were destroyed. In order to strengthen the flood prevention, the farmers in 1936 organized a special dredge company. Besides the dredging activities this company also took initiatives for flow control, using Lake Vomb as a regulation reservoir.

At this time a lucky coincidence appeared: the metropolitan Malmö-Lund area in the neighbourhood, earlier supplied with ground water, was prospecting for additional water. In other words: when the farmers wanted to get rid of surplus water, the urban areas needed more water.

Phase II: 1939–1969 (Water export for urban water supply)

The 20th century was characterized by a rapid redistribution of population, the urban growth was 2–4% per year and Malmö township more than doubled its population from 127,000 inhabitants 1930 to 265,000 inhabitants 1970. The water supply for the growing urban population was urgent. Further dependence on ground water supply was not possible because of risks for overexploitation of the aquifer and a resulting salt water intrusion from the sea. The solution looked for was long-distant surface water transfer. The geographical relations between distribution of the population and the Kävlinge river basin are illustrated in Fig 3.

In the Kävlinge river basin, the central natural water reservoir, the Lake Vomb, was selected as water source (total area 12.4 km^2, max.depth 6 m). Construction of the Vomb intake water-works, pumping stations and transfer lines started in 1939. The permitted withdrawal was maximized to 500 l/s. The development of the expected and the actual water consumption in the urban areas 1910 to 1980 are illustrated in Fig 4. By the end of the 50's the specific and the total water demand increased more than expected. In 1969, the discrepancies between actual and expected water consumption were close to 25%. To satisfy this demand, the withdrawal increased from 500 l/s to 1,500 l/s, i.e. came close to 40% of the mean in-flow. In order to meet these new requirements the downstream low flow was reduced by water-court verdict. During the 60's the flow control was improved by expanding the reservoir capacity of Lake Vomb.

250

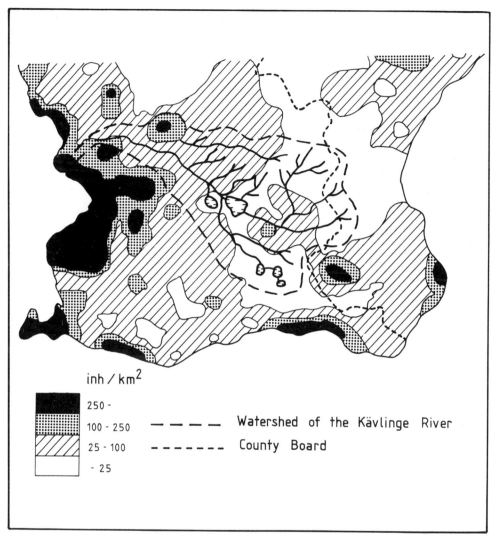

Fig 3 The demand areas and the water resources areas. Kävlinge river basin and the population density (in h/km²) in sourthern Sweden (from Castensson & Thelander 1982).

Phase III: 1970—1987 (Ecological disturbances)

The considerable expansion of the reservoir capacity of Lake Vomb and the increased flow control in combination with other water demands caused ecological disturbances in the lake. By this time, the river basin development had expanded from a single urban water supply problem to a more general multiple regional water supply problem. The most

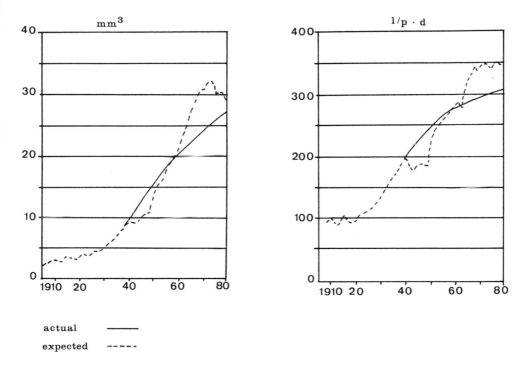

actual ────────

expected ─ ─ ─ ─ ─

Fig 4 Expected and actual urban water consumption in Malmö township 1910 -1980.
a) total consumption (mm³/yr), b) specific consumption (1/pd).

severe damages have been registered on the fishing, for example destroyed fish reproduction sites because of high water level regulation amplitudes. Water-logging on surrounding farmlands, deteriorated water quality and eutrophication problems are other examples.

Since the mid 1960's, when the flow control operation manual had been decided many important water related changes took place. Some of the mentioned ecological disturbances were to be reduced by a more adaptive water management. The rapid changes in water supply and water use during this phase were, however, not fully considered by the water management authorities. Therefore the water allocation rules got obsolete in relation to the changes among the water related activities during the last decade:

— more than doubled water withdrawals for irrigation
— reduction of municipal waste water discharge
— increased leakage of chemical substances from farmlands
— reduced industrial water withdrawals
— deeper environmental consciousness among citizens.

The lack of flexibility and adaptation to water use changes is historically based on the fact that the water-court verdicts in principle remain valid until a new verdict is given. Changes are not monitored and the water courts per se are not expected to initiate changes in water allocations prescriptions.

At this phase the Kävlinge river basin reached a development level of high complexity. Separate changes could no more be viewed isolated (Dávid 1980); instead the water management should be considered in a basinwide perspective.

Phase IV: after 1987 (Multiple river basin integration)

The future development of the Kävlinge river basin is expected to follow two main strategies. The strategies have been adopted from a comprehensive study by the International Institute for Applied Systems Analysis (Kindler et al. 1982). The first alternative implies further integration of the existing water supply systems. Gains from that strategy would be more reliable deliveries in periods of water shortage. The second one implies additional water supply by transfer from distant water sources. Since 1975 a 200 km tunnel from Lake Bolmen is under construction. The tunnel will have a delivery capacity of 6 m³/s and is planned to be in operation from 1987 with an initial delivery of 0.5–2.0 m³/s. The future integrated water supply system, illustrated in Fig 5, implies that three different river basins will at that time become integrated in the same water supply system.

Concluding Remarks

The results from a historical river basin development process in the form of various water related measures, technical devices, etc., are the responses to complex socio-economic processes, explained by changes in production and economic superiority etc.

Concerning the Kävlinge river basin development, the following concluding remarks can be made.

— The urban growth and consequently the rising demand for urban water supply has been the strongest development factor. Originally small and isolated withdrawals have expanded drastically. Isolated single-purpose water development schemes have expanded to basinwide and multi-basin water management problems.

— Besides the development of the river basin, new water demands and changes in existing demands have taken place. More adaptive operational rules would have reduced many of the ecological balance problems and contributed to a more efficient water use.

— Many of the existing basin-specific water conflicts will be reduced by additional water supply from an adjacent river basin. It is expected that the water conflicts will expand to an interregional level. This calls for a stronger national conflict resolution capacity.

— A river basin development with high emphasis on urban water supply is an expensive and long-term activity. Considering the rapid changes taking place in environmental and societal conditions, a reasonable high level of flexibility would be a cheap guarantee for future miscalculations and unforeseen damages.

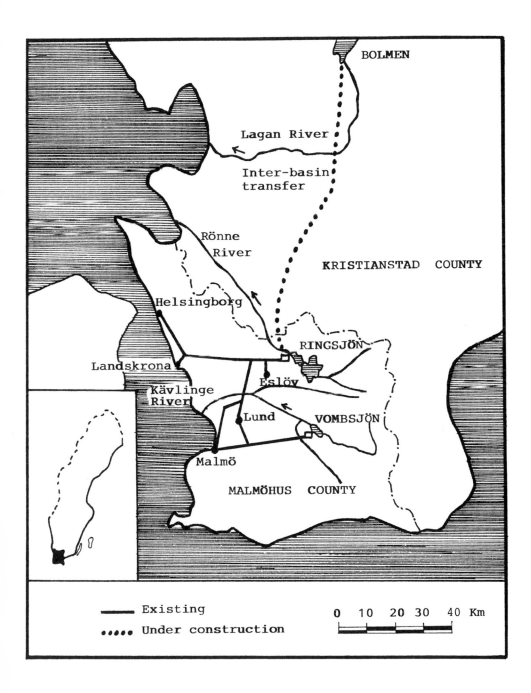

Fig 5 General scheme of the integrated water supply systems (from Kindler et al. 1982).

254

References

Castensson, R.; Thelander, A.-L.: Mark-vattenkonflikter (Land and water use conflicts). Rapport R 99, Byggforskningsrådet, Stockholm 1982 (In Swedish).

Castensson, R.; Lönegren, H.; Schaar, C.: Mark- och vattenanknuten naturvård — analyser av översikts-planer for FRP-sjöar och skötselplaner för våtmarker (Land and Water Conservation — analysis of land and water conservation plans for great lake basins and wetlands). University of Linköping, Department of Water in Environment and Society, Tema V Report 8, 1984 (In Swedish).

Dávid, L.: Multiattribute analysis of decision problems in regional water management. Workshop on Conflicts in Regional Water Management, IIASA, Laxenburg 1980.

Falkenmark, M.: Vatten — resurser, användning, problem. Ett försök till hydrologisk helhetssyn (Water — resources, use, problems). Jordbruksdepartmentet, Ds Jo 8, 1979 (In Swedish).

Hjorth, P. et al.: Vattenresursplanering för Kävlingeåns avrinnings- och influensområde. Del 1. Inledande inventering (Water resources planning for Kävlinge river basin and its influence area). Rapport 3021, Institutionen för teknisk vattenresurslära, Lunds tekniska högskola, Lunds universitet, 1979 (In Swedish).

Kindler, J. et al.: Issues in Regional Water Resources Management — A case study of southwestern Skåne, Sweden. IIASA, Laxenburg 1982.

McKinley, C.: The Valley Authority and Its Alternatives. In: Friedmann & Alonso (eds.), Regional development and planning. The MIT Press, Cambridge, Mass. 1964.

Wengert, N.: A Critical Review of the River Basin as a Focus for Resources Planning, Development and Management. In: North, Dworsky & Allee (eds.), Symposium Proceedings Unified River Basin Management. AWRA, Minneapolis, Minn. 1980.

White, G. F.: Strategies of American Water Management. — Ann Arbor Paperbacks. The University of Michigan Press, 1969.

Wolf, P.: Utdikad Civilisation (Drained Civilization). Svenska Lax- och Laxöringsföreningen Malmö u.p.a. 1956 (In Swedish).

Lundqvist, J.; Lohm, U. and Falkenmark, M. (eds.):
Strategies for River Basin Management, pp. 255–264
© 1985 D. Reidel Publishing Company

3.3 Political interference and the role of traditions in management

The Making and the Breaking of Agreement on the Ganges

Crow, B., The Open University, Milton Keynes MK7 6AA, England

Abstract: The paper describes some of the immediate political circumstances which contributed to the making of the Ganges Waters Treaty in 1977 and to its demise five years later. Then the paper identifies the social group which appears to be most influential in the setting of the agenda for river development in South Asia. Finally the paper concludes that conceptions of the 'national interest' and expressions of class interests require careful analysis if the timing and nature of river development decisions are to be understood.

Introduction

This paper is an attempt to describe some of the political circumstances which have influenced the making, in 1977, and the breaking, in 1982, of agreement over the sharing and development of the River Ganges. It focusses on elements of two political battles: one, when agreement was reached, over the minimum flow Bangladesh would accept, and a second, when the area of agreement was reduced, over the guaranteeing of that share. Then, the paper identifies the social group which appears to be most influential in determining the nature and setting the time scale for river development in South Asia.

Most scholarly accounts of river basin development concentrate on technical and legal structures and carefully avoid consideration of political and socio-economic conditions which influence river development. For example, one recent collection of academic contributions on the development of the Ganges (Zaman et al. 1983) omits any political description or analysis but nevertheless makes political recommendations. This paper sketches some of the political battles and political forces which have determined the nature of development of the River Ganges. It is a plea for a genuinely multi-disciplinary approach to the subject — including not just engineers, lawyers and the occasional reticent diplomat, but also political economists and political sociologists.

The Making, 1977

In March 1977, the Congress Party led by Indira Gandhi was defeated in a general election and the Janata Party formed the new government of India. One of the first acts of the new government was to reach an 'understanding' with the government of Bangladesh over the sharing of the River Ganges. This was achieved in April of 1977 and six months later the understanding was translated into the first treaty laying down the principles for sharing the waters of the Ganges.

The 1977 treaty had two functions: to share the existing dry season flow of the Ganges, and to establish negotiating procedures (and a deadline) for discussions on how the flow could be increased in the future. By comparison with Indo-Bangladesh agreements before and since the treaty was a comprehensive and flexible instrument. Nevertheless it lacked neither loopholes nor critics and generally reflected the particular circumstances which allowed its emergence.

In both its functions, the treaty contains clauses enshrining the outcome of complex, hard-fought battles. These battles have not been documented but they can occasionally be glimpsed in the more revealing accounts.

In its first function the treaty has three central features: a schedule laying down the proportion of water allotted to each side, a definition of the period during which that schedule should apply and a clause determining the practice to be followed should the river flow be less than anticipated. All three features have been the subject of public and private debate and the exercise of political power between the two governments. Some flavour of these battles can be obtained by examining how the first — the schedule of water shares — emerged.

The key to the schedule was the share released to Bangladesh during the driest period of the year. This figure was the central and almost the sole content of the April 'understanding' which preceded the November 1977 treaty. All the rest of the sharing schedule and the treaty were constructed around it.

At the time of the understanding, newspaper accounts reported tense and enigmatic comments as the Indian negotiators prepared to leave Bangladesh. Whilst the senior negotiator for Bangladesh, Rear Admiral M. H. Kahn, temporised:

> "I shall have to give considered thought to many things; there are many ramifications; we will have to think about the modalities and mechanism."　　　　　　(Bangladesh Observer 19/4/77)

The senior Indian negotiator, Defence Minister Jagjivan Ram, denied agreement had been reached:

> "Take it from me, nothing has been settled."　　　　　　(Far Eastern Economic Review 6/5/77)

and his colleague, Foreign Secretary Jagat Mehta, contradicted him with a single-sentence, formal statement:

> "An understanding has been reached, the details of which are to be worked out ... as soon as possible."　　　　　　(Far Eastern Economic Review 6/5/77)

The most thorough account so far published of the Ganges waters dispute (Abbas 1982) regurgitates these enigmatic statements (pp. 77—78) without explaining why they took place. This omission prompted one of the participants, Rear Admiral Khan, to write to Abbas:

> "I thought you would come out with a vivid penpicture of the Bangabhavan discussions. The descriptions of those encounters could have given the idea of decision making at that stage. You also did not mention the administrative coup that the then CMLA (Chief Martial Law Adminis-

trator) Major General Zia organised with the Foreign Office, which compromised our demand by agreeing to a figure below 35,000 cusecs. This is one of the important historical facts of the Dispute and the public would have found that an interesting reading." (Khan 1982)

According to Khan (interviewed by the author in 1982), agreement only emerged, belatedly and, in his case, reluctantly, when he was outranked and outflanked by the then military leader of Bangladesh, Major General Ziaur Rahman. Several meetings between Khan and Ram were fruitless. The pair had met on several occasions during the previous year (Jagjivan Ram was also Mrs. Gandhi's chosen negotiator on this issue). Ram had been present on two occasions when Mrs. Gandhi had engineered a diplomatic humiliation against Khan (in September and December of 1976). During his meetings with Ram in April 1977 Khan held out for a Bangladesh share of the Ganges water close to 1130 m^3/s. At that figure there was no agreement and the Indian delegation packed their belongings and prepared to leave empty-handed. In accordance with usual diplomatic procedure, a meeting with the head of the host state was scheduled as a courtesy just prior to the departure of the Indian delegation. It was at this meeting that Khan was 'outflanked'. Without prior consultation with Khan (but on ground prepared by the two Foreign Secretaries) Ziaur Rahman proposed that Bangladesh's share could be set at 980 m^3/s. Jagjivan Ram asked for a further small concession of 14 m^3/s, "for my electors". The deal was struck, and the Indian delegation left for the airport.

Khan quietly fumed but could not oppose his senior colleague in the military junta (Khan 'represented' the Navy whilst Ziaur Rahman had the numerically and politically stronger army behind him). When asked if there was agreement he temporised but could not deny it. This reluctance to confirm the arrangement annoyed Ram and prompted his explosive "in that case, nothing's settled". But agreement had been reached on a key figure (which remains in force even today, whilst much else has been repudiated).

A minor quarrel, it might appear, between two individuals over at most 140 m^3/s of water; hardly the stuff of history, less still a guide to the construction of legal and administrative structures for ensuring equitable water development. Acting on such perceptions, most 'serious' and 'academic' writing on water disputes simply ignores the political battles from which treaties emerge. Unfortunately for such accounts, the exertion of political power is one, possibly the most important, determinant of when, how and in what form treaties are made. It was the stuff of history. The two individuals represented important factions and ideologies which bear directly on this and similar disputes.

The Breaking, 1982

For the five years during which the 1977 treaty held sway its first section, the sharing of the existing flow, worked admirably and without conflict. Sadly, its second, the negotiation of longer term development of the river, was stillborn.

There were two obstacles to progress in agreeing ways of augmenting the dry season flow (which was the primary agreed criteria for river development):

— Bangladesh's sovereignty and
— Nepal's sovereignty.

India proposed to increase the flow in the Ganges by building a barrage on the Brahmaputra and transferring water through a 320 km Brahmaputra-Ganges link canal (of up to 2850 m^3/s capacity). The barrage is to be sited within Indian territory and therefore would give India a measure of control over the flow of Bangladesh's second major river. Since the barrage over the Ganges at Farakka was used to control that river against the expressed wishes of Bangladesh, the government of Bangladesh is reluctant to countenance another within Indian territory.

Bangladesh proposed joint action to build storage reservoirs at a dozen major sites and many smaller ones in the Nepalese Himalayas. The Indian government was, however, unwilling to recognise any Bangladesh right to a share in the increased flow from such storage. India therefore refused (though with some wavering on at least two occasions) to back any joint or separate approaches by the two governments to seek the cooperation of the Nepalese government.

Without compromise on these two issues, the carefully drafted legal instruments and the expensively maintained administrative, technical and diplomatic institutions were utterly wasted.

In the absence of joint progress, the two governments have set up separate technical staffs to investigate the feasibility of alternative schemes. In the case of Bangladesh, the World Bank has provided a credit facility to enable British consultants to be employed on this work. Inevitably, however, such work proceeds on a stunted technical and political base.

When the time came, after the three years laid down in the treaty, to review progress and contemplate what should happen at the expiry of the treaty, a series of acrimonious meetings took place which recapitulated old conflicts the treaty had apparently resolved. During the five years of the treaty the Janata government in India had been replaced by a new government headed again by Mrs. Gandhi. Throughout her term of opposition Mrs. Gandhi had been critical of the treaty and she made no attempt to renew it. When Nov 5, 1982 came the treaty was not renewed but allowed to expire.

In its place Mrs. Gandhi negotiated a 'memorandum of understanding' with the new leader of Bangladesh General Ershad. Although this was heralded as a resumption of the earlier terms (The Hindu, International Edition, October 16, 1982), it differed from the 1977 treaty in at least one important respect affecting the terms of sharing.

During the review meetings in 1981, a senior Bangladesh Foreign Office official told me: 'The primary aim of the Indians is to knock off the 80% clause'. The new Memorandum of Understanding omits this clause, which reads as follows:

> "... if during a particular 10-day period, the Ganga flows at Farakka come down to such a level that the share of Bangladesh is lower than 80 percent of the (agreed share) the release of waters to Bangladesh during that 10-day period shall not fall below 80 percent of the (agreed share)."

This clause was the legal device which ensured that Bangladesh would not suffer from ongoing development of the Ganges in India and Nepal. The waters reaching the Farakka Barrage (in West Bengal) would dwindle as irrigation was extended in India and Nepal, but Bangladesh's share was guaranteed at 80% of the expected flow. Without that clause, the schedule of sharing contained in the new Memorandum becomes increasingly irrelevant with the passage of time: Bangladesh has a legal hold on a rapidly diminishing asset.

Even some of General Ershad's aides were surprised by the precipitate decision to drop the 80% clause. According to one senior member of his technical staff (interviewed in 1983), the Bangladesh team anticipated that the clause would be part of the new agreement. During meetings between the two heads of state and their immediate advisers it was dropped. According to Bangladesh's new negotiator on this issue, Agriculture Minister Obaidullah Khan (interviewed in 1983), the 80% clause is replaced by 'informal agreement' not to go below the previously guaranteed minimum (and in the 1983 dry season Bangladesh's share was generously exceeded). Another member of the team reports that Bangladesh approached the negotiations in a 'positive attitude of accommodation' saying, in effect, 'let us share the burden' of low flows rather than imposing the burden (of primarily Indian development) upon India.

On this occasion, as happened in 1977 with the Janata Government and the intervention of General Ziaur Rahman, a new government, and negotiators less hampered by the history of negotiation, have significantly changed the circumstances of the negotiation. Mrs. Gandhi, however, held fast to her stand of five years earlier. In May 1984, the Memorandum of Understanding was allowed to lapse. Since then a sequence of sensitive political events — election in West Bengal, the assassination of Mrs. Gandhi, an impending general election in India and negotiation about elections in Bangladesh — have restrained the Bangladesh government from commenting publicly upon their failure to renew the agreement. The result: the pressure of time on Bangladesh is increased. But from whom does this pressure come?

The Rich Peasant Thesis: Who Wants the Water?

Governments do not require water; it fulfils no inherent purpose for the institution of a state. The Indian government does not want the water of the Ganges for itself; it only acts to secure that water because it perceives demands for that water expressed by various groups of its constituents. Such demands are articulated in a variety of ways, one, but not the only one, being through the ballot box at elections.

In parts of South Asia the demand for water is being most consistently, almost inexorably, articulated by a particular group of agriculturalists. This is the group of farmers who have been able to corner and make good use of the 'package' of new agricultural technologies known as the Green Revolution — high yielding seed, fertilisers, pesticides, cheap credit, tubewells and tractors. This group of 'Rich Peasants', who use significant quantities of hired labour and produce not only for their own consumption but primarily for sale, are one of the successful examples of economic development in South Asia. They have begun to accumulate capital and to take on some of the characteristics of capitalist entrepreneurs. With increasing economic power have come the ability and desire to wield political power. The current agitation in the Indian Punjab is one expression of this desire and it is significant that one of the central demands of the agitation is for more water. In the words of one eminent political economist, the Rich Peasant 'has marched boldly through the door of politics... Indian state power has been exercised on his behalf' (Byres 1981, 445).

The Rich Peasants are 'masters of the countryside' in the most dynamic areas of India — Punjab, Haryana, Western UP, parts of Andhra Pradesh and of Tamil Nadu (and it is in these regions that irrigation has been extended most rapidly — see Tab 1). As a class they have been able to gain preferential access to the productive assets they require. Characteristically access to water has been provided by private, small command area, diesel pumpsets — diesel to avoid the vicissitudes of India's erratic and scarce electricity supply, private to circumvent the difficulties of state controlled irrigation schemes, and small because it is flexible ('optimal exploitation of the potential of HYV seeds ... requires a qualitatively new level of water control' Chattopadyaya 1977). Almost all of the marked acceleration in expansion of irrigated area since 1965 has come from wells. A large proportion of these pumpsets tap ground water rather than rivers or canals but they do so generally in areas where irrigation canals have raised the ground water level.

Tab 1 Irrigation ratios — India, Bangladesh 1967—1968 and 1975—1976

	Percentage of net sown area irrigated		
	1967—1968[1]	1975—1976[2]	Change 1968—1976
India			
Ganges and Brahmaputra basin states			
Assam	n.a	n.a	—
Bihar	24.3	29.8	5.5
Harayana	32.2	50.3	18.1
Madya Pradesh	6.4	8.8	2.4
Rajasthan	12.4	17.1	4.7
Uttar Pradesh	32.4	40.2	7.8
West Bengal	26.5	n.a	n.a
Other States			
Andhra Pradesh	27.2	34.9	7.7
Gujarat	11.3	15.0	3.7
Himachal Pradesh	16.4	16.6	0.2
Jammu + Kashmir	41.2	40.2	−1.0
Karnataka	10.8	15.3	4.5
Kerala	19.3	9.2	?
Maharashtra	8.1	11.0	2.9
Orissa	n.a	19.2	?
Punjab	58.4	73.8	15.4
Tamil Nadu	43.2	46.7	3.5
Bangladesh[3]	n.a	11—15	—

Sources: [1] India, 1972, Table 4.9
[2] Bharadwaj, 1982, Table 8
[3] Edwards et al., 1978, Table 1 + App. 1

Outside the Punjab (where 85% of the net sown area was irrigated in 1980--1) there is considerable scope for expansion of tubewell irrigation from ground water. Within the Ganges basin, Bihar, Madya Pradesh and West Bengal all irrigate less then one third of net sown area. A rapid extension of the area irrigated from ground water would, however, soon encounter localised water shortages. If concerted effects were not made to replenish the ground water on a large scale (through extension of canals and rainfall runoff controls), water table lowering would soon be generalised. Even with complementary development of canals and ground water irrigation, the availability of water in the Ganges basin could soon become a critical restraint on the progress of the Rich Peasantry.

How soon? It is difficult to tell. The Indian Government is understandably coy about circulating estimates of water demand or supply. Such estimates as exist are frequently contested, inevitably when national development plans are threatened by an international dispute and estimates themselves become subjects of dispute.

In its proposal for augmenting the dry season flow of the Ganga (India 1978, 47) the Indian Government notes: 'even the irrigation projects in hand can consume the entire fair weather flow at Farakka'. To avoid that happening, India has planned the development of the Ganges basin — a development similar in purpose to that proposed by Bangladesh but smaller in scale, quietly implemented and controlled by the Indian government. India has started developing all possible storage reservoirs within the Indian part of the Ganges basin and has quietly embarked upon bilateral negotiations to construct a series of major reservoirs in Nepal (at sites identified in the Bangladesh proposal for the augmentation of the dry season flow of the Ganges (Bangladesh 1978) but fewer of them and on a more modest scale). The confidential estimates of the Indian irrigation department indicate that

"the storages in the Ganga basin in India are utterly inadequate for meeting even the minimum local needs."

and

"the needs for Ganga waters are so urgent and so large that any additional storages that may be created in Nepal will not be able to effectively meet the requirements in time as well as in quantum" (India 1978, 54 & 56).

We therefore encounter the following problem: Rich Peasants can, and have already started to, extend the irrigated area in the less advanced states of Bihar, MP, UP and West Bengal. As they do so the Government will attempt to provide the infrastructure of canals to support their efforts but in due course (one estimate from a senior Indian negotiator in 1978 was 'within fifteen years') water shortages will become a severe constraint.

Two possible ways of resolving this problem have been suggested. A group at Harvard has suggested a massive attempt to increase the amount of water stored in the ground. This proposal they termed the Ganges Water Machine (Revelle & Lakshminarayan 1975). Whilst the scheme has many characteristics which might lend itself to implementation by and for a Rich Peasantry, the Indian Government does not believe it would work. A second way of meeting the demand of India's Rich Peasants is the Indian Government's Brahmaputra-Ganges link canal proposal (India 1978) which is now the subject of detailed feasibility studies by the Indian Government. Alternative solutions such as increasing the efficiency of water use or changing cropping patterns have not been seriously discussed in Indian strategy documents (such as India 1982).

As the Indian estimates in Tab 2 indicate, the flow per unit cultivable area of river basin is ten times larger in the Brahmaputra than the Ganges, and the flow per unit of river basin population is seven times. Those estimates are subject to the severe practical reservation that they are annual flow estimates and unless a river's flow is evenly distributed throughout the year or is readily tameable, annual estimates represent a somewhat utopian view of what is available. These rivers are highly seasonal and not readily tamed.

Supposing that detailed investigation – week by week – of the Brahmaputra flows shows (as I think it does) that there is 'spare' water in the Brahmaputra, how could that help India's rich peasants?

India's Proposal (India 1978) outlines a canal to be built from Jogighopa in Assam (110 km up the Brahmaputra, North from the Bangladesh border) to the existing Farakka barrage (18 km up the Ganges, West from the Bangladesh border). Transfers of water from the Brahmaputra will not, therefore, be of any direct use to Indian agriculture. But at the size (tentatively) mooted by India of 2850 m³/s this canal could replace all the minimum flow of the Ganges as it enters Bangladesh plus all the requirements of Calcutta Port, with something to spare in addition. Thus all of the current Ganges flow and any additional flow obtained from reservoirs in India and Nepal could be utilised in India. This could provide plenty of water to irrigate all the cultivable areas of the Indian Ganges basin and probably some surplus which could be transferred south to drought areas (plans for a National Water Grid or for regional interbasin transfers to equalize water availability are gradually rising up the Indian Government's development agenda).

Conclusion: Politics Rules

The point of this paper is to argue that legal tools, geographical concepts and engineering technology are insufficient to control either rivers or governments. People control both, but the way in which they do so, through their collective institutions and their differences as classes and nations in conflict, require careful analysis.

Tab 2 Water resources of the Ganges and Brahmaputra basins – Indian government estimates (1978)

River basin	Mean annual flow $10^9/m^3$	Population 1977 million	Annual per capita flow 10^3 m^3/yr	Cultivable area 10^6 ha	Annual flow per unit area (m^3/m^2)
Ganges	379	291	1.3	65	0.6
Brahmaputra and Meghna	715	78	9.2	12	6.0

Source: India, 1978:26 (converted to SI units).

I have tried to show that conceptions of the 'national interest' and expressions of class interest bear upon the nature and timing of the development of the Ganges and Brahmaputra rivers. I cannot pretend that these insights lead to readily applicable guidelines either for engineers or negotiators. In general, it is useful to know the origins of demands for river development and to be on the look out for the sorts of influences which will direct the outcome of negotiation.

Amongst the very preliminary conclusions which I would draw from the analysis are the following.

The timing of negotiations is crucial to their success. In 1977, both the Indian government and the Bangladesh government in different ways brought a will to settle to the negotiation. Bangladesh brought the defeat of Khan's conception of national interest. India brought the new governments zeal to achieve and be seen to achieve objectives which had eluded its predecessor. The elements of good timing seem to include

— the resolution of political dispute (i.e. progress may be made in the wake of state or national elections or of power struggles within governments)
— favourable third party relations (i.e. Bangladesh is more likely to make headway when India's relations with Pakistan are bad and with China good).

Successive governments of Bangladesh have believed that time is on their side, so long as they do not negotiate away (too cheaply) India's access to the waters of the Brahmaputra. Increasingly members of the Bangladesh negotiating team have perceived their hold on the waters of the Ganges as somewhat tenuous whereas the Brahmaputra is seen as within their control. This perception has been reinforced by the success of Indo-Nepalese negotiations over the construction of reservoirs on the Nepalese tributaries of the Ganges. By the middle of 1984 Bangladesh negotiators were reluctantly concluding that Bangladesh's claim to a share of the augmented flow of the Ganges was slight. The Bangladesh government's bargaining position had been seriously weakened by Nepal's decision to proceed with the reservoirs on a bilateral rather than a trilateral basis. Both of these perceptions are likely to hold in the medium term, but the strength of the current Punjab agitation and the importance it places upon water should be heeded. The Rich Peasants of North India are an increasingly powerful political force, their economic strength has been built upon access to abundant water. As that abundance becomes scarcity the Indian government will feel impelled to take even drastic measures. Within ten years, a Brahmaputra-Ganges link canal circumventing Bangladesh could be put seriously upon the agenda. At that point Bangladesh's control over the Brahmaputra, and thus a major part of its leverage in the negotiations, will dwindle rapidly.

These tentative conclusions suggest that at an appropriate time, within the next 6—7 years, the government of Bangladesh should present India with a package which secures Bangladesh's dry season supplies for all time and maximises the substantial secondary benefits which India will offer (irrigation facilities from a link canal, flood mitigation on both rivers, cheap hydroelectric power). Such a time may be presented in the immediate aftermath of the next Indian general election.

References

Abbas, B. M.: The Ganges Waters Dispute. University Press Ltd., Dhaka 1982.

As you were at the Ganges talks. Far Eastern Economic Review (May 6, 1977)

Bangladesh: Proposal for the augmentation of the dry season flow of the Ganges. Dhaka 1978 (unpublished).

Bangladesh Observer, April 19, 1977, quoted in Abbas (1982)

Bharadwaj, K.: Regional differentiation in India. Economic and Political Weekly, Annual No. (April 1982)

Bindra, S. S.: Indo-Bangladesh relations. Deep and Deep, New Delhi 1982.

Byres, T.: The new technology, class formation and class action in the Indian countryside. Journal of Peasant Studies, 405—454 (July 1981)

Chattopadyaya, B.: Water, Cereals and Economic Growth. People's Publishing House, New Delhi 1977.

Crow, B.: Appropriating the Brahmaputra: the onward march of India's rich peasants. Economic and Political Weekly, 2097—2102 (December 25, 1982)

Edwards, C.; Brigg, S.; Griffith, J.: Irrigation in Bagladesh. University of East Anglia, Discussion Paper 22, 1978.

India: Report of the Irrigation Commission, New Delhi 1972.

India: Proposal for Augmenting the Dry Season Flow of Ganga. New Delhi 1978 (unpublished).

India: National Perspectives for Water Resources Development. New Delhi 1982.

India & Bangladesh: Memorandum of Understanding. Bangladesh Times (8.10.82)

Khan, Rear Admiral M. H.: Letter to B. M. Abbas (1 July 1982)

Revelle, R ; Lakshminarayan, V.: The Ganges Water Machine. Science, 188, 611—16 (May 9 1975)

Zaman, M. (ed.): River Basin Development. Tycooly International Publishing Ltd., Dublin 1983.

Lundqvist, J.; Lohm, U. and Falkenmark, M. (eds.):
Strategies for River Basin Management, pp. 265–277
© 1985 D. Reidel Publishing Company

The Mahaweli Strategy of Sri Lanka — Great Expectation of a Small Nation

Madduma Bandara, C. M., Dept. of Geography, University of Peradeniya, Peradeniya, Sri Lanka

Abstract: Sri Lanka, a small nation reputed for its ancient 'hydraulic civilization', is at present locked in a gigantic struggle to develop the resources of her mightiest river system — The Mahaweli. The Mahaweli Project is not merely another river development programme for the Government and people of Sri Lanka. In terms of the extent of land involved, the amount of financial and other resources invested, and the number of people directly affected or shifted, it pervades almost all aspects of the economy and life of the people of the Island. The visionaries of this massive project expected major achievements in the spheres of irrigated agriculture, hydro-power production, employment generation and population redistribution. The paper summarizes the salient characteristics of the Mahaweli Project, reviews its progress and discusses some of the challenges it has faced.

Introduction

The Mahaweli Project is the largest river basin development programme undertaken by the Government of Sri Lanka. The project is based on a Master Plan produced by an UNDP/FAO team during the years 1964–68 (Fig 1). An assessment of the possible benefits led the present regime under President Jayawardhana to make a decision in 1977 to accelerate the activities of this project and to telescope its time horizon for a major part of the Project to six years from its original thirty years in the Master Plan. The large capital outlay and the expertise needed for heavy construction work involved in the implementation of the project was too massive for a country of the size of Sri Lanka. Therefore, the Government had to invite foreign aid and investment from a host of friendly countries for the accomplishment of this enormous task.

As the construction and engineering programmes gradually taper off and wind up, socio-economic, management and enrironmental issues are beginning to surface and dominate. The most prominent among these are the questions of irrigation water management, human settlements, environmental degradation and cutural regeneration. The visionaries of the Mahaweli Development Project expected much more from it than merely providing irrigation water, or increasing hydropower generation and employment opportunities, which of course form three of the major benefits that make the project viable. More than anything else they expected the 'catalysing effect of the Mahaweli to work centrepetally, to improve the quality of life of the rural interior'. How these great expectations would be realised, within the envisaged time horizon under the existing and proposed systems of management is still too premature to be assessed.

266

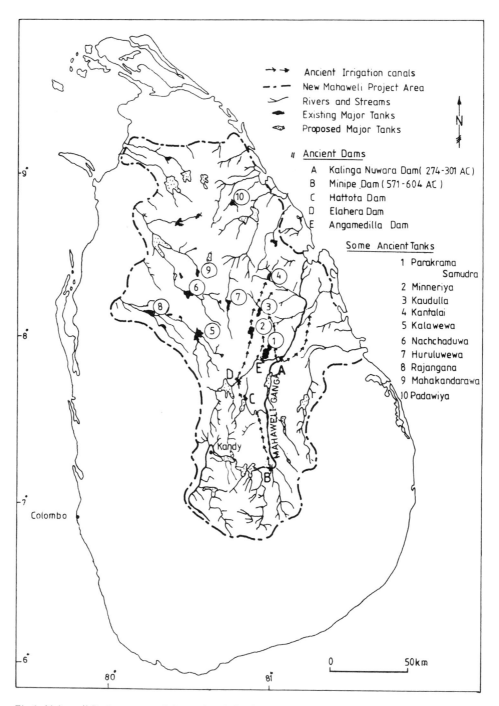

Fig 1 Mahaweli Project area and the ancient irrigation system based on Mahaweli Ganga

Historical Sketch

Sri Lanka has been associated with a 'hydraulic society' since the early periods of her history (Leach 1959). The magnificient ancient irrigation works of Sri Lanka (Brohier 1935) indicate a high standard of professional skill in the sphere of water resources development, utilization, and management. Thus even the mighty river of Mahaweli did not escape the attention of the ancient hydraulic engineers. As Fig 1 shows, since the time of King Mahasena (274–301 AD), a great builder of ancient irrigation works in Sri Lanka, significant attempts have been made to dam the main stream of Mahaweli. Some of the canals originating from these dams carried water to distances of over 75 km. The civilization based on these irrigation systems had however, collapsed around the 12th century AD, for reasons which are still not fully comprehended.

The modern interest in the development of the Mahaweli river basin has originated from the vision of Hon. D. S. Senanayake the first Prime Minister of Independent Sri Lanka (Senanayake 1935). The Canadian Hunting Corporation resource survey (1962) provided the first comprehensive assessment of the resources of the Mahaweli Ganga basin. Then the UNDP/FAO study published in 1969 led to what is now called the Mahaweli Master Plan. The recommendations of this report began to be gradually implemented since early 1970s with assistance from the World Bank. Several comprehensive feasibility studies were conducted before the implementation of the Project notably by Sogreah (1972) and Nedeco (1979). The current Accelerated Mahaweli Programme was started in 1978 and a substantial part of its main projects have been completed while a few others are still in progress.

Catchment and its Water Resources

The Mahaweli Ganga forms the largest and the longest river system in Sri Lanka. It is 315 km long and is longer than the combined length of any other two rivers of the Island. The basin covers some 10,750 km^2 of land or nearly 1/6 of the total land area of Sri Lanka (Fig 2).

The head streams of the Mahaweli Ganga orignate at altitudes of over 1340 m above sea level in the wettest area of the Island where the annual rainfall often exceeds 5000 mm. Thus one of the major tributaries of the river, i.e. the Kotmale Oya rises at an altitude of 2250 m above sea level and joins the main river after passing through an area of high relief and rainfall. This upper portion of the catchment due to its year-round high discharge and steep gradients had become the main theatre of hydropower development in Sri Lanka. The first major reservoir in the Mahaweli cascade, the Kotmale reservoir is now being constructed with the support of the Government of Sweden. The annual discharge of the Kotmale Oya below this point is estimated to be around 953 million m^3 (Tab 1).

The river, flowing in a zig-zag course winds its way around the city of Kandy and enters the picturesque Dumbara valley near Polgolla. The main diversion structure of the Mahaweli is located at this point from which a 8 km long tunnel carries some 60 m^3/sec

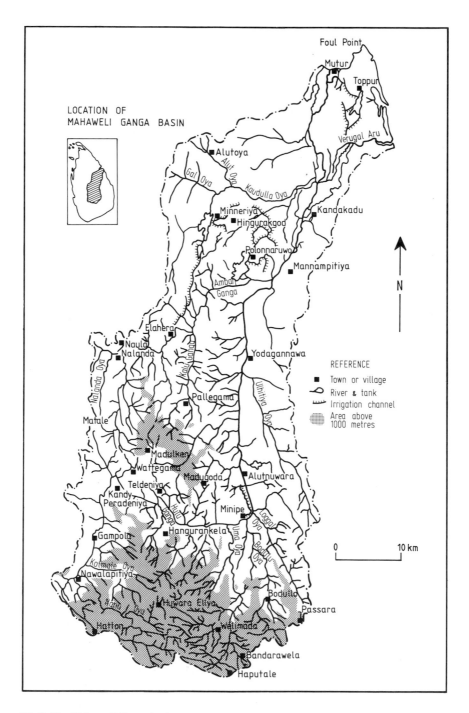

Fig 2 The Mahaweli Ganga basin

	J	F	M	A	M	J	J	A	S	O	N	D	Year
Kotmale Oya (Morape)	47	26	22	38	72	107	125	117	104	121	98	73	953
Mahaweli Ganga (Peradeniya)	97	56	56	90	174	261	252	250	215	276	252	190	2164
Mahaweli Ganga (Randenigala)	306	204	123	184	262	350	352	331	307	421	459	470	3843
Mahaweli Ganga (Weragantota)	666	398	269	324	343	375	391	384	333	504	600	762	5339
Mahaweli Ganga (Manampitiya)	1096	656	375	389	354	366	394	365	339	583	816	1306	7152

Sources: NEDECO — Hydrological Crash Programme Mahaweli Development Project 1981.

Tab 1 Monthly flow of Mahaweli (in million m³)

of water. The Ukuwela power station located at the end of the tunnel generates 40 MW of power. The main north-bound irrigation canal begins from this point.

After passing the Polgolla Diversion structure, the river enters a valley with steep slopes, rapids and gorges. The next major dam in the Mahaweli cascade — the Victoria Dam, located in the vicinity of Victoria rapids, has just been completed with the support of the Government of the United Kingdom. In addition to the discharge of the main river, the Victoria reservoir is fed by Huluganga — a tributary originating in the Knuckles range which is another super humid area that receives over 5000 mm of rainfall.

Immediately below the Victoria dam another major reservoir in the Mahaweli cascade — the Randenigala reservoir, is now under construction with the support of the Government of the Federal Republic of Germany. The annual discharge of the river here is estimated to be in the region of 3843 million m³. This reservoir will be the largest reservoir in the Accelerated Mahaweli Programme.

The river after its descent from the highlands, swings sharply northwards to settle on a straight course upto Manampitiya. Upstream of the point of inflexion the ancient wier that diverted water to the left bank of the river could still be seen. On the right bank, the new Minipe transbasin canal begins its long journey which finally ends at the Maduru Oya reservoir, which was completed last year with the support of the Government of Canada. Over 30 km of this canal had been concrete-lined. At the end of the straight stretch of the river which falls well within the 'dry zone' of Sri Lanka, another ancient dam across the river — the Kalinga Nuwara dam was located. The annual discharge of the river at Manampitiya which is not far from this point, is over 7000 million m³. The river develops a host of distributaries below this point, before reaching the sea near the Trincomalee Bay.

The Mahaweli basin which covers an area of over 10,000 km² spreads over all major ecological regions of Sri Lanka, namely, 'the Dry Zone', 'Wet Zone' and the 'Intermediate

Stations	J	F	M	A	M	J	J	A	S	O	N	D	Year
						(1870–1980)							
Anuradhapura	99.5	45.7	80.0	177.0	90.2	22.7	31.2	43.7	74.5	248.2	259.1	226.1	1397.8
Trincomalee	171.6	72.1	47.6	56.8	64.7	27.2	52.4	101.0	103.1	223.0	356.4	357.8	1633.6
Kandy	121.5	67.3	104.8	182.0	157.5	205.5	170.7	141.8	146.1	287.9	274.8	218.4	2078.4
Bandulla	218.0	94.4	118.7	196.8	114.5	48.5	55.5	87.9	97.2	232.9	270.4	299.1	1833.9
Nuwara Eliya	136.9	63.1	87.0	150.3	194.0	274.1	256.6	188.8	195.2	258.0	222.4	206.8	2233.2

Source: Yoshino et al. 1982, Climatological Notes, 31, Institute of Geoscience, University of Tsukuba, Japan

Tab 2 Rainfall in the Mahaweli river basin

Zone'. Thus the rainfall varies widely within the basin area (Tab 2). The regime of the river is governed mainly by the seasonality of rainfall, which constitutes the most important factor in the exploitation of the water resources of the Mahaweli Ganga. Thus the wettest months in the upper catchment area are the driest months in the lower basin of the Dry Zone. Therefore, harnessing of upper Mahaweli waters to irrigate the Dry Zone Lowlands which comprise of over 60% of land in the total basin area had become one of the main water resource utilization strategies since ancient times. The upper 12% of the basin actually contributes some 27% of the total volume of discharge from the catchment. This upper catchment is now predominantly under tea estates while the lower basin is covered by a deciduous type Dry Zone forest which still persists where it is not cleared for agriculture.

Main Features of the Project

The Mahaweli project is essentially a multi-purpose river valley development programme which involves the creation of a chain of large reservoirs, canals, irrigated areas and new settlements. Thus irrigation and hydropower generation remain two of the most significant immediate benefits of the Project. However, the resettlement of people in new development areas and the generation of new employment opportunities constitute some of the other important benefits. The magnitude of the task of Mahaweli River Development is clearly presented in the UNDP/FAO Master Plan.

According to the recommendations of the Master Plan the project area spreads over an extent of 0.61 million ha of which 0.36 million ha were to be irrigated. The plan also envisaged generation of a total of 508 MW of hydropower from its varied projects. The whole plan was phased over a period of 30 years for stepwise implementation of about a dozen projects.

The Accelerated Programme

The Accelerated Mahaweli Programme, started in 1978, concentrated its attention on a few major reservoir projects of which some have been already completed and the work is progressing on schedule in others (Tab 3).

Reservoir	Capacity million m³	Hydropower Development (Installed capacity) (MW)	Extent of land benefited (ha)	Date of completion	Supporting country	Cost of Project (in million)
Kotmale	174	134	36,500	1985	Sweden	SEK 1964
Victoria	128	210	45,000	1984	UK	£ 146
Maduru Oya	467	1.5	46,750	1983	Canada	C$ 80
Randenigala	860	122	30,200	1986	FRGermany	DM 500

Tab 3 Major reservoirs of the Mahaweli Project

In addition to the four major reservoir projects, the Minipe Transbasin Canal and the Ulhitiya-Ratkinda reservoir system had already been completed. Much work has also been done on the front of human settlements and environmental protection and conservation. It is estimated that the Accelerated Mahaweli Development Programme will involve the construction of some 362 km of main canal, 3622 km of distributary canals, and over 400 km of road. On the settlement front, it will involve setting up about 350 townships, 1600 village centres and some 8000 hamlets. All this was planned to be completed in a period as short as 6 years. Inspite of the enormity of the exercise, the tremendous amount of work that had been accomplished during the last 5 years is a tribute to the new Ministry of Mahaweli Development created particularly to handle this task.

Shift of People to new Settlements

It has been estimated that nearly one million people will be accommodated in the Accelerated Programme area in addition to its present inhabitants. The settlers fall into several categories including, (a) farm family units settled on new lands, (b) non farm support family units (engaged in services and trade), and (c) farm or support family units settled on improved lands.

In the older colonization schemes based on major irrigation projects, the migration of people into them was largely a matter of individual choice. Those migrants, mostly belonged to poorer landless families from more congested parts of the Island. In the case of the Mahaweli Project, in addition to these usual settlers, certain other groups who had to make a real sacrifice for a greater national cause has also been included. Some of the biggest evacuation operations in the history of Sri Lanka had to be performed within a short period of time from the areas submerged by large reservoirs such as Victoria and Kotmale.

Most evacuees from reservoir project areas had to leave a congenial environment in the hills and many of them had to part from fertile lands occupied by their ancestors for centuries. The shift of these people proved to be an equally challenging task to the Mahaweli authorities, as that of the challenge faced by engineers in taming the river through the construction of a series of colossal engineering structures. Although the Government could compensate for the economic losses sustained by the evacuees, the very act of uprooting large numbers of families and resettling them in new areas involved many more things that could not be quantified or assessed in monetary terms alone. Nevertheless, in terms of the anticipated economic returns all the Mahaweli projects were considered justifiable.

One of the main theatres of the drama of evacuation of large numbers of families was in the area submerged by the Victoria reservoir built with assistance from the British Government. In the history of Sri Lanka it is difficult to find as large an exodus of people as what had been witnessed from the Victoria reservoir bed. A survey conducted by the writer (Madduma Bandara 1982) revealed that the project affected some 283 villages while submerging a major township in the area. The number of families whose residence was located within the submerging area stood at 3687 comprising a total population of 18,382 persons. Similarly, the Kotmale resevoir built with assistance from the Government of Sweden, affected some 66 villages and 4 tea estates leading to the evacuation of some 2691 families from the submerging area (Bulakulama 1979).

Many existing old settlements in the downstream Mahaweli Development areas also had to undergo considerable transformation in the process of their integration with the new plan of land development. Some of the old minor irrigation reservoirs had to be drained for expanding the cultivable extent, and a number of old nucleated villages had to be dispersed and their residents moved to new settlements. Similarly, several old colonization schemes based on major irrigation reservoirs had to be integrated into the new settlement plan.

Settlements Based on the "Cluster" Principle

The new settlement plan of the accelerated Mahaweli Project is based on a 'cluster system' of hierarchical settlements. The smallest cluster of houses is called a 'hamlet' which is located at a maximum distance of one to two kilometres from the respective irrigated allotments. A hamlet will comprise of about 100 family units. Each family unit gets 0.4ha of highland as a housing lot and a home garden and one hectare of irrigable land for a rice field.

About four or five hamlets are consolidated to form a village and four village centres form a township. These 'cluster' settlements were different in their layout from the 'ribbon' settlements which were developed along irrigation canals and road fronts in the older colonization schemes. In the selection of new settlers too, some consideration was given to social cohesion and previous experience in agriculture, in addition to landlessness and family size.

All the necessary infrastructural facilities such as roads, shools and hospitals etcetera are provided by the Government. However, it is expected that the settlers will establish other facilities such as shopping centres, and community centres through their own efforts. It is an interesting experience to observe the gradual emergence of these hamlets and townships such as Galnewa, Bulnewa, Medagama and Giranduru Kotte in the new Mahaweli settlement areas.

Ecological Impact

In a large river basin development programme such as that of the Mahaweli Project, some disturbance to the existing ecosystems is inevitable. Environmental impact assessments were made in 1976 and in 1980 (Tams, USAID 1980). The annual soil losses in the upper catchment are estimated to be in the region of 10 tons per hectare per annum. Although, this rate of erosion is not a direct result of the Mahaweli Programme, it is computed that it can lead to a total sediment yield of a little more than a million cubic metres per year. Compared with the storage capacity of major reservoirs, this rate of sedimentation is not expected to impair the effective operating life of the reservoirs for at least 50 years. Nevertheless, if left unchecked the rate of erosion will continue to increase. Therefore, it was felt that there is an urgent necessity to adopt an effective watershed management policy in the upper catchment areas.

It is with this intention that concerted efforts are now being made to reforest the reservoir areas of the upper catchment. Thus on one hand the state agencies such as the Forest Department have started upper Mahaweli Forestry Projects based on the forest plantation model. On the other hand voluntary organizations such as Nation Builders Association are now experimenting with new concepts such as 'community forestry'.

The implementation of the Accelerated Mahaweli Programme involves the clearence of over 80,000 ha of undeveloped land for agricultural and other development needs. The demand for timber and fuel wood is expected to increase significantly in the next 25 years. In particular, it is estimated that nearly one million settlers will have a fuel wood requirement of around 570,000 m^3 annually and the existing forests can probably supply fuel wood needs for only two or three years. This indicates a need for new thinking in the sphere of developing forest systems that cater for the energy needs of the people in both the upper catchment as well as in the lower development areas.

The clearance of large tracts of forested land would also significantly reduce the natural wild life habitats, thereby decreasing the carrying capacities of animal populations. In recognition of this fact, four national parks are now been developed in the downstream Mahaweli Development areas.

Issues in Irrigation Water Management

At the time of the preparation of Mahaweli Master Plan, water management does not seem to have been a serious major concern. As Chambers (1974) observed the UNDP/ FAO Report of 1969 is 'remarkable for its failure to see and grapple with the problems and opportunities of water management'. In this three volume report some direct reference to water management is made only in one page. Thus Chambers goes on to say that "this report may be quoted as an example of the persistent failure to recognize the complexities of water management and of the organizational requirements for it".

The situation has significantly changed since the writings of Chambers 10 years ago. Several research and development programmes were carried out in the Mahaweli Development areas, particularly in the Kalawewa and Minipe Irrigation systems.

A Water Management Secretariat was set up in the premises of the Ministry and attempts are now being made to optimize the distribution of available water among different irrigation systems within the Mahaweli development areas (Fig 3).

The water management secretariat is responsible for the management of the complex water distribution system both at the macro-level as well as at the micro-level. Accordingly, two computer models have been developed to handle this task. Of these two, the macro-model uses historical streamflow data and demand for irrigation water and hydropower to evaluate alternative policy options. The micro-model on the other hand helps to assess the irrigation demand and diversion flows within the complex system of tanks and irrigation areas (Tab 4).

At the field level a special group of water controllers *(jala pālaka)* have been appointed to assist the farmers. Attempts are also being made to increase the community participation in water management. Nevertheless, the sociological dimension of water management does not appear to have been fully realised. It is hoped that the International Irrigation Management Institute, which is now in the process of being established within the Mahaweli basin would significantly contribute to the understanding of irrigation water management problems in the Mahaweli development areas.

Many hydrological implications of the Mahaweli Development activities will become observable only after the main reservoir systems are made fully operational. The changes that may come cannot be quantified until the reservoir operation policies are finalised. It is however, clear that the overall volume of flow of the river would be reduced by about 50 %, mainly due to the diversion of Mahaweli waters to irrigation systems outside the basin. Thus the flood peaks are likely to be subdued and the frequency of flood occurrence will be reduced. The reduction of water flow in the main river is likely to affect the lower flood plain *villu* (i.e. marsh) areas below Manampitiya. It is estimated that at least half the present *villus* will be permanently lost. In certain areas brought under irrigation, problems of water logging and salinity can already be observed.

The reduction of river flow, particularly below the dams, while leading to downstream degradation, may also increase the incidence of malaria. The creation of small pools of clear water provides ideal breeding grounds for mosquitoes. This situation was experienced in the Tennekumbura area following the diversion of the Mahaweli at Polgolla in 1976. It had become necessary to flush out these mosquitoe breeding grounds by occasional water issues from the Polgolla dam. There may also be several micro-climatic changes in the vicinity of the three large reservoirs which are now being built around Kandy. All these indicate, that there is a necessity to establish an effective system of ecological monitoring.

Great Expectations

For most Sri Lankans the Mahaweli Project is not merely another river valley development programme. In terms of the extent of area covered, the amount of financial and other resources devoted and the number of people shifted or affected, it pervades almost all aspects of the economy and life of the people of Sri Lanka. It is now clear that, the Mahaweli Project is significantly changing the geography of Sri Lanka at an unprecedented

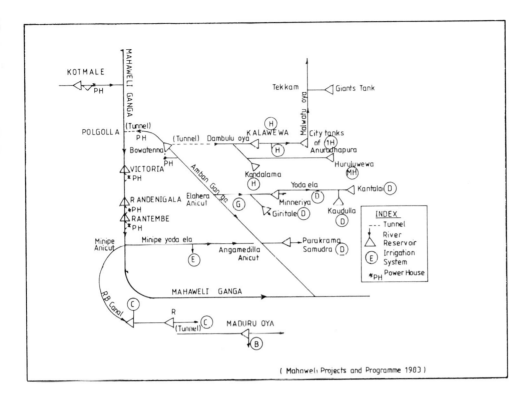

Fig 3 Water distribution diagram of Accelerated Mahaweli Programme

Tab 4 Water distribution in the Mahaweli system (in million m³)

| Season | (Long-term averages are given in brackets) | | | | | |
	Polgolla diversion	Amban Ganga flow	Bowatenna	Diversion to H area	Elahera diversion	Angame dilla
1982/83 Maha (Oct.—March)	567 (682)	383 (402)	950 (1084)	364 (441)	342 (614)	133 (206)
1983 Yala (April—July)	263 (499)	31 (136)	294 (635)	153 (321)	149 (335)	33 (74)

(Mahaweli Projects & Programmes, 1984)

scale. The benefits that are expected from the project particularly in the sphere of food production, hydro-power generation and employment, if realised, are bound to bring a considerable degree of stability to the political economy of Sri Lanka. The strategy of the Government, clearly is to make Mahaweli a 'lead' Project from which a wide range of economic and social benefits would radiate in to the entire nation; It is a project that is expected to give life to an ailing economy of the rural interior. The great men who conceptualised the Mahaweli Development Programme 'did not launch it merely to ensure a plenteous agriculture to service the supply lines to the cities or to merely provide hydro-power to keep the wheels of industry moving' (Ministry of Mahaweli Development 1983).

Historically, much of the development areas of the present Mahaweli Project had an advanced civilization as far back as twenty centuries. These areas were abandoned for various reasons and the people moved into the wetter interior around the 12th century AC. Now, the return journey had begun and the new settlers are in a way returning to their ancient homeland in the Dry Zone plains. It is therefore, felt important to ensure that the same factors that led to the collapse of the famous hydraulic civilization of these regions will not become operative again, be they natural causes or ones of human origin.

For visionaries of this great river development project, Mahaweli was only a means to an end and not an end in itself. The type of life and agrarian society that would emerge in the new settlements of the Mahaweli Development Project is considered the end product of the entire exercise. As the Hon. Minister of Mahaweli Development aptly pointed out: "what the men of vision envisaged was the 'dawn of a Mahaweli Era' with the return of the people to their ancient homeland in the plains. They envisioned that on returning to their traditional homelands the people would reawaken to their cultural ethos and build a new civilization like that which flourished in the ancient kingdoms of Anuradhapura and Polonnaruwa." This 'new society' would inevitably be based on old 'cultural roots' which still survive in these areas in the form of traditional systems of social values.

References

Brohier, R. L.: Ancient Irrigation works in Ceylon. Part I, II and III. Government Press, Ceylon 1934.
Bulankulama, S.: A Study on the Socio-economic conditions of the People and Area to be submerged by the Kotmale Reservoir, Settlement Planning Division, Mahaweli Development Board, Polgolla 1979.
Canada-Ceylon Colombo Plan Project. A Report on a Survey of the Resources of the Mahaweli Ganga Basin Ceylon. Hunting Survey Corporation Limited, Toronto, Canada 1962.
Chambers, R.: Water Management and Paddy Production in the Dry Zone of Sri Lanka. ARTI Occasional Publication, Colombo 1974.

Leach, E. R.: Hydraulic society in Ceylon. Past and Present, 15, 2—25 (1959)

Madduma Bandara, C. M.: Report of a Survey of the People and Private Property Affected by the Victoria Reservoir Project. Victoria Project Office (A, S, & D), Kandy 1982.

Ministry of Mahaweli Development: Mahaweli Projects and Programmes — 1984.

Nedeco:Mahaweli Ganga Development Programme Implementation Strategy Study. Ministry of Mahaweli Development, Colombo 1979.

Nedeco: Hydrological Crash Programme, Mahaweli Development Project. Hydrology Division, Irrigation Department 1981.

Senanayake, D. S.: Agriculture and Patriotism. 1935.

Sogreah: Mahaweli Ganga Development. Feasibility Studies, Grenoble 1972.

TAMS/USAID: Environmental Assessment: Accelerated Mahaweli Development Programme. Ministry of Mahaweli Development 1980.

UNDP/FAO: Mahaweli Ganga Irrigation and Hydro-Power Survey. Final Report, Vol. I, II & III, FAO/SF: 55/Cey-7, Rome 1969.

Lundqvist, J.; Lohm, U. and Falkenmark, M. (eds.):
Strategies for River Basin Management, pp. 279–285
© *1985 D. Reidel Publishing Company*

Coordinated Management of Land and Water Resources; Regionalization and Institutions

Kindler, Janusz, Institute of Environmental Engineering, Warsaw Technical University, ul. Nowowiejska 20, 00-653 Warsaw, Poland

Abstract: At the background of an overview of some basic interrelationships between land and water, the paper adresses two related issues of regionalization and institutional arrangements. Based on few examples it is shown that administrative regionalization that almost precisely follow the watershed boundaries does not necessarily lead to better planning, more effective management and coordinated use of land and water resources. Even the best planning schemes often fail if there are no adequate institutional arrangements governing land and water use. The difficulties in interaction among many entities and rules concerned are examined. To achieve coordinated management of land and water resources, the river basins, which are coherent hydrological units relevant to water control, should be seen in the broader societal contexts.

Introduction

The coordinated management of land and water resources is an increasingly serious concern throughout the world, in countries of widely differing economic and political philosophies, whatever the level of development. The growth of population and the urban sprawl, industrialization, the pressure on agriculture at both the extensive and intensive (fertilizers, pesticides and herbicides) margins, water pollution, devastating floods and droughts, desertification and deforestation are only a few of the major headings under which one could catalog the reasons for the urgency being felt in this field. It also seems to be becoming increasingly understood that the problems related to coordinated land and water use and management extend far beyond the endowed physical resources base; to ignore the existence of different sets of economic, social, cultural, and institutional factors results in a failure to appreciate the social objectives and aspirations of the society. The arrangements theoretically possible in physical terms may be socially unacceptable.

This paper addresses briefly some problems related to coordinated management of land and water resources, taking into account that many uses of land and water keep changing, sometimes into undesirable directions and at the alarmingly high rates. Many of these problems have been discussed for decades and by many, they reemerge periodically in different forms and now in the context of the relevance of the river basin approach for such coordination.

Land and Water

The processes involved in the transfer of moisture from the sea to the land and back to the sea again form what is known as the hydrological cycle consisting of two principal phases, the atmospheric and the continental one. No simple figure can do justice to the complexities of this cycle which involves evaporation from the water (foremost oceans) and land surfaces, transportation of clouds and water vapour over continents, condensation of moisture available in the atmosphere, precipitation which reaches oceans or the land surface and ultimately evaporates back to the atmosphere. The global hydrological cycle is certainly not indifferent to the land use changes, with deforestation of the equatorial belt, which receives almost half the globe's terrestial precipitation, being probably the most disruptive threat. Discussion of possible changes in the global hydrological cycle due to deforestation, carbon dioxide-induced atmospheric warming, and other land use-related factors lies outside the scope of this paper. But before one moves to the basin scale of enquiry it's good to remember that things may change on the global scale, including spatial and temporal distribution of precipitation which is, after all, the prime source of all our fresh water.

In terms of water resources management in the areas of either water deficiency or excess, three components of the hydrological cycle that merit the most attention are precipitation, evapotranspiration, and runoff. The general pattern of annual runoff corresponds to that of precipitation modified by soil and geological characteristics and other natural and antrophogenic factors prevailing in the given area. Among all these factors, the use, treatment and management of land affect both evapotranspiration and the quantity and quality of runoff, as well as the location and magnitude of water uses. A high percentage of precipitation and stream withdrawals is used on agricultural land for production of food and fiber. Intensive land use for urban and industrial purposes creates concentrated water supply, flood control, and waste disposal problems. Changes in land use cause changes in demands for water. But as shown by several appraisals of watershed experimentation sponsored by UNESCO/IHP around the world "there is no sufficient understanding of basic relationships among precipitation, evapotranspiration, water movement in the soil, and water movement in the stream to permit clear-cut generalizations about the effect of altering a parameter of land use or vegetation" (White 1971).

The land resources hold potential for many different uses. Forests, wood-land and related wildland are important regulators of hydrological cycle. These land also provide habitat for wildlife and recreation opportunities. Much of the meat production comes from pasture and range land. Land requirements for cities, industries, highways, solid waste disposal and many other non-agricultural land uses of expanding urban societies are to large extent satisfied at the expense of the above mentioned land uses. But land shifting to non-agricultural uses includes often much of the cropland which is the fourth general land use category. Each of the land uses is subject to change, in terms of both quantity (increasing or diminishing area) and quality (e.g. cropland changes in response to irrigation, fertilization and new agronomic technologies). Management and use of land need to be planned to minimize conflicts between uses, to reduce the economic and social costs of land use changes, to increase and made more efficient production of goods and services and, among others, to enhance and protect water resources. With full recognition of the importance of water for all types of economic and social activities, it would be incorrect to presume that under all circumstances water resources play an overriding role in land use planning and management. This is especially evident in more advanced economies as demonstrated, for example, by Howe (1968, 1976) who studied the effects of water resources development on economic and regional growth. But on the whole specific forms of land use and management undoubtedly affect supply, availability and distribution of water resources and vice versa — water resources development and water use (or misuse) affect the land resources base.

The engineering schemes of water resources development, mostly storage reservoirs and water transfer facilities, most visibly affect the available land resources base. For example, the present area of storage reservoirs in South-East Asia which is about 3 million ha is expected to increase in the year 2000 to 15 million ha (Saha & Barrow 1981). Even if there is no direct loss of land due to its inundation, elevated ground water tables in the areas adjacent to water impoundments, as for example in the lower Vistula basin in Poland, may negatively affect agricultural productivity. The loss of natural environmental conditions due to water resources development is an issue often subject to heated debates. The answer to this question can be very different in different countries and contexts. A loss of wild river may be lamented upon by trout anglers, while in other situation it may be greeted with enthusiasm by population seeing new reservoir as a base for development of a new recreation center. There are no easy answers to such questions. What is good in one situation at a specific time may be bad in another, no general statement can be offered.

The negative aspects of water use (contrary to development) are mostly rooted in the unwise utilization or overuse of this resource. Soil salinity, waterlogging, water quality degradation due to uncontrolled wastewater discharges, soil erosion leading to sedimentation of reservoirs, are just few consequences of unwise water use affecting badly land resources. As a matter of fact such negative effects usually extend far beyond land resources, affecting human health, sanitary conditions, and ultimately reducing productivity and overall quality of life in a given region.

Managing the competing interests of various types of land use, and relating them to the requirements for successful management of water resources, are a complex task recognized for centuries. The foregoing discussion has concentrated mostly on the physical relationships between land and water. With this background, let's examine briefly the notion of a region and institutional arrangements which must provide some way of reconciling the differing objectives so that decisions can be made about how land and water will be utilized and who will use them.

The Region

Planning and management of land and water use is a regional excercise. But the regions may be defined and delineated in several different ways. For example, in the history of geographical thought, one of the basic concepts discussed for years is that of a "geographical region". There seems to be a general agreement, however, that at the very base of this concept there is a notion of human organization. The geographical region is usually understood as an areal unit where humans organize their life in some specific way, not only adapting themselves to the natural conditions prevailing in that unit but also changing them according to the actual needs. If we follow this definition carefully, economic regions representing the areas dominated by a certain type of economic activity and administrative regions (provinces, counties, municipalities, voivodships, etc.) represent nothing but certain types of geographical regionalization. But the term "region" is used as well to describe the areas distinguished on the basis of other factors than human organization. Among them are natural regions, e.g. river basins, which are homogeneous from the point of view of some physiographic characteristics. The boundaries of such regions, contrary to the boundaries of administrative or economic ones, are usually stable and their change, if at all possible, cannot be accomplished rapidly.

Very often the drainage basin which provides a coherent hydrological unit relevant to water control is not coincident with the administrative or economic regions within which society functions. Many people tend to see this fact as one of the major impediments to coordinated planning and management of water and land resources. There is empirical evidence, however, that this is not necessarily true. There are regions in the world where rivers, at least in history, played exceptionally important role in organization of local populations and this fact is reflected in administrative regionalization following almost precisely the watershed boundaries. What is even more striking, this regionalization takes into account the usually different physiographic characteristics of the upper, middle and lower parts of the river basins (Janiszewski 1959). For example, the upper part of the Amazon basin makes the Loreto Department in Peru. The middle and lower parts of this basin constitute the Amazonas and Para states in Brazil respectively. Similar situation may be observed in the Po river basin in Italy. The provinces of Piemonte, Lombardy and Venice (called in Italian "regioni") correspond to the upper, middle and lower parts of this basin. Two other examples of this nature are the Ganges and Nile river basins. The states of Uttar Pradesh and Bihar in India follow the upper, middle and lower parts of the Ganges river, while the similar subdivisions of the Nile river basin correspond to three independent countries of Uganda, Sudan and Egypt. The current situations in all these basins unfortunately proves that administrative regionalization that follows the watershed boundaries does not necessarily lead by itself to better planning, more effective management, and coordinated use of land and water resources. It seems that something else than delineation of boundaries of a planning or management areal unit is responsible for the difficulties in achieving coordinated use and management of these resources. Drawing upon the experience with water and land use planning in Europe, Asia and the Americas, it seems increasingly clear that many of the related problems should be above all attributed to the institutional arrangements which govern land and water use in specific regions of the world. Without taking adequately into account what these arrangements are and what are the available options for their eventual modification, planning of coordinated use of land and water may often prove to be not much more than a paper excercise.

The Institutional Arrangements

Although the importance of institutional arrangements for wise and sustainable use and management of natural resources is vigourously voiced by many, the term "institution" is used in a variety of ways. One of the best definitions of this term is given by Fox (1976) who explains that it refers:

> "... either to an entity; an organization or an individual, or a rule, a law, regulation, or established custom. An institutional arrangement is defined as an interrelated set of entities and rules that serve to organize societies' activities so as to achieve social goals. Each nation has an institutional arrangement for managing water resources. This arrangement establishes the conditions under which water resources can be developed and used and provides organizations and individuals with certain resources and authorities to carry out prescribed tasks ...".

There are no ways to describe all possible institutional arrangements for management of land and water resources since they differ widely from one nation to another. In most cases, however, land use planning and management for agricultural, urban, recreational, and wildlife preservation purposes is conducted by other (and usually quite numerous) agencies than the one responsible for water resources management (if such exists at all). Even if a river basin authority is established, the water management decision process cannot be completely unified "because it is impossible to establish a single organisation that can embrace all facets of water management and use" (Fox 1976).

Thus, decision process concerning either land or water or both involves interactions of may entities and rules. How difficult these interactions could be is well illustrated by an example taken from White (1971). This example refers to the American practice but its main features are quite typical for many other nations and contexts as well:

> "... The question of whether planning for the main streams needed to go hand in hand with soil and forest management upstream was debated in theory over three decades... Discussion of the principle of integration was heated, but glossed over several difficulties which would arise when theory would be put to practical test... From the outset, there was disagreement between the Department of Agriculture and the Corps of Engineers as to the practicability of doings so (harmonizing measures of land use improvement and treatment with measures of engineering control of stream and channel in the lower reaches of the same drainage area), and a tacit understanding that each agency would go its own way. But administrative troubles showed themselves promptly in contending efforts between the Forest Service, with its woodland interests, and the Soil Conservation Service, with its concern for cultivated lands, to carry responsibility for dealing with land use components in upstream areas. Strains with the Department of Agriculture became acute. Review during 1940 by interested federal angencies of the Trinity River Reports raised agonizing doubts as to the practicability of producing integrated reports ...".

The institutional arrangements involve not only difficult interactions among many entities and rules, but these entities and rules themselves may be often subject to quite drastic changes. For example, in 1960 in Poland, the National Water Authority (NWA) was established. Few years later the NWA was charged with responsibilities for air pollution control as well. But in 1972, in the process of some governmental reorganizations, the NWA was abolished and its functions were taken over by three ministries of Agriculture, Regional Administration and Environmental Protection, and Communication. Three other ministries of Mining and Power, Construction, and Geology remained intimately involved in the decisions concerning hydropower generation, construction of hydraulic engineering works, and ground water utilization respectively. In the meantime (1975) the system of territorial administration was changed increasing the number of voivodships (provinces) in the country from 17 to 49. The number of regional water authorities (among others charged with the issuance of water-use permits) attached to the territorial administration changed accordingly. Finally in 1983, in response to several difficulties concerning availability of water and its quality in some parts of the country, a new Ministry of Environmental Protection and Water Resources Management was established (the former Ministry of Regional Administration and Environmental Protection remained in charge of regional affairs only). Once again it should be emphasized that such institutional changes, sometimes to the better sometimes to the worse, can be observed in many countries (e.g. changes in the organization of water industry in England and Wales), and they are of crucial importance for coordinated land and water management schemes.

One additional remark seems to be in order. Considering institutional arrangements which achieve given policy objectives, it is also important that they are capable of motivating all involved in land and water management decisions towards effective implementation of these arrangements. There are situations that administrative structures seem to be ideal, laws and rules look perfect, but the things do not work as expected. Quite often this is caused by the lack of adequate motivating mechanisms. This difficulty has been most evident where authorities concerned have relied chiefly upon a regulatory process to achieve policy objectives as in regulation of land use and in the allocation of water supplies among users (Fox 1976).

Concluding Reflections

In this paper two aspects of coordinated management of land and water resources were discussed, namely regionalization and institutional arrangements. Of course, there are many other factors of importance which could have been addressed as well. For example, variability of water supplies from season to season and year to year, the immobility of land resources, the structure of property rights, long times needed for rectification of such related processes as soil erosion, intertemporal fairness (distribution of burdens across generations). It seems to be particularly important, however, that planning for coordinated management of land and water is not too narrowly concerned with land and water per se. Planning should relate as much to changes in human preferences, habits, desires, aspirations, and abilities to manage land and water as to the land and water themselves. There should be also more reliance on sequential decision making, learning feedbacks and experimentation, than on detailed and usually inflexible policy schemes. To increase chances for coordinated management of land and water resources, the river basins, which are coherent hydrological units relevant to water resources planning and management, should be seen in this broader context.

References

Fox, I. K.. Institutions for water management in a changing world. Natural Resources Journal 16, 743—758 (1976)

Howe, C. W.: Water resources and regional economic growth in the United States, 1950—1960. The Southern Economic Journal 34, 477—489 (1968)

Howe, C. W.: The effects of water resource development on economic growth: the conditions for success. Natural Resources Journal 16, 939—955 (1976)

Janiszewski, M.: Regiony geograficzne Polski. PZWS, Warszawa 1959 (in Polish).

Saha, S. K.; Barrow, C. J. (eds.): River basin planning: theory and practice. John Wiley and Sons, London 1981.

White, G. F.: Strategies of American water management. Ann Arbor Paperbacks, The University of Michigan Press, Ann Arbor 1971.

Lundqvist, J.; Lohm, U. and Falkenmark, M. (eds.):
Strategies for River Basin Management, pp. 287–292
© *1985 D. Reidel Publishing Company*

Equity Issues in the Implementation of River Basin Planning

Goulter, I. C., The University of Manitoba, Dept. of Civil Engineering, 342 Engineering Building, Winnipeg, Manitoba R3T 2N2, Canada

Abstract: River Basin planning is concerned with the development and management of the water resource in the basin. As with any planning process, the ease of implementation of the chosen strategy is extremely important in the success of the plan. One of the issues that has a significant effect on the case of implementation is equity in the way in which benefits and/or costs are distributed among the water resource users. The unique nature of the river basin, however, makes it very difficult to develop 'equitable' strategies. In many cases there is very little reciprocity in the way in which benefits or costs can be allocated among upstream and downstream users. In situations where such reciprocity does exist, uniform treatment of the resource users appears most suitable for the development of equitable solutions. In those cases where such reciprocity is not present equity can only be achieved through governmental intervention in the form of taxes and subsidies.

Introduction

The whole concept of river basin planning is based upon the selection of a plan or sequence of decisions which best fulfills the predetermined objectives of the planning process. This principle is true whether the planning objectives are economic, such as basin development to maximize national revenue or environmental/social such as improvement in recreational opportunities within that river basin.

An implicit feature in any effective planning process is the assumption that the planning body has the authority to implement a chosen plan. Since river basins generally cover relatively large areas, the dominant planning authority is either a government office or some form of semi-autonomous government agency. "Strong-arm" measures can always be used by these government authorities to implement these plans. A more desirable scenario, however, would be the achievement of similar, but not necessarily exactly the same, levels of the objectives through a co-operative effort among present and potential users of the basin resources and the appropriate government authorities. One of the best ways to achieve such a consensus is to ensure fairness to all users of the basin resource, i.e., achieve equity among users. In fact, it has been asserted that equity has historically been a prime concern of any planning procedure because of its importance in determining basin strategies which are politically acceptable and therefore enforceable (Brill 1976).

Relative to other types of planning units, river basins have unique physical characteristics which often make it difficult, if not impossible, to distribute the overall benefits and costs of certain types of basin development among all parties within that basin. This paper discusses the issues involved in the equitable implementation of river basin strategies. Those planning objectives which are most likely to experience these problems are identified. Previously reported approaches for the resolution of some of these problems are reviewed and analyzed with respect to the suitability for resolution of other types of problems. Using the results of earlier approaches a set of general criteria for the equitable distribution of benefits and costs is established.

Physical Parameters of River Basin Planning

River basin planning is almost always directed towards some form of allocation of the water resources in the basin. The river itself is either the primary source of the resource or is the link between the useable portions (e.g. storage in lakes or reservoirs) of that resource. The features that make allocation of the water resource in a particular river system so interesting are the relationships between upstream and downstream uses of water.

'Use' in this context means not only use for the direct development of benefits at the point of utilization but also use in the sense of accepting damages or costs at an upstream point in order to improve or produce benefits at some downstream point. In this second situation an upstream user may incur 'costs' associated with either reduced production or inability to expand without receiving any of the benefits that accrue as a result of those costs.

This non-reversible link between uses at different locations in the basin differentiates river basin planning from other types of planning units where the trade-offs associated with the allocation of the resource among competing users generally have a semi-reciprocal basis. The lack of a semi-reciprocal trade-off means that it is very difficult to achieve any great degree of equity in the way in which benefits and costs are distributed throughout the basin. Under these conditions a reasonable degree of equity is normally only possible with either significant degradation in the level of attainment of the original objectives or considerable governmental intervention. The first alternative is obviously counter productive to the whole objective of the planned development. The second alternative, which is contrary to the assertions of the introduction, is generally a last resort, especially in those countries which do not have a tradition of governmental intervention.

Equity

Without even considering the additional difficulties associated with its implications in river basin management, the definition of equity in relation to any development process is a complex issue. In its most simple form equity is a condition where all resource users benefit or are penalized to the same degree. Obviously, it is very unlikely that this ideal will ever be completely achieved. The improbability of reaching the true optimum should not, however, prevent the consideration of the issue in the planning process.

In developing mathematical models, (especially those involving systems analysis), for application to the planning or management of river basins, the analyst is faced with attempting to incorporate some form of equity in the optimization process. The incorporation of equity into the modelling process may be severely complicated by the lack of a clearly stated and well accepted definition of equity.

The accepted definition of equity will change depending on the issues under consideration. In some situations equity is best achieved by minimizing, according to some predetermined measure, the changes experienced by individual users within the basin. This situation assumes of course that the current situation is in itself equitable. In other cases, equity might best be achieved by ensuring that the individual users contribute by the same amount to the basin development or benefit to the same degree from the basin development. This approach penalizes those resource users who are already using or protecting the resource efficiently. In both cases a more refined approach may result from different measures of equity being used for different subsets of users. These different measures permit some degree of recognition of variation in equity needs or flexibility to change which may exist for the different types of users, c.f., municipal waste dischargers vs industrial waste dischargers. While neither approach is completely satisfactory, they both hold some attraction in that they are easily incorporated into mathematical planning models.

It should also be noted that equity is normally achieved at the expense of increased overall basin costs or reduction in overall basin benefits. For this reason equity issues have, at times, been considered multi-objective problems with the trade-off between equity and costs or benefits being the parameter of importance.

Review of Equity Issues in River Basin Planning

In the following discussion, only those river basin development plans which have benefits and costs internal to the basin are considered. If the bulk of the benefits or costs are external to the basin, the equity issues for that basin no longer relate to the river basin itself but rather to the planning strategy for the whole region or country. These national issues are beyond the scope of this paper.

Water quality planning models were the first of the river basin planning models to encounter and recognize equity issues. A number of early theoretical models had as their primary objective the minimization of total cost over the whole basin to achieve a specified level of water quality in the river system, e.g., Liebman & Lynn (1966), ReVelle et al. (1968) and Thomann (1972). Equity consideration related to the minimization of this cost ranged from uniform treatment requirements across the basin, uniform treatment requirements within groups of dischargers and minimization of the range between maximum and minimum levels of treatment (Water Pollution Control Administration 1969; Thomann 1972; Brill et al. 1976 and Brill et al. 1979). In an attempt to minimize governmental intervention and to permit market forces to determine the equity issues related to achieving a given level of water quality, the concept of transferable discharge permits was introduced (David et al. 1980; Eheart et al. 1980; Eheart 1980).

A characteristic feature of this type of analysis is that only costs (or in other cases only benefits) have to be distributed equitably. A degree of reciprocity exists in this case. Increased expenditure by upstream discharges on pollution control may permit reduced expenditures at downstream points. The fact that the reverse relationship holds, if not to the same degree, shows the existence of the reciprocity. The contrast in equity considerations between this situation and those situations in which reciprocity does not exist will be made evident in the following section.

Over recent years there has been considerable effort and research directed towards the mitigation of flood damages on river basins through the use of non-structural measures. These non-structural measures were considered in isolation or in combination with structural measures. Earlier studies addressing this issue were performed by Lind (1967), James (1967) and Whipple (1969). In these early studies the most common non-structural measure is the application of land use management to restrict future development in or remove existing development from the flood plain areas of the basin (Day 1970; Bialas & Loucks 1978; Hopkins et al. 1978, 1981; Goulter & Morgan 1983).

Restriction or removal of development in the flood plain under these land use managemant policies is equitable as long as all regions of the flood plain are subjected to the same level of adjustment and existing users have access to non-flood plain lands. However, the studies by Hopkins et al. (1978, 1981) also addressed the concept of reducing the increases in flood peaks normally associated with increased urbanization or development within the basin. In this way it was considered possible to consider tradeoffs between the costs in the upstream regions of the basin (costs due to lost development opportunities, and costs required to reduce or eliminate the contributions to the increased flood peaks normally associated with urbanization) and savings due to decreased flooding and increased land values downstream. The reciprocal trade-off, i.e., the trade-off between increased revenue/benefit or reduced flood costs in the upstream regions due to reduction in the contribution to peak flows in the downstream reaches does not exist. Since the objective in these approaches generally is to maximize the net economic value of the basin, equity measures in these situations are particularly difficult to formulate and analyze. Furthermore, unlike the water quality management situation, benefits and costs can accrue simultaneously to individual users within the basin. Due to the one-way relationship that characterizes these situations it is very difficult to distribute both the benefits and costs equitably across all basin uses. Benefits in a particular portion of the basin are normally achieved through costs in another part of the basin. It is these kinds of non-reciprocal relationships that were presented in the introduction as making the equity issues in river basin planning so difficult to resolve.

General Approaches to Equity Considerations in River Basin Planning

In addressing equity considerations it is suggested that the objective of the planning process be analyzed closely to assess the degree of reciprocity that exists between upstream and downstream users. The problems can be divided on this basis and treated differently as described below.

In those cases where a reasonable level of reciprocity exists, guidelines of uniform treatment of individual users appear appropriate. The uniform treatment need only apply to individual groupings within the basin rather than relating to all users in the basin. Equal treatment may be a general level of activity, i.e., level of discharge treatment for each user, or equal starting conditions such as that used in the transferable discharge permit concept. The crucial issue is that each user be treated equally initially. How he responds after that is determined by each user individually. Procedures such as the transferable discharge permits concept hold some attraction as they minimize direct government action on the individual users, thus giving the sense of a co-operative action. Consideration of range of treatment levels as a measure of equity is not recommended as it does not allow any opportunity for free market conditions, such as those described above, to resolve the equity problem without a considerable degree of often unwanted government intervention.

The above approaches to equity resolution are the most suitable when either benefits or costs, but not both, are to be distributed across the basin. The equity problem is then simply involved with the distribution of the single parameter of basin performance. The provision of equity is more difficult when both benefits and costs are to be considered. In these situations there is generally a strong relationship between costs in one area and benefits in other areas. In fact the planning process may be directed to the problem of allocating benefits and costs across the basin to achieve overall optimal basin performance. The provision of equity is even more complex when there is a lack of reciprocity, i.e., cost and benefit relationships between two locations on the basin are not reversible. By its very nature this condition makes any solution to basin planning inequitable.

There are, however, a number of generalizations that can be made. If the objective of the planning exercise is a 'fine-tuning' of the basin, e.g., the case of flood damage reduction on rural watersheds discussed by Goulter & Morgan (1983), then an equitable solution may be to attempt to minimize the deviation from the existing conditions that any user is required to undergo. In the rural watershed case, this means minimizing the changes in agricultural practice that must occur on any land parcels in the basin. By attempting to make the required changes relatively small the degree of government intervention may also be reduced.

If the objective of the planning process is control of large scale development in the basin, e.g., control of urbanization, to maximize overall basin revenue or benefits, the equity problem is extremely complex. Large scale development implies that considerable deviation from existing conditions will be experienced. The non-reciprocity of the benefit and cost relationships makes a true trade-off between benefits and costs almost impossible. The only feasible response to equity requirements in these cases appears to be government intervention to tax, in some form, those receiving benefits from basin development. These tax funds would then subsidize or reimburse those who suffer as a result of the benefits accruing to others. This solution is contrary to the recommendations given in the introduction but appears to be the only alternative available to handle equity issues of this type.

Conclusions

While equity considerations in river basin planning have been investigated in some specific areas, the overall question of how to define or implement equitable planning remains essentially unresolved. The major problem appears to be non-reciprocity in the relationships between benefits and costs in the downstream and upstream portions of the basins. Equity issues where either benefits or costs (but not both) are to be considered can generally be resolved through some form of uniform treatment of basin users. Where benefits and costs are to be considered simultaneously and where strong relationships exist between these benefits and costs, planning problems are very difficult to resolve equitably. In these cases either substantial government intervention is required or the residual inequities accepted.

References

Bialas, W. F.; Loucks, D. P.: Non-structural Floodplain Planning. Water Resources Research 14, 67–74 (1978)

Brill, E. D.; Liebman, J. C.; ReVelle, C. S.: Equity Measures for Exploring Water Quality Management Alternatives. Water Resources Research 12, 845–851 (1976)

Brill, E. D.; ReVelle, C. S.; Liebman, J. C.: An Effluent Change Schedule: Cost, Financial Burden and Punitive Effects. Water Resources Research 15, 993–1000 (1979)

David, M.; Eheart, W.; Joeres, E.; David, E.: Marketable Permits for Control of Phosphorous Effluent into Lake Michigan. Water Resources Research 16, 257–262 (1980)

Day, J. C.: A Recursive Model for Non-structural Flood Damage Control. Water Resources Research 6, 1262–1271 (1970)

Eheart, J. W.: Distribution Methods for Transferable Discharge Permits. Water Resources Research 16, 833–843 (1980)

Eheart, J. W : Cost-Efficiency of Transferable Discharge Permits for the Control of BOD Discharges. Water Resources Research 16, 908–986 (1980)

Goulter, I. C.; Morgan, D. R.: Analyzing Alternative Flood Damage Reduction Measures on Small Rural Watersheds Using Multiple Return Period Floods. Water Resources Research 19, 1976–1982 (1983).

Hopkins, L. D.; Brill, E. D ; Kurtz, K. B; Wenzel, H. G.: Analyzing Floodplain Policies Using an Interdependent Land Use Allocation Model. Water Resources Research 17, 469–477 (1981)

Liebman, J. C.; Lynn, W. R.: The Optimal Allocation of Stream Dissolved Oxygen. Water Resources Research 2, 581–591 (1966)

Lind, R. C.: Flood Control Alternatives and the Economics of Flood Protection. Water Resources Research 3, 345–357 (1967)

ReVelle, C. S.; Loucks, D. P.; Lynn, W R.: Linear Programming Applied to Water Quality Management. Water Resources Research 4, 1–9 (1968)

Thomann, R. V.: Systems Analysis and Water Quality Management. 286 pp. Environmental Research and Applications, New York 1972.

Water Pollution Control Administration: Delaware Estuary Comprehensive Study. US Department of the Interior, Philadelphia, Pennsylvania, USA 1966.

Whipple, W.: Optimizing Investment in Flood Control and Floodplain Zoning. Water Resources Research 5, 761–766 (1960)

Lundqvist, J.; Lohm, U. and Falkenmark, M. (eds.):
Strategies for River Basin Management, pp. 293–298
© 1985 D. Reidel Publishing Company

4   RIVER BASIN RESOURCES ADMINISTRATION AND ANALYSIS

4.1 Approaches in river basin administration and development

Different Types of River Basin Entitites — A Global Outlook

Burchi, S., FAO, Via delle Terme di Caracalla, 00100 Rome, Italy

Abstract: This paper reviews the various kinds of river basin entities established in different regions of
the world. These entities are classified in three main groups, namely, (a) valley authorities, with responsibility for broader economic and social development basin-wide; (b) basin entities, responsible
in varying degrees for water resources management at the basin level; and (c) co-ordinating commissions or committees, established with the aim of bringing harmony in water policies and plans at the
basin level. Various cases illustrating the three classes of basin entitites are drawn from the experience
of various countries. The considerable diversity of the cases illustrated allows for very tentative conclusions only to be drawn.

Introduction

The river basin, sub-basin or group of basins have come to be regarded as the focus of
administrative water management responsibilities as a result of the growing influence,
since the turn of the century, of the concept that the river basin should be treated as a
unit of planning. Governmental entitites at the basin level concerned with one or more
aspects of the water resources management process can be grouped into three main classes: (1) the valley authorities, which are in charge of water resources management as a
component part of the broader development process for which they are also responsible;
(2) the basin entities, which are in charge of one or more aspects of water resources management only; and (3) coordinating basin commissions or committees, with functions
generally limited to planning of water resources development, and coordination. It is the
purpose of this paper to review the river basin entities established in the various regions
of the world according to the above model.

Three Main Classes of River Basin Entities

Valley authorities for broader development

The idea that the unit of planning and management should be the river basin was strongly
associated with multi-purpose projects from the beginning of this century. Economic

development by river basin units as opposed to other types of economic region, however, did not come to fruit until the 1930's. The pioneering institution was created in the United States in 1933 in the Tennessee Valley Authority (TVA). This areally and functionally consolidated, autonomous agency was the prototype for a number of others in various parts of the world whose task would go far beyond the management of water resources *per se*. The TVA is a government corporation with powers to plan, construct and operate multi-purpose projects, and to achieve economic and social development goals (Teclaff 1967, 127).

Other valley authorities subsequently established in other countries reflect a comparable emphasis on basin-centred regional planning and administration, and on general economic and social development. In India, the *Damodar Valley Corporation* was created in 1948 with the task of promoting public health and agricultural, industrial, economic, and general wellbeing in the Damodar valley (Teclaff 1967, 132). Similarly, the *Gal Oya Development Board* in Sri Lanka was set up in 1949 as a government corporation, and was charged with broad development responsibilities in the Gal Oya river basin (Teclaff 1967, 135). In Brazil, the San Francisco Valley Commission (Comissão do Vale do São Francisco) established in 1948, is charged with planning, coordinating and executing works for the development of the basin water and related resources (Teclaff 1967, 137– 138). Colombia's Regional Autonomous Cauca Corporation *(Corporaciòn Autònoma Regional del Cauca)* was created in 1954, and subsequently re-organized in 1960, with the task of promoting agricultural and industrial development, mining, and social welfare (Teclaff 1967, 138).

Another entity with far reaching powers was established in Afghanistan in 1953 as the *Helmand Valley Authority*. The plans drawn up contemplated power generation, land reclamation, irrigation, industrial development, resettlement of nomadic tribes, and the provision of educational institutions, public health centres, and modern housing (Teclaff 1977).

The distinguishing feature of the valley authorities, apart from their broad mandate for economic and social development, has been their administrative structure. The enabling legislation envisaged them as highly autonomous entities, corporate in form, separately funded, and responsible to the central government rather than to the water administration or any sector of it. The TVA for instance was made responsible directly to the President of the United States and endowed with greater independence and flexibility than perhaps any other department or agency of the federal government. The Comissão do Vale do São Francisco likewise was set up as a purely federal agency, directly responsible to the President, and with only token representation of the basin states. The Helmand Valley Authority was made to report only to the Ministry of Finance.

This type of administrative structure, which provides little scope for coordination with the regular government departments in charge of water resources management, with political entities below the national level, or with user interests, has met with considerable resistance. As a result, the valley authority concept has rarely been implemented more than once in any individual country. However, the limited acceptance of the valley authorities did not detract from the general acceptance of the river basin concept, as basin organizations of more modest scope can be found in many countries.

Basin entities for water resources management

Among the earliest basin entities of more limited scope than the valley authorities are the basin associations *(Genossenschaften)* of the Ruhr, brought into being at the turn of the century for the orderly management of water supply and pollution abatement programmes in a densely populated and highly industrialized area. These associations control six river basins — those of the Wupper, Ruhr, Emscher, Lippe and Linksniederrhein (several streams), all of which are tributary to the Rhine, and the Niers, which is tributary to the Maas. They are all alike in structure and function, but they vary in the scope of the tasks assigned to them by the enabling legislation. The Lippeverband has the most extensive powers, ranging from pollution control, to drainage, conservation, and even regional development. The Emschergenossenschaft and the Linksniederrheinische Entwässerungs-genossenschaft are concerned with pollution and drainage, the Ruhrverband with pollution only, and the Ruhrtalsperrenverein with water supply. Composed of local government units and private corporations which use river facilities, the associations are funded by assessment of the members, government grants, and loans. They are governed by: (a) an assembly consisting of the elected representatives of the membership; (b) a board of directors which conducts day-to-day business and represents the association; and (c) a board of appeal, which hears appeals from the decisions of the board of directors. The autonomy of the associations is not so great as might appear at first glance from the composition of their membership. The state (Länder) governments exercise a close supervision and all new projects, as well as the regulation of operation and use of existing ones, have to be approved by the competent ministers (Teclaff 1967, 126--127).

France's *Compagnie Nationale du Rhône* was established in 1933 for the development of power, irrigation, and navigation in the Rhône river. The Compagnie, however, is not a truly basinwide institution, for it does not encompass the Swiss section of the river nor its tributaries. Unlike the Tennessee Valley Authority, the Compagnie is a stock company, whose shares are held by public organizations interested in the development of the Rhône, and by chambers of commerce representing private interests. Close government control is ensured through the right to appoint a majority of members of the board of directors of the company (Teclaff 1967, 129–130).

The Valley Authorities, the German *Genossenschaften*, and the French Compagnie National du Rhône are all basin or sub-basin entities established by *ad hoc* legislation as the need for such institutions arose. That kind of basin entities is contrasted by the basin entities which are envisaged by legislation of nationwide scope, and which form part of the national administrative framework for water resources management at an intermediate level of government. In very broad outline, the functions of this kind of basin entities are not so wide-ranging as those of the valley authorities, yet they tend to be quite comprehensive and to include regulation, allocation, and management of water resources. This comprehensive type of basin entities is characteristic of developments in Europe.

In *England and Wales* by the Water Act of 1973 the then 27 River Authorities were consolidated into 10 Regional Water Authorities, under the joint direction of the Secretary of State for the Environment and the Minister of Agriculture, Fisheries and Food. The

Water Authorities are responsible for virtually all aspects of water resources management, including water allocation, pollution control, water supply, drainage, fisheries and recreation, at the level of groups of basins. The membership of the Water Authorities is made up of appointees of the central and local governments (FAO 1975).

A somewhat similar administrative structure has evolved in the *German Democratic Republic*. Basin authorities *(Wasserwirtschaftsdirektionen)* had been established there in 1958 in seven basins or groups of basins. A national Water Management Agency was created in 1969, but within three years was absorbed, together with the basin authorities, into a new Ministry of Environmental Protection and Water Management (Teclaff 1977, 22).

In *Hungary*, basin administration enjoys wide powers, though with considerable more direction from the central government than their British counterparts. The 12 regional water authorities of Hungary, which roughly correspond to hydrologic units, have not only regulatory but also construction responsibilities. They are supervised by a single national entity of cabinet rank, the National Water Authority (Teclaff 1977, 15, 23).

In *France*, the six basin financial agencies *(Agences financières du bassin)* are more limited in scope, as their primary function is to help finance the control of water pollution. They do not have regulatory authority of their own, but are nevertheless associated in the discharge of the regulatory functions of the responsible government authorities. Representatives of the central and local governments, and of water users, sit on the board of directors of the basin agencies in equal numbers (FAO 1975, 75; Chardon 1976).

In *Spain*, the basin Confederations *(Confederaciones Hydrogràficas)* are technically an integral part of the government administration (Ministry of Public Works), but enjoy considerable decision-making and financial autonomy. They have jurisdiction over water allocation and waterworks in respect to individual basins or groups of basins. The Board of Directors of the Confederation is composed of representatives of the central and local governments, and of water users (FAO 1975, 197–198).

Coordinating commissions and committees for water policy harmonization

Federal countries have often resorted to basin commissions and committees to effect coordination of the water policies of the constituent states or provinces sharing parts of the same river basin. The majority of the commissions created in the United States by interstate compacts are of this type. Some are strong commissions, empowered to develop plants, policies, and projects and to allocate waters, such as the Delaware River Basin Commission. Others are entrusted merely with water apportionment, which has already been spelled out in detail in the compacts, and are further limited in their powers by the requirement of unanimity or near-unanimity of decisions. An example of this kind of commissions is the Upper Colorado River Basin Commission. Another example is the planning organization which was set up in 1956 for the Rio Colorado in Argentina by formal agreement between the five basin provinces. In India control boards for several basins or parts of basins (e.g. the Kosi, Rihard, and Chambal) were set up by informal agreement between the states and the central government. They could give only recommendations, which need the sanction of the governments concerned, and actual construction of works is carried out by engineers of the participating states.

Coordinating bodies have been established also by interdepartment — as opposed to interstate or interprovincial — agreement. A coordinating body of this type is represented by the committee for the Marikina river in the Philippines, which was created in 1953 and included representatives of the National Power Corporation, Bureau of Public Works, and Metropolitan Water District (later of the National Waterworks and Sewerage Authority), each of which carried out different phases of project development in the Marikina Valley. An example from Mexico is embodied in the *Comisiòn Hidrològica de la Cuenca del Valle de México*, with functions limited to study and planning, and with representation from the central and local government as well as private associations.

The trait which distinguishes this group of basin entities is that they have been created on an *ad hoc* basis for individual river basins as the need for coordination arose, and not by a generalized devolution of authority. Commissions and committees whose powers are confined to coordination and planning have been criticized as ineffectual. However, since they represent less of a threat to the powers of government departments engaged in water resources development and at the same time satisfy the need for some representation of basin interests, they have met with less opposition in practice than the valley authorities.

Conclusions

The river basin, sub-basin, or group of basins as an administrative unit of water management responsibilities is well represented today virtually in all regions of the world. There is a great diversity of approaches, however, ranging from the powerful, general development-oriented Tennessee Valley Authority to the weaker interstate or interprovincial coordinating commissions, through a panoply of intermediate basin entities whose functions are neither as broad as those of the former type of basin authorities, nor as narrow as those of the latter type of commissions. Of this last category of basin entities, some have been established by *ad hoc* legislation, but others have resulted from general legislation devolving governmental water management responsibilities to the basin level of administration.

In view of the great diversity found, it is not possible to draw any general conclusion. It is worth recalling, however, that the United Nations Interregional Seminar on Water Resources Administration held in New Delhi in 1973 endorsed a two-tier type of water resources administration with co-ordination at the national level and decentralization at the regional or basin level (UN 1975) — a model which seems to correspond to the approach followed in England and Wales, Hungary, or the German Democratic Republic.

References

Chardon, J. M.: Legal and Institutional Criteria Utilized for Establishment of the Inventory, Development, Conservation and Rational and Integrated Use of Water at the Basin, Sub-basin and Regional Levels. Annales Juris Aquarum, 2, p. 850, 855, 864, Caracas (1976)

FAO: Water Law in Selected European Countries. FAO Legislative Study 10, Rome 1975.

Teclaff, L.: The River Basin in History and Law. M. Nijhoff, The Hague 1967.

Teclaff, L.: Legal and Institutional Responses to Growing Water Demand. FAO Legislative Study 14, Rome 1977.

UN: Proceedings of the Interregional Seminar on Water Resources Administration, New Delhi 1973. UN Doc DP/UNT/INT 70, 1975.

Lundqvist, J.; Lohm, U. and Falkenmark, M. (eds.):
Strategies for River Basin Management, pp. 299–305
© *1985 D. Reidel Publishing Company*

The River Basin Concept as seen from a Management Perspective in USA

Wengert, N., Political Science Department, Colorado State University, Fort Collins, Colorado 80523, USA

Abstract: In the USA interest in river basins as units for water planning began in the late 19th century as technological progress in engineering provided data for such broader approaches. The emphasis on integrated, comprehensive river basin planning peaked in the 1930s and 1940s as epitomized by the Tennessee Valley Authority experience and New Deal emphasis on rational decision making. Well into the 1950s efforts to intertwine the water concepts of river basin development with regional economic and social planning continued to receive support. But for various reasons in the 1960s till today river basin development and regional planning have pulled apart, river basin management being limited largely to hydrological aspects of river systems engineering. Regional planning and development, in turn, has ceased to focus on the river basin, becoming primarily concerned with inter-related urban socio-economic and political factors and forces with some attention to environmental values and constraints. Thus regions have become foci for urban socio-economic and political analyses, and river basins continue as significant hydrological planning units for efficient control of water resources.

Introduction

While ancient history gives pragmatic examples of water management in large rivers, truly comprehensive basin management had to await engineering and technological advances as well as ability to handle ecological, socio-economic and political interrelationships for large areas or regions. On the scientific and technological side precipitation, soils and geologic data and a wide range of water measurements over time were needed. On the ecological, socio-economic and political side, competence was needed to analyze action interrelationships and consequences, to tabulate benefits and costs of development, to assess developmental alternatives, and thus, finally to choose desired consequences.

In most early instances of awareness of rivers as systems stress was on water and only incidentally on the causal interrelationships of water management to ecological factors, to socio-economic and political choices — all affecting large geographic areas or regions properly designated "River Basins" (Holmes 1972,1979). River basin management required reliable data to provide foundations for cause and effect projections and (in present day terminology) for stream and eco-system modelling. Significantly, the evolution of river basin management in the United States — with its numerous large rivers, large in length and large in water flow — was often a response to the crises and necessities of floods and flood control (Holmes 1972).

Taking advantage of what nature had provided, early settlers in the United States found its many Eastern Rivers useful for navigation. Some dredging and clearing of obstructions was necessary, but these early efforts can hardly be characterized as river management focused on the basin. As navigation became more important, it was the US Army (ultimately the Corps of Army Engineers) which undertook tasks of navigation development in order to reach backcountry forts to protect white settlers from native American Indian (Holmes 1972).

In a similar fortuitous way, early location of urban settlements on the shores of rivers, streams, bays, and lakes was a response to opportunities provided by the natural environment. Location of towns and cities reflected the needs of export and import trade, a major economic activity almost from initial settlement. Even the early use of smaller streams for water power to drive flour mills, saw mills, and ultimately textile mills and other manufacturing establishments involved primitive, single purpose management of the water resource. These early water power mills were clearly site specific, enterprise specific, and project oriented and hardly reflected basin management.

River Basin as a Focus for Management

An active awareness of the river basin as a focus for management developed first out of often futile attempts to manage flood waters. Only as economic development increased did the costs of flood losses become severe. As a first step, flood risks on the lower Mississippi river stimulated organization of special districts to finance and build local (not basin-wide) dykes, levees, flood walls, diversion ways and other structures to attempt to reduce the destruction of crops and fertile bottom lands of the so-called "Mississippi Delta" which extended several hundred miles North from the Gulf of Mexico to Memphis, Tennessee. These early flood control works were often unsuccessful just because they sought to cure basin problems by engineering works locally-conceived and implemented. But without quantitative data on the flow and flood patterns of the larger Mississippi river system it was impossible to design effective controls. The great floods of the 1870s developed an awareness that local works would not be sufficient and led the US Congress to create the Mississippi River Commission in 1879.

Early approach to rivers as systems

For constitutional reasons the mission of the Commission was stated as improvement of navigation on the lower Mississippi. Devasting floods often did destroy existing channels, building sandbars were channels had been before the flood. So it was logical as well as constitutional for the Commissions's concerns to include flood control. The Corps of Army Engineers, assigned to assist the Commission, combined its concern for navigation with a concern for flood control. As a federal agency, moreover, the attention might and ultimately did extend beyond the very local and beyond even state boundaries. By the turn of the century headwaters began to be included in the planning perspective. Thus, gradually concepts of planning navigation and flood control for a major river as a system (River Basin Management) began to emerge — although the phrase "systems analysis" would not enter the English vocabulary until much later (Hall & Dracup 1970).

The National Inland Waterways Commission appointed by President Roosevelt in 1907 brought to public attention evolving concepts of planning watersheds as single, total entities (basin planning) stressing headwater storage as well as inputs from tributaries along the way. Effectively integrating water development with the socio-economic opportunities in the basin was yet to come.

Twenty years later a major project with a basin and regional orientation was the Hoover Dam project begun in 1928 on the Colorado river to provide water and electric power for the city of Los Angeles. The large Mississippi flood in 1927 led Congress to request the Corps of Engineers to undertake so-called "308" studies (named after the section of the law which authorized them) as the foundation for flood control, navigation and power development on major US rivers. The "308 Report" for the Tennessee river provided the initial base for development of that river by the Tennessee Valley Authority after 1933 (Holmes 1972).

River basins as geographical regions

River basin management, thus, emerged from a melding of the technical realization that rivers could best be understood and perhaps managed as systems with concepts of the region as a logical and proper geographic focus for integrated resource planning and development investment (Hunt 1974; Krutilla 1969). In many respects, river basin management reached its fullest articulation in the late 1930s and early 1940s in the application of the so-called "TVA Idea" to the Tennessee river. The combination of river basin management with regional resource development became a popular set of concepts supported by many Americans. Ultimately, proposals were introduced in the Congress authorizing river development authorities for many of the major rivers including the Columbia and the Missouri. In 1950, a President's Water Policy Commission urged the Comprehensive development of "Ten Rivers in America's Future" (Pres. Water Res. Policy Comm., 1950, Vol. II). Basic to these proposals to extend regional authorities were:

> First, the idea of dealing with a river and its tributaries on a comprehensive, integrated, multi-purpose basis in order to maximize flood control, navigation, irrigation, hydro power production, water supply for cities and farm irrigation, for recreation and other water benefits;

> Second, development of the river on a systems engineering basis including attention to regional development — the region and the river basin being assumed to be a single, coterminus planning and development unit.

Although agriculture and forestry as principal land uses in the Tennessee valley were not entirely neglected in the TVA program, and TVA maintained a fish and wildlife staff in its organization, stress (in the context of the Great Depression) was on development — on getting the economy moving. Ecological concerns were seldom voiced and environmental impact remained for future articulation.

Melding river basin planning and regional development

In proposing more river basin authorities, advocates assumed that the nation could logically be divided into major planning and development regions combining in each region a number of States, and boundaries for each region being determined by major watersheds. Thus, planning and management of resource development in each of the regions was to be guided by a river basin authority modeled on TVA.

This restructuring of the US never occurred, but numerous river basin studies were undertaken, many from 1933—43 by the National Resources Planning Board (Smith 1956), and many books and articles, both pro and con, on valley authorities were written (Wengert 1967). After the construction hiatus of World War II, the US Army Corps of Engineers began again to develop the nation's rivers, increasingly applying a systems approach in planning and constructing river projects. Similarly, the Bureau of Reclamation in building storage reservoirs for irrigation, water supply and hydropower in the 17 western States also approached rivers as engineering systems. And the Soil Conservation Service, entering the water development field much later than the other federal construction agencies, promoted small upstream reservoirs which probably went farther in integrating construction with other resources than did other federal activities.

At the national level a variety of devices were created to coordinate and integrate construction of river works with economic development objectives. The Department of the Interior created a net-work of Field Committees and both federal and state agencies collaborated in regional plans consistant with basin development goals. Ultimately, a sub-cabinet level coordinating committee was converted into the Water Resources Council with a statutory mission to coordinate, if not control, river basin commissions and committees and generally to facilitate development of viable river basin plans (Wengert 1980). The Council drafted a "Principles and Standards" volume to guide programmatic construction agencies in determining benefits and costs of specific river development projects. During the Carter Administration environmental values were included in the draft document. But for various reasons the Reagan Administration put an end to the effort to formulate guiding "Principles and Standards", indicating opposition to most water projects.

New concepts of regional aggregation

By the 1970s comprehensive, integrated approaches seeking to meld river basin planning and regional development began to disintegrate as a new concern with water and air pollution moved to the center of the political stage (Wheat 1973). Among causes contributing to the split between the application of a systems approach to the nation's rivers in combination with regional development are the following (not necessarily in the order of their importance):

1.   First, the river basin was not, with few exceptions, a logical and effective basis for regional planning and economic development. For example, TVA's power marketing area

never coincided with the natural watershed of the Tennessee river (Hunt 1974; Fain 1976; Jackson et al. 1981; Levis 1973).

2. In its polemics and lore, "The TVA Idea" especially as formulated by David Lilienthal was in fact not based on a community of interests among valley residents, although over time some common interests developed. The Tennessee valley (and most other large river basins in the US) could not be considered sociological communities. Many Americans have only vague ideas about the river basin in which they happen to reside. Community roots tend to relate to symbols other than a river basin (Newman 1972). Even rural people are more influenced, if not dominated, by the urban culture which surrounds them and draws their children away from farming to urban job opportunities (Mosely 1974).

3. With the Great Depression buried in the prosperity generated by World War II, Korea and Viet Nam, general development policies and programs under government auspices seemed less important. But as a result of the activities of interst and pressure groups government involvement in development of the private economy continued. It is worth recalling Harold Lasswell's definition of politics in the US: "Who gets what, when, how". I have often added "where" to his list, but perhaps his omission of a spatial concept implies the low level of importance he attached to spatial, including regional benefits.

4. Thus regional development and even natural resource development are no longer popular catch phrases of American politics. A new vocabulary which includes such words as: *environment, social welfare, gross national product, urban redevelopment, air and water pollution control, hazardous chemicals management, acid rain* and many other essentially urban oriented terms are now important in the language of American water politics (Price 1982).

5. Throughout the several decades during which regional river basin development aroused enthusiasm, construction agencies of the Federal Government continued building river works, in the process what in political slang had come to be called "politcal pork". It was thus not considered gauche for Senator Kennedy, like most of his congressional colleagues, to stress the slogan "He can do more for Massachusetts". Seeking geographical benefits for the home state or the home district remains a characteristic of American politics, tending to militate against rational, regional planning, to minimize envrionmental values, to ignore ecological concerns, and to skew benefit-cost analyses.

6. While river basin management in a hydrological sense remains important, it has effectively been severed from concepts of basin planning for a region and remains simply a systems approach to river engineering.

7. At the same time, regionalism is not dead (Chadwick 1978) but has changed rather significantly. Today's regions are rarely watersheds but rather urban, metropolitan agglomerates. A recent monograph, published by the Joint Center for Urban Studies of MIT and Harvard University, entitled "Regional Diversity: Growth in the United States, 1960—1990" (Jackson et al. 1981, p. 5) has suggested that the term "region" is ambiguos in present usage. The authors indicate that four levels of aggregation can probably be identified in today's literature: 1. those areas which the US Bureau of the Census simply calls "places"; 2. specifically identified economic areas, as in agriculture or industrial production; 3. regions referred to in ordinary speech as New England, the Mid-West, the High

Plains, etc. which have fairly vague boundaries; and 4. what might be called "major" regions, based on general usage, such as The North, The South, The West.

It needs no emphasis that in this categorization of regions there is hardly room for River Basins. Nor are regionally defined environmental concerns recognized. A few cultural regions (e.g. Appalachia, the Florida Gold Coast) may still be identified, but radio, television, national advertising and information media are bringing about a homogenization which erodes regional uniqueness, and minimizes any influence the river basin QUA region may have had.

Conclusions

To restate my conclusions: Where during the 1930s to the 1950s the intertwining of river basin concepts with ideas of regional planning and development were creative, even stimulating, (Kuderna 1974; Leven 1970; Lewis 1971), today they have pulled apart, river basin and river basin management being primarily synonyms for river systems engineering. Regional planning and development, in turn, has become primarily a description of an aspect of interrelated urban socio-economic and political factors and forces, with more-or-less casual attention to environmental values and naturally determined geographic boundaries, such as river basins. Today's regionalists in fact are not interested in fitting these "old fashioned" concepts into their models. Thus, regions are foci for urban socio-economic and political analysis and river basins are significant as hydrologic units, defining water system boundaries in order to plan efficient control and use of the water resource.

References

Chadwick, G.: Regional Planning: A Systems View of Planning Towards a Theory of the Urban and Regional Planning Process. 2nd edition. Pergamon Press, Oxford, New York 1978.

Fain, R. Jr.: Ecological Systems and the Environment. Houghton & Mifflin Company, Boston, Massachusetts 1976.

Hall, W. A.; Dracup, J. A.: Water Resources Systems Engineering. McGraw-Hill Book Company, New York 1970.

Holmes, B. H.: A History of Federal Water Resources Programs, 180—1960. Economics Research Service, US Department Agriculture, Miscellaneous Publication No. 1233, Washington, DC 1972.

Holmes, B. H.: History of Federal Water Resources Programs and Policies, 1961—70. Economics, Statistics, and Cooperative Service, US Department of Agriculture, Miscellaneous Publication No. 1379, US Government Printing Office, Washington, DC 1979.

Hunt, C. B.: Natural Regions of the United States and Canada. W. H. Freeman and Company, San Fransisco, California 1974.

Jackson, G.; Masnick, G.; Bolton, R.; Barlett, S.; Pitkin, J.: Regional Diversity, Growth in the United States, 1960—1990. Auburn House Publishing Company, Boston, Massachusetts 1981.

Krutilla, J. V.: Multiple Purpose River Development Studies in Applied Economics. Resources for the Future, Inc., The Johns Hopkins University Press, Baltimore, Maryland 1969.

Kuderna, F.: Survey of River Basin Planning Techniques. Illinois Institute for Environmental Quality, Chicago, Illinois 1974.

Leven, C.: An Analytical Framework for Regional Development Policy. MIT Press, Cambridge, Massachusetts 1970.

Lewis, W. C.: Regional Economic Development: The Role of Water. Prepared for the National Water Commission, Utah State University Foundation, Logan, Utah 1971.

Lewis, W. C.: Regional Growth and Water Resources Investment. Lexington Books, Lexington, Massachusetts 1973.

Mosely, M. J.: Growth Centers in Spatial Planning. Oxford, New York 1974.

Newman, M.: The Political Economy of Appalachia: A Case Study in Regional Integration. Lexington Books, Lexington, Massachusetts 1972.

Price, K. A. (ed.): Regional Conflict and National Policy. Published by Resources for the Future, Inc., Washington, DC 1982.

Smith, L. W. (Compiler): Record Group No. 187: Records of the National Resources Planning Board: Preliminary List of Published and Unpublished Reports of the National Resources Planning Board, 1933—1943. The National Archives, Washington, DC 1956.

Wengert, N.: A Critical Review of the River Basin As a Focus for Resources Planning, Development, and Management. In: Unified River Basin Management. American Water Resources Association, Symposium Proceedings, Gatlinburg, Tennessee, May 4—7, 1980. Published by AWRA, Minneapolis, Minnesota 1980.

Wengert, N.: The Politics of Water Resources Development as Exemplified by TVA. In: Moore, John R. (ed.), The Economic Impact of TVA. The University Press, Knoxville, Tennessee 1967.

Wheat, L. F.: Regional Growth and Industrial Location. Lexington Books, Lexington, Massachusetts 1973.

Lundqvist, J.; Lohm, U. and Falkenmark, M. (eds.):
Strategies for River Basin Management, pp. 307–310
© 1985 D. Reidel Publishing Company

The Thames Basin, England — Legal and Administrative System

Sinnott, C. S., Consultant; 86, Wensleydale Rd., Hampton, Middlesex, England

Abstract: The paper outlines the legal and administrative systems for land use and for the control of water resources in England. It discusses their application, particularly in terms of the interaction between land use and water use, noting that in a developed river basin most new developments are relatively small and incremental to the already heavily exploited resource system. The impact of two major land use proposals is discussed to illustrate the issues which arise.

Introduction

In England there are comprehensive planning powers over land use and total control of water management including withdrawals, discharges and pollution. These powers derive from Acts of Parliament extending over many years and in most cases are administered in the last resort by the Secretary of State for the Environment. However, the principal decision-making bodies are the local authorities (County and District Councils) for land use and the Regional Water Authorities for the control of water in all respects. The Secretary of State for the Environment can "call-in" any land or water use application and determine it himself and any rejected applicant can appeal to the Secretary of State to reverse a local refusal.

This background might suggest that the compatibility of proposed land use and water use is examined by the national government's Department of the Environment but this is not the usual practice. Before coming to describe that practice and demonstrating it with regard to the basin of the river Thames it must be emphasised that all the relevant powers are incremental and permissive. They deal only with externally-generated proposed new uses of land or changes of use and with requests for new water withdrawals or discharges. Each proposal is therefore judged against its impact on the existing situation in the locality or river basin. Moreover, if a proposal is consented to, there is no power to enforce its realisation, although the period during which the consent is valid can be limited.

Operation of Land and Water Use Powers

There are two basic ways in which land use and water use may interact. Firstly, a proposed new use of land may impact upon water management by leading to a demand for water (thus also giving rise to waste water), by interfering with surface or sub-surface drainage and/or by giving rise to pollution. Secondly, a proposed change in water management may impact upon land-use either directly through the required structures or indirectly through consequential environmental and ecological changes (real or imagined).

Land use planning

In an historically well-developed area such as the Thames basin most land use planning proposals will be relatively small in scale with respect to the existing conurbations. Such proposals will emanate from either private developers or public authorities acting in the role of developer. The application will be made to the local authority (District Council) for the area concerned. The local authority is required to consult the water authority on any proposal to carry out works on the bed or banks of a river and on specified developments which might pollute water resources.

For other developments, including those which might impinge upon the capacity of water services, the local authority is strongly advised, but not required, to consult the water authority. If appropriate, the water authority will comment upon its ability to supply a proposed development with water, to remove its waste water and on any expected impact on drainage. Should the water authority believe that it cannot meet the new demands that would be made for its services then it can ask the planning authority to refuse the application.

The water authority, particularly if the proposed development impinges upon the flood drainage of existing properties, may advise against or object to the development. The planning authority need not act upon the water authorities' view and can consent to a development despite objection from the water authority. Should a planning authority refuse an application for development the applicant can appeal to the Secretary of State for the Environment to reverse the decision. If the reasons for refusal cite the water authorities views then the water authority will be expected to give evidence to support its opinion, possibly at a public inquiry.

Where an industrial or commercial development is concerned a water authority can require the developer to contribute to the costs of enhancing the water supply and/or sewage disposal system to meet the needs of the development. In such cases the granting of land use planning consent may be linked with the completion of such financial agreements. In practice this situation also applies to residential development since a developer can be required to finance the consequent extension of existing services. With a large development this could include the financing of a significant increase to the general infrastructure of water services.

Water planning

In England and Wales the regional water authorities are the administering bodies for water resources and pollution control. Since the water authorities also have public water supply and sewage disposal responsibilities they are major users of water resources and polluters themselves. In addition water abstraction licences and pollution discharge licences are held by industry and by other public water supply organisations.

The position on water resources control is relatively simple because Parliament has laid down that a new licence to abstract water must not be granted if the abstraction would derogate the rights of preexisting licence holders, and by implication should be granted if it does not so derogate. Thus when presented with an application (from itself or elsewhere) for a licence to abstract groundwater or surface water a water authority must determine whether the new abstraction would harm existing rights to abstract. In practice this relatively simple hydrologic concept is greatly complicated by objections to the possible environmental impact of new water resources, particularly if it is thought that river flows will be affected. Objections on these grounds are often put forward or supported by local authorities.

Water authority applications to themselves for water resource licences are determined by the Secretary of State for the Environment who also hears appeals from other applicants who have been refused by the water authority. This is a fruitful area for conflict. Pollution control legislation is also administered by the regional water authorities who advise the Secretary of State for the Environment on the issue of consents for discharges to watercourses. These consents specify quality and quantity and are required for the water authorities discharges from thir sewage treatment works as well as from industry. Whereas water resources management is inevitably based upon the finite availability of water, pollution control management has not such a natural basis since there is no limit to the amount of pollution a watercourse can receive. It is the practice of water authorities to set desired River Quality Objectives based upon public water supply abstraction needs or on a political assessment of environmental desires and then to determine individual discharge consent conditions so as to achieve overall the River Quality Objectives. This approach raises problems in the allocation of the permitted total pollution load amongst dischargers and the attitude to be taken to applications from potential new dischargers. As with land use planning and water resources management there is a right of appeal.

The water authorities have absolute control of industrial discharges to their sewerage system and in addition levy a charge on the quality and quantity that is accepted.

Application

In the Thames river basin most issues of land-use planning and water service planning arise through applications for new developments which are small relative to the existing infrastructure. They thus present few technical problems in the provision of water and removal and treatment of waste-water. Nevertheless even small developments can impinge upon surface water drainage. The high density of population in the Thames basin has led to considerable urban development in the flood plain of the river to such an extent that runoff of flood waters is restricted. The Thames Water Authority vigorously opposes plans for further development when the consequent obstruction to flood flows would aggravate flooding upstream. All local land use planning authorities are provided with maps of the historic floodplain and are strongly advised to refuse consent for further development in critical areas.

Major Developments

Although the usual situation involves the handling of many thousand applications for small developments occasionally large issues arise which require a special approach. Two examples will show the conflicts between land and water planning which can arise and their resolution.

In the years following the Second World War it was the policy of the British government to establish new towns to counter the continuing expansion of existing conurbations. This was particularly the case with the London area where a number of sites some 50 km from central London were considered for designation for such special development. The water authority and its predecessors were consulted on the availability of water and waste-water services and on the costs of providing them. At each stage in the expansion of the chosen new towns the water authority is consulted, and advises upon the problems and costs it will encounter if the expansion proceeds.

Issues of a similar kind have arisen from proposed new airport development. An airport generates much local employment and hence, planned or otherwise, considerable population expansion in its neighbourhood. The ability to provide the infrastructure to support this population is a major consideration and the water authority will be pressed to provide evidence on costs and environmental consequences of the provision of water services.

Although in the Thames region it is technically possible to provide water services anywhere the cost of doing so will vary from place-to-place, being dependent upon the availability of water resources and the capacity of the local rivers to receive waste-water effluent and increased surface water run-off. These factors are often seized upon by opponents to the new town or airport development so bringing the water authority to the forefront of the political debate. The water authority therefore needs to have firm technical and financial evidence to present on the consequences on water services of the alternative developments being examined, often through the medium of a Public Inquiry.

Major developments which are generally welcomed in the locality concerned present a somewhat different problem. The water authority can require or seek financial contributions from developers where there are significant new costs in the provisions of water services. In general an industrial developer can be required to contribute but housing development has a right to water services albeit with financial obligations. It will often be in the interests of a developer to finance the provision of water services to achieve a timing of their availability which meets his business needs. In practice the developer, local authority (land-use planning) and water authority may jointly consider all aspects with land-use planning consent conditional upon an appropriate arrangement covering finance and timing between the developer and the water authority. This may cover the provision not only of water mains and sewers but also extension to water and sewage treatment works.

Lundqvist, J.; Lohm, U. and Falkenmark, M. (eds.):
Strategies for River Basin Management, pp. 311–315
© 1985 D. Reidel Publishing Company

The Relevance of River Basin Approach for Coordinated Land and Water Conservation and Management

K. V. Krishnamurthy, Hydroconsult International, B-1, 2nd floor, LSC, J. Block, Saketh,
PO Box 1100017, New Delhi, India

Abstract: The historical evolution of the concept of the river basin development, in general, and in particular reference to India, is briefly traced in the introduction. The basis of river basin development in India, particularly since 1947, is then rapidly described as a key concept in regional development both in agriculture and industry. The Damodar Valley Corporation is cited as an example. Then a number of conflict situations are outlined — conflicts of a regional nature, important upstream modifications having downstream repercussions; work on one bank affecting the other bank etc. The nature of the functional conflicts are briefly alluded. The limitations of the river basin approach are seen against the recent emergence of an approach of inter basin transfer of water. Details are described of a national perspective recently evolved by Government of India. In conclusions, it is pointed out that while a strict river basin approach is not now found entirely possible, necessitating a consideration of interbasin transfer of water, at the same time it cannot be stated that the concept is out-dated because the basis for the determination of the possibilities of large scale transfer of water from one 'surplus' region to another 'deficit' region is in itself dependent on the river basin approach.

Introduction

Ancient human civilizations were known to be nurtured on the banks of rivers. The Babylonian civilisation on the banks of the Euphrates, the historic Nineveh on the banks of the Tigris, the Harappa and Mohenjodaro on the banks of Indus and the Nalanda and Rajgir on the banks of the Ganga are well-known examples. When social production was predominantly agricultural, resulting in the form of settled communities in early human history, it is not surprising that such agricultural settlements were founded on river banks and thus gave rise to what subsequently came to be known as hydraulic civilizations.

Not merely did man use river waters by gravity and later by pumping by primitive means, but storages were constructed on rivers and their tributaries and water diverted for agricultural purposes, for many centuries in India. Not only was river water stored but there were innumerable small storages of even rain water captured in shallow depressions and used subsequently for agricultural production. Thus artificial irrigation was known in India for a very long time although modern irrigation is relatively recent. Extensive canal systems have been built in India, connected to major rivers like the Ganga in the 18th and 19th centuries, like the Upper Ganga Canal in North India and the famous Mettur Canal System in the South. However, all this did not specify a river basin approach. It represented a development concept based on the river itself as the main artery of development.

The River Basin Approach

The concept of river basin as the unit for development received a new impetus in India particularly after the advent of national independence in 1947. The years immediately preceding the Second World War i.e. the late thirties and those that followed the War i.e. the wid-forties, represented a period of stagnation, in terms of agricultural develpment, or for that matter of economic and social development, because the national energies were channelised for the prosecution of the War. Agricultural development received a new impetus as a result of postwar development and more particularly after the advent of independence in 1947.

In this period, the concept of the river basin as a unit for river development assumed a basic role in organising the development, both agricultural as well as industrial. Agricultural from the point of view of utilisation of river water either by diversion or from artificial storages, and industrial because power was considered as an indespensable element in the essential industrial infrastructure. Moreover, hydroelectric power assumed a fundamental importance because water was considered as a replenishable natural national resource without the necessity of importing. By comparison coal was considered a fixed natural resource which would disappear with constant depletion.

The example of the Tennesse Valley in the United States was considered worthy of evolution and the early postwar years in India witnessed the establishment of a similar experiment when an autonomous corporation was established called the Damodar Valley Corporation. The DVC was supposed to be the Indian TVA — an example of autonomous institution for the integrated and comprehensive development of water resources of a river basin.

Conflict Situations

Although the river basin provided a natural unit for development, both for agriculture and industry, an increasing use of the land and water resources of a basin gradually pointed to the potential conflicts within the river basin. These unfolded themselves as a result of different administrative units, provinces or what came to be later called as States and subsequently also between different water uses.

The conflicts between different geographic units of the basin were such that an excessive off-stream utilisation in the upper reaches resulted in the depletion of the available water resources in the downstream areas. In addition to national conflicts between upstream and downstream areas, there were also conflict situations between lateral riparians. The prevention of inundation on the right bank by the construction of a long line of embankment was found to be transferring the problem of inundation on to the left bank and vice versa. Straionght-jacketing of a river for purposes of flood control on both banks would accentuate the problem upstream by afflux caused by the embankment system and produce flood situations in areas which were not hitherto flooded by afflux and therefore the entire river basin had necessarily to be dealt with as a unit for planning development. There is no escape from this.

Also in a functional sense, there were potential conflict situations. If flood water had to be absorbed behind a dam, a certain amount of storage had to be reserved for this purpose. If storage space is kept empty to cover the flood risk, that much of water would be lost both for irrigation and power generation if that space was not filled up with water towards the end of the rainy season. This problem became particularly acute in India because over a great part of the country the rainy season is confined to a period of four months, the remaining eight months being totally dry. Similarly, in the case of a multiple purpose project involving hydroelectric power and irrigation, if water is drawn through the turbines for power generation that much of water is lost for irrigation in the subsequent dry season. Such potential conflicts among water uses made it necessary to integrate the water use within a river basin. Thus the river basin presented itself as a natural unit for solving the potential conflict situations both in spatial sense as well as in a functional sense and the concept of a river basin as a unit for planning comprehensive and integrated water development therefore reigned supreme for a long time i.e. during the last three to four decades.

The conflict situation made it necessary to institute adequate legal and administrative tools or instruments for facilitating solutions. A number of inter-state tribunals were established for the allocation of water resources. Many inter-state control boards were established for facilitating joint construction, operation, regulation and maintenance of major control structures in river basins.

Limitations of a River Basin Approach

It cannot be stated that the river basin approach has been exploited in full in the conditions of storages and diversions on many rivers. In fact, the long-term target is the irrigation of about 113 million ha by the turn of the present century. This represents the doubling of the efforts in the last two decades of the present century over what has been accomplished during the preceding three to four decades since the advent of independence, a colossal effort in comparison to what has been done in the past in India or in comparison to what has been accomplished in many other countries in the world. But even so, there is increasing recognition that planning has to transcend the concept of river basin as a unit. There have been a number of proposals under the consideration of the Government of India for a large scale inter-basin mass transfer of water resources.

In the early seventees, proposals were formulated for transferring the waters of the Brahmaputra into the Ganga and from the Ganga into the Cauvery. This proposal involved a series of links between the various river systems in India from north to south. It envisaged a transfer of about 25 billion m^3 of water during the monsonn season by pumping from the Ganga river over a head of about 600 metres to irrigate about 4 million ha. This scheme has recently been modified and a national perspective has been evolved comprising two main components, namely, the Himalayan rivers on the one hand the Peninsular rivers on the other. The development of the Himalayan rivers envisages the construction of storage reservoirs on the main Ganga and the Brahmaputra and their principal tributaries in India and Nepal. The scheme will benefit not only the States in the Ganga and Brahmaputra basins in India but also neighbouring countries. The implementation of such a scheme will naturally depend upon international cooperation.

The Peninsular river development is envisaged in four parts;
1. Interlinking of Mahanadi, Godavarai, Krishna, Pennar and Cauveri.
2. Interlinking of west-flowing rivers north of Bombay and south of Tapi.
3. Interlinking of Ken with Chambal.
4. Diversion of west-flowing rivers.

This perspective has been accepted for a further study and investigation of the potential for large scale transfer of water resources amongst the Peninsular rivers.

Although the principle of transfer from surplus to deficit areas is accepted on all hands, the existence of a surplus is to be established in the first instance as the basis for effecting such a transfer. A surplus or a deficit is to be established within a State and within a river basin. Thus the current experience shows that although the river basin has a certain limitation in planning development of land and water resources it cannot be entirely dispensed with as being irrelevant. In fact, the river basin is the basis for the determination of the surplus and for the diversion of water into an adjoining river basin. Nowhere in the world can it be stated that the river basin approach has spent itself as a logical approach for development. At the same time, the approach does not permit any rigidity. Herein lies the need for resilience. Planning should be done on the basis of a basin as the unit, because there are influential social and political pulls for transferring or not transferring water from one area to another. It must be recognised that the requirements of the basin have first priority before surplus or deficit can, in fact, be established. Legal and administrative tools must take this into account even within the same county. The problems become very complex in the case of an international river where conflicts have to be resolved to the satisfaction of all the riparian countries. The conclusion of the treaty between India and Pakistan for the utilisation of the waters of Indus is an example which shows that satisfactory solutions can be found for resolving conflict situations even within an international river basin, given the necessary political goodwill. But these legal and administrative tools cannot in themselves be considered either as incentives or disincentives. The incentive is common political goodwill engendered on the basis of a joint recognition of a common economic destiny.

It has been the experience in India that the growing urbanisation has undoubtedly created additional demands on the use of surface and ground waters but it has not impinged adversely on the perspective of river basin as a unit for development. There has been increasing recognition that not only land and water resources should be planned in an integrated manner but the entire river basin should be considered as an integral ecosystem, calling for management of various resources in upper regions which have an effect on the hydrological regime of the river basin. The approach of the eco-system transcends the concept of land and water management, the integrated development of which has always been recognised as being essential in the interest of both. The approach of the eco-system implies the development of the forests on the one hand and also the prevention of pollution on the other as a result of growing urbanisation and industrialisation. The latter aspects have brought up the question of water quality more than the quantitative aspects implicit in the allocation of water resources.

Conclusion

In conclusion, it can be pointed out that the river basin approach for coordinated land and water conservation and management is found not entirely adequate within itself because the transference of water from one river basin to another is increasingly felt necessary in the conditions prevailing in India. At the same time it cannot be stated that the concept is outdated because the transference of water from one area to another in India is to be effected on the basis of sufficiency of the fulfilment of the needs of the river basin itself. To that extent the concept of the river basin approach is still valid and serves as the basis for transcending the limitations imposed on it as a result of increasing demands for accelerated development.

Lundqvist, J.; Lohm, U. and Falkenmark, M. (eds.):
Strategies for River Basin Management, pp. 317–328
© *1985 D. Reidel Publishing Company*

317

Rationale for Establishing a River Basin Authority in Tanzania

Mascarenhas, Adolfo, IRA, PO Box 35097, Dar es Salaam, Tanzania

Abstract: In a passive way, river basins have been used as a focus of activity since prehistoric times. With greater technical and engineering skills, some modern river basin development projects have stood out as triumphs of human skills. As developing countries, such as Tanzania, try to maximize their resources, care has to be taken, that the river basin authorities which are created, take into consideration both engineering as well as social and ecological factors. This article attempts to illustrate that river basin authorities offer good opportunities for an integrated approach to planning. An authority should be able, among other factors, to balance the long term from the immediate benefits, reduce environmental risks and maximize social benefits.

Introduction

It is first of all necessary to put the issue of River Basin Development or to use the more recent terminology, Integrated River Basin Development, within a historical perspective. Some river basins have been cradles of great civilizations. The Indus, the Euphrates and Tigris in Asia and the Nile in Africa stand out as good examples. In most cases, it is the construction of a dam which makes it possible to efficiently use a river basin. Dams then are "among the greatest human triumphs, constructive enterprises conducted with machinery, manpower and efficiency that, regrettably are usually reserved for war". (Stanley; Alpers 1975). The first known case of a dam across a river, the Sadd el Kafara, is more than 5,000 years old. The use of the river basins in these early cases was passive. It required the emergence of technological and engineering skills for people to be able to undertake construction activities on a scale which one could admit that human intervention was both active and dramatic. However, as several authorities point out, control of river does not mean mastery over nature.

Water management of rivers demands a great deal of social control and at least one authority has argued that autocratic control was a feature of the early kingdoms (Wittfogel 1957). The authority had to be invested with powers to regulate water use, control the amount of land cultivated, extract surplus by tax collectors and demanded that people acknowledge the suzerainty and authority of the religious or political leaders.

River basin development has not taken place merely because of man's quest for mastery over nature. Rivers have been dammed or the waters in the basin otherwise managed because people needed to benefit from the water resources. The development of river basins requires collective action of many people and institutions. The resources needed are large even for rich countries. Consequently, even in the United States, the haven of the free enterprise system, it was not the private sector but the Federal and state governments which were behind the massive Tennessee Valley Authority (TVA) project.

The TVA was undertaken in the early 1930's at a time of a chronic economic depression with its attendent socio-economic upheavals. However, it still stands out as an example of massive mobilization of the people: unskilled, jobless, artisans, professionals and bureaucrats. Disastrous floods were controlled, soil erosion reduced and, above all, many were given a new lease to be productive people. The project worked because of the commitment of those in it and a degree of honest service seldom seen in the public sector. Even after 50 years the TVA continues to function and dams are built or their construction is halted, on account of ecological issues.

In the general context of the above, the Rufiji Basin Development Authority (RU-BADA) in Tanzania has a relatively short history, but the work ahead is no less formidable. The basic planning work for the engineering works, especially for the dam at Stieglers, is well advanced and while it is easy and perhaps rightly so, to be proud of human engineering skills, it is equally imperative that some attention be given to the developmental aspects and rationale for establishing a river basin authority. In Tanzania, as in many parts of Africa, the rational use of river basins forms an important strategy for development.

Rationale for a River Basin Authority for the Rufiji Basin

Strategies for development

A useful summary on river basins development in Eastern Africa now exists in an abridged form for 6 African countries — Somalia, Kenya, Sudan, Tanzania, Ethopia and Uganda (Mascarenhas 1982). The dominant theme is that river basin development schemes and the damming of parts of the rivers is commonly associated with large scale projects and are the most visible and obvious strategies for development on two fronts: to make agriculture less dependent on the vagaries of climate and on the energy front to find alternatives for fossil energy.

In all cases, there is need for an audit of the schemes in terms of human well-being. A critical appraisal is needed of the dams and the schemes. The price has to consider what the people and the nation have to pay and the benefits that are realized. Audits of river basins are unpopular and can seldom be given a clear certificate. They would frequently reveal a high financial price, human suffering and show that opportunities for other equally viable investments were forefeited.

Tab 1 shows the size and Fig 1, the areal extent of the different river basins in Tanzania. In relative terms, the river basins which have received considerable attention are, the Pangani, the Wami, the Ruvu and the Rufiji. Systematic efforts at gathering information on these four basins go back to the 1950's. Among these four rivers, the waters of the Pangani and its tributaries are used most intensively. There are 9 reports covering the entire basin for the period 1934—1968 and 26 specific project reports. The Nyumba ya Mungu dam was built as a result of these reports but it still required yet another FAO study to investigate the areas suitable for irrigation development. An important recom-

Tab 1 River basins of Tanzania

(In square kilometres)	
Rufiji River Systems	178,085
Internal Drainage Basins	153,802
Lake Tanganyika Basin	149,935
Lake Victoria Basin	79,569*
Lake Rukwa Basin	77,340
Pangani	56,303
Ruvuma	50,900
Wami	45,340
Lake Nyassa Basin	27,105
Ruvu	26,865
Matandu	18,565
Mbwemkuru	16,255
Lukuledi	12,950
Mavudji	5,600

*Hydrological Year Book.

Source: Atlas of Tanzania

mendation to establish a school for training irrigation technicians was not followed. Still one can hardly say that the issue of the condition of human settlement have been adequately covered.

The Rufiji basin, which covers nearly a fifth of the area of Tanzania, was neglected until the 1950's. Once the huge water resources and other potentials became evident, the Rufiji basin was selected for a comprehensive survey. The work was organized around an FAO consultant team though officers from the Tanganyika government were involved in most aspects of the study and in particular in geological investigation. The terms of reference for the FAO study was:

> "to assess the irrigation potential which exists in the basin as well as the probable cost of exploitation of that potential. To this end the irrigation areas are to be defined reasonably closely and their relative merits ascertained" (FAO 1962)

Studies were carried out over a seven year period, involving 12 FAO members, 2 government officers worked full time while 15 others were involved for shorter periods. The cost of the FAO study was £ 750,000; almost a third of which were related to topographic surveys and mapping: another substantial proportion was used for the setting up of a comprehensive network of gauging stations. (Berry et al. 1970). The appraisal revealed the extent of the physical potential for irrigation of 600,000 ha.

Limited by its term of reference the team did not place too much emphasis on the alternative uses of water for HEP. Since 1961 there has been a major reorientation and at least 2 large studies have been commissioned and completed on the HEP potential especially of Stieglers Gorge. The NORAD funded study which is the latest, also emphasized the hydropower aspects. The Act establishing the Rufiji Basin Development Authority while comprehensive, also emphasizes the HEP generating role. (Tanzania Act 1975) Notwithstanding that many people now regard the development of the Rufiji basin as a multiple

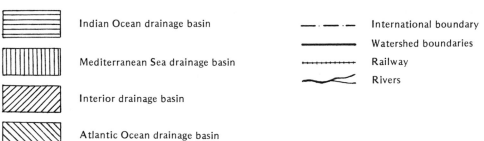

Fig 1

purpose project, we obviously are dealing with a problem which changes dimensions and which requires attention.

The evolution of authorities in Tanzania

Before discussing the rationale for establishing a river basin authority it is perhaps instructive to note the evolution of other forms of authorities. The Rufiji Basin Development Authority (RUBADA) is not the first to be established in Tanzania and one would venture to say not the last. Many are familiar with the various crop authorities created for the promotion of specific crops like the Cashewnut Authority of Tanzania, Coffee Authority of Tanzania and the increasingly diminishing Tanzania Sisal Authority.

It is perhaps easy to understand and to justify the creation of such authorities for they all aim to promote and maximize benefits accruing to all parties concerned, including the Government. However, we may wish to ponder as to why, of late, there has been so much of criticism against authorities, within and from outside Tanzania. One hesitates to ask whether RUBADA also falls in this class but given that the authority recently itself chose to ask this question, we should attempt an answer.

There are other 'Authorities' than those dealing with crops. For instance, the Capital Development Authority (CDA) was created soon after RUBADA, in order to implement a major Party and Government policy, namely, to shift the capital from Dar es Salaam to Dodoma. The composition of the board of the CDA was not too unlike the first RUBADA board. It was composed of the more powerful ministries including Finance, Planning, Water and Energy, Transport and Communication. With the exception of three board members, all the rest were ministers. Clearly, the Government wanted to back the CDA to the hilt.

In contrast to the above, there is the very early authority established in the country — The Ngorongoro Conservation Area Authority (NCAA). The equivalent position as of a Director General is held by the Conservator. NCAA also has a board of directors whose members are not ministers. Distinct as the two boards are, they have one common feature of interest: They are both excised from the regional administration and exercise special control over the area demarcated for their special attention. In the case of the NCAA, it covers about 8,000 km², of some of the most priceless country in the world, (Mascarenhas 1983); the CDA covers a much smaller area and the intention is to integrate man with his environment (Kahama 1975).

Interstate river basin authorities

Large river basins can cover several countries. In such situations agreement among several countries is necessary to maximize on the resources of the river basin. The agreement between the riverine states of the Nile goes back to 1929 and there have been subsequent revisions. The formation of commissions with executive powers facilitates the interstate operations and implementation of the use of river basins.

The latest example of such collaboration between states is the Kagera Basin Development Authority. The main justification was that:

"... the creation of a Kagera River Basin Authority appears essential to manage and coordinate activities at a regional level."
and
"the development of the project area's land and water resources can and should be conceived as a coordinated whole to create a larger economically active and trading area involving the three countries" (Consulting Engineers Consortium 1973).

Differences in resources and management between Burundi, Rwanda and Tanzania are so great that the political will and backing of the three governments and an Authority seemed the only logical way of expediting development. The problems and constraints are so complex and yet inter-related that they can meaningfully be discussed only in a holistic context.

Nature as a unifying force

Nature is a great unifying force and the need for a river basin authority stems from this central fact. Two of the most obvious resources found in a river basin are the land and water. The extent of land covered by some of the large river basins in Africa have already been mentioned. In the Tanzania context the fourteen river basins found in Tanzania range in size from the mighty Rufiji basin covering 178,085 km² to the relatively small Lukuledi basin which only has 12,950 km². The Rufiji basin is large and covers more than 20% of the country's area. The volume of water and of other resources are by any standards very considerable, only if they could be harnessed.

The need for an authority does not arise on grounds of the sizeable real estate alone. On the basis of area, it could be argued that the large area of the basin makes it unwieldy. After all even the regional administration breaks up the country into twenty units. Such an argument is untenable. The basin is a system which is unified — this unity of resources cuts across administrative regions wholly or partly.

Minimize conflicts of interest

The rationale for a river basin authority is also based on the professional need of the planners and the tools they use for analysis and implementation. During the 1960's the government sought assistance from the Food and Agricultural Organization for realizing the potential benefits from the Pangani and the Wami river basin. One of the recommendations in both studies was that a River Authority be set up so as to coordinate development. The justification in the case of the Pangani was that there was a close interrelationship between flood control, irrigation development and the ensuing competition for water. In the case of the Wami an authority was justified on the basis of the multipurpose aspects of the scheme (Berry et al. 1970). Unfortunately these recommendations were not followed. It resulted in competition for water between HEP and agricultural needs, problems with silting of the Nyumba ya Mungu dam, the salinization in the Kahe area, etc.

The idea of an authority is promoted to minimize the apparent conflict of interests. A hypothetical example from the Rufiji basin and preoccupation of the officers from two regions (Dodoma and southern Morogoro) will illustrate how such situations could be resolved by an authority. In the case of Dodoma with its semi arid environment, the amount of water required for human and livestock consumption is a priority, while with southern Morogoro, the availability of water is not a problem. However, unplanned water extraction for livestock numbers in excess of the carrying capacity of the land can cause soil erosion and environmental degradation. The silting of dams will cause floods, reduce power and transfer the problems from Dodoma and create new problems in Morogoro. Ultimately it will lead to a reduction in the national benefits. In sum, river basin authority justifies itself if it can minimize the apparent conflict of interests.

Secure continuity in planning

The development of river basins is a long term process in most cases because of the sheer magnitude of the tasks to be undertaken. In most cases just collecting data is time consuming and expensive without really showing any tangible results. The lack of data and poor documentation is one of the major barriers in prolonging the gestation period of RUBADA. The process of gathering data, which is comprehensive, is made even more difficult in Africa, because in many parts, the population density is so low that these areas are the least known. Thus, within the Rufiji basin, we have one of the largest game parks in the world but there is little additional information about the area. Are there minerals? What is the soil quality like? But the story is not too different even for the smaller river basins such as the Wami basin which covers 45,340 km^2.

Briefly, the western part of the Wami Basin is semi arid, lacking adequate water, is overgrazed and eroded; near Kilosa, flash floods occur. Going back to the early 1920's proposals were made to regulate and control land and water resources. In the late 1950's three dams were built at Ikowa (1957); Hombolo (1958) and Dabolo (1958).

Since there were only some correspondence and safari reports of railway engineers the FAO was requested in 1964 to carry out a general reconnaisance survey. Despite useful but rudimentary work in the upper and central parts of the Wami which contain the Magole and Dakawa plains, the recommendation of the FAO experts was to focus attention on the coastal areas of the Wami. The FAO produced at least 5 other reports between 1968–1969 but the social and agricultural considerations "are not quite so useful because of their dealing with generalities" (Berry et al. 1970). Can a dam built on the Wami solely for irrigation purposes pay for itself? There are spots on the Wami which were identified for HEP as far back as 1945. Yet this aspect was overlooked. One cannot but help agreeing with the recommendation made by FAO for the creation of a Wami River Basin Authority. Out of necessity such an authority would perhaps bring about an integrated development for the benefit of Dodoma, Morogoro and Coast Regions. Shifting priorities without attributing reasons for doing so does not help in imparting the much needed consistency. The Wami basin has certainly suffered from a lack of continuity of work. Similarly the piecemeal efforts like on the Pangani are not too useful either.

Minimizing side effects through planning

Despite the fact that at least after 1972, ecology and environmental issues are increasingly becoming topical, in reality there is surprise and the response comes *post facto*. Yet with a little more scrutiny the writings are in many cases clear on the wall. Changes in river basins cause fundamental alteration to human well-being and without deliberate planning, whatever benefits calculated earlier were frequently lost. This is particularly true of the health aspects of dams, irrigations and man-made lakes. A literature review reveals that the side effects in many parts of Africa have actually worsened after dam construction (Mascarenhas 1978). As has pointed out again recently the:

> "Negative impacts have taken many forms: exotic diseases or plagues; increased incidence of indigenous diseases; impairment or reduction of industrial production based upon an area's resources; an increase in social problems related to population migration; a weakening or over-extension of social services; loss of diversity and stability; exaggeration of both the degree and incidence of disasters occasioned by natural phenomena, such as earthquakes and hurricanes resource impoverishment; foreclosure of future developmental alternatives; and the need for a large, unanticipated addition of financial and other resources to keep the project operational."
>
> (OAS 1979)

A properly functioning authority is in a far better position to see that a situation is created where all the precautions are taken. In this respect while the state of the art is fairly well understood there is still a danger that disciplinary bias or excessive attention to the engineering aspects so dominate the scene that ecological and environmental issues are only incidentally considered.

The emphasis on ecology is not merely to keep environmentalists or university professors happy but basically because disease, loss of sustained resources and most forms of destruction are the antithesis of development. Ecologists, politicans, educators have more in common than is usually considered.

The timing of the ecological consideration is important. Environmental interest have to be considered in the planning stages. Consideration after signing of the construction activities both prejudice the case against a proper assessment and modifications are generally expensive. Finally on the ecological aspects it is worth remembering that:

"Since the primary concern of the environmental movement is improvement of the long term quality of life for human beings, a context is created in which the environmentalist and the developer as politican, economist, social activist, planner, or technican, can work together"   (OAS 1979)

Some Concluding Issues on Forms of Authority

The nature of an authority

The word 'Authority' has an unnecessary disciplinarian overtone rather than reflecting, the nature of being disciplined. True development needs discipline, even though other measures including the cultural revolution, fascism and nazism have been attempts to bring about a so called national development. A river basin authority frequently seems to have powers to be very dictatorial. Part of the reason why governments give power to authorities is to get things done. However, there are very many ways to use power.

One could vest an authority in a single institution as in the case of RUBADA or it could be broken into components. While it is preferable to have one authority, to simplify communication and integration, this may not always be possible. An authority could have autonomous components, in order to give maximum attention to each basin. In Nigeria, the River Basins Development Authority has 10 sub-authorities to manage the river basins. In Ghana there is a single authority.

Ability to urgently recast priorities

It is to be expected that the major concern of an authority will reflect the preoccupation with the present. Thus in keeping with national planning concerns in Ghana in the late 1950's, the Volta River Development Act of 1961 empowered the Authority to be responsible for the planning, execution and management of the Volta river. In Nigeria, conscious of their oil wealth, the main functions were to comprehensively develop surface and underground water resources for multiple uses, as well as control floods and soil erosion, etc.

In Tanzania, concern with the dependence on importation of crude oil the preoccupation has been on the energy sector. Most of the planning and use of finances has been geared towards this aspect. However, it is all the more important that an authority adjusts itself to the times. It becomes both necessary and expedient to have in a portfolio short and long term priorities. For instance most of the technical aspects of the Stieglers Gorge dam are fairly well covered. However, it is unrealistic to expect anything dramatic in construction efforts within the next 5 to 10 years. Should not this be the time, for RUBADA to recast its immediate priorities. A first step has already been undertaken in the directions of promoting and regulating developmental activities (RUBADA 1983). But a great deal remains to be done.

The balance between legalistic and participatory aspects

The enabling act, which created RUBADA is both comprehensive and far reaching. The application of the act, in realistic terms is open to a good deal of interpretation. One notes that a noble attempt is being made by RUBADA to balance between the legalistic/bureaucratic and the participatory aspects. For instance it has been stated that: "in order to fulfil this function we need the cooperation of the regions and other institution carrying on projects in the basin ..." or the "Implementation of these recommendation was the task of different ministries. The role of RUBADA remained that of coordinating the implementation ..." (Kabuzya 1984).

At the moment in Tanzania there is no shortage of the legal or administrative powers. It will be noted that participation is not easy. For instance in the case of RUBADA it was noted that: "The secretariat received little official reply from the implementing agencies." (Kabuzya 1984). Establishing a working relationship takes time.

The need for a strategy for development

A strategy among other aspects would require monitoring of the big forces and trends. Essentially, we are dealing with a national issue, in which collaboration becomes a key word. For instance one should be struck by the fact that both Maliasili (Ministry of Natural Resources) and RUBADA have the same goals in tree planting. Can this common goal be implemented? Destruction of trees is taking place on a big scale. How does one cope with this — for we know that the law in itself is not enough.

If we do not ask questions on development we should not expect answers. Therefore, monitoring and evaluation must always be there as part of a strategy.

RUBADA has now come to a stage when it needs to have a strategy for development. Development can be illusive. It would require in many cases to reexamine the options which are presently being proposed. For instance how many peasants or villages in Tanzania have 100,000 shillings to spend per hactare on irrigation? If irrigation is the option being suggested should the more fundamental question not be whether there are water management practices among peasants or equally important whether the traditional forms of irrigation could not be improved? The costs could well be lower and the benefits quicker to realize.

There are contradictions which have to be resolved. Thus why do pastoralist move into the Usangu plains in numbers that seem to cause concern? The Kilombero District with less than 13 persons per km^2 would absorb a far larger population than it presently carries. Yet, to transport about 180,000 people, to double the population density would require at least 36 million shillings for transport alone. Therefore, what would be required to mobilize spontaneous migration into the Kilombero valley?

We may at this juncture give a secondary role to the assessment of the resources of the Rufiji basin because we know that they are considerable — more than 600,000 ha of land suitable for irrigation, the largest game park in Africa, nearly 100 forest reserves, several major sites for HEP development, etc. However, the more important issue is how the resources can be utilized for the well being of people and the nation. Therefore, it becomes necessary to concentrate on harnessing the ingenuity of the peasants and the creativity of people. The Rufiji basin may then live up to the expectation of being Tanzania's major frontier of expansion and development.

Concluding Remarks

RUBADA is a new type of authority in Tanzania. If the main focus since its creation was HEP, construction and generation, and this function is not immediately feasible, the issue of identity now becomes vague and yet it is now more crucial for RUBADA to have a new focus.

Some of the key issues which should be evident from the first part of this paper is the need for a time horizon and for priorities to be assessed; for the integration of the major components with each other; for maximizing of benefits; for deployment of re-cources and for the working out of modalities — the list is long. A short elaboration on modalities would be useful. For instance does an authority actually get involved in research on chemical oceanography or become an engineering firm or should the authority only manage and coordinate?

The use of river basins as a focus of planning is in the final analysis undertaken to bring about development. In the delegation of authority, bodies charged with river basins must have enabling powers to bring development. This authority in collaboration with others can: improve the use of resources rather than be at the mercy of floods or drought, better control productive resources such as soils and water, perhaps even increase the resource base by opening up new land, which hitherto had not been used or doubling its use by irrigation.

As has already been stated, the ecological considerations are emphasized to make the environment a better place. Briefly all these efforts are aimed to bring improvements in the quality of life. It is impossible to visualize how all these factors could be resolved without an authority.

References

Berry, L.; Kates, R. W.: Planned Irrigated Settlement: A study of four villages in Tanzania. BRALUP Research Paper No. 10, 63 pp., 1970.

Consulting Engineers Consortium: Planning the Development of the Kagera River Basin. Vol. 1 General Report, Prague, pp. G2, 1973.

FAO: Rufiji Basin Report. Food and Agricultural Organisation, Rome, Vol. 1–7, 1962.

Kabuzya, B. J.: A Review of the First Annual Conference of the Development of the Rufiji Basin held at Iringa in October. RUBADA, 17 pp., 1984.

Kahama, C. G.: A Tailor Made City. FAO CERES 48, Rome, Nov–Dec 1975.

328

Mascarenhas, O.: Health Repercussions on Man-made Lakes, a selected annotated bibliography. BRALUP Research Report No. 29 (New series), 40 pp., 1978.

Mascarenhas, O : River Basin Development in Eastern Africa: A preliminary appraisal. Worcester, Mass., Clark University International Development Program, East Africa Regional Studies, Regional Paper, No. 10, 36 pp., 1982.

Mascarenhas, A.: Ngorongoro: A Challenge to Conservation and Development. AMBIO 12, 3—4, Stockholm (1983)

OAS: Environmental Quality and River Basin Development: A Model for Integrated Analysis and Planning. 107 pp. Organisation of American States, Washington DC 1979.

RUBADA: Promotion and Regulation of Development Activities in the Rufiji Basin. 15 pp. The Rufiji Basin Development Authority, Dar es Salaam 983.

Stanley, N. F.; Alpers, M. P.: Man-made Lakes and Human. 47 pp. Health Academic Press, London 1975.

Tanzania Act: The Rufiji Basin Development Act. 19 pp. Republic of Tanzania, Dar es Salaam 1975.

Wittfogel, K.: Oriental Despotism: A Comparative Study of Total Power. Yale University Press, New Haven 1957.

Lundqvist, J.; Lohm, U. and Falkenmark, M. (eds.):
Strategies for River Basin Management, pp. 329–335
© *1985 D. Reidel Publishing Company*

4.2 Models for resource analysis and policy implications in a river basin context

Methodologies for Water Resources Policy Analysis in River Basins[1]

da Cunha, L. V., NATO Scientific Affairs Division, B-1110 Bruxelles, Belgium

Abstract: The special nature of water resources problems determines the specification of adequate technical, economic and institutional measures, the implementation of adequate sets of these measures in so-called water resources strategies and the consideration of coherent combinations of these strategies referred to as water resources policies.

The role of models in the definition of these policies is described, with particular emphasis on the development and application of policy oriented models to problems involving conflicts in water resources management in river basins, in particular resorting to series of linked strategy/impact-prediction models for evaluating alternative management policies.

Introduction

Water is a natural resource whose availability strongly affects economic development and social welfare. For most water uses a significant part of the water withdrawn in a river basin is returned to the environment, usually to the same river basin and with degraded quality, which may affect the existing water reserves causing, for instance, the pollution of rivers, the eutrophication of lakes and reservoirs or the salinization of aquifers. On the other hand, economic development also influences water resources both in quantity and quality, particularly in relation to water pollution due to waste water discharges.

An integrated water management policy for each river basin or set of river basins has to be developed intersectorially, including the various water using sectors (agriculture, industry and domestic uses) and the institutions responsible for water resources data acquisition processing and dissemination, for the planning, construction and operation of water resources infrastructures, and for environmental protection.

Water resources management implies certain procedures that can be considered as involving three steps:

— identification of the system to be managed;
— prediction of the behaviour of the system;
— management of the system.

---

1 The ideas presented in this paper were partially developed at the launching stage of a project established at the Laboratorio Nacional de Engenharia Civil (LNEC), Lisabon, Portugal, in co-operation with other Portugese institutions. This project was partially sponsored by the "Science for Stability" Programme which is a part of the NATO Science Programmes. The author was a member of the LNEC staff and Director of the above-mentioned project in 1982 and 1983 until assuming his present functions.

The *identification* of the system involves the specification of the characteristic features of the water resources system relevant to the different management problems that have to be faced, these features being of a physical, economic, social or environment nature.

The *prediction of the behaviour* of the system corresponds to finding out the response of the system to certain actions taken by man that excite the system. These include pollution discharge in the water bodies, urbanization, change in agricultural practices, or building of works which condition the behaviour of water resources within the system.

The *management* of the system involves the selection of the best alternative to attain certain objectives. Management decisions are based on the previous steps of identification and prediction.

At the identification and prediction stages adequate monitoring of the system is essential. At the prediction and management stage modelling is a first rate tool.

What has been mentioned previously justifies that water resources problems should always be considered as intersectorial problems. In addition, they are also interdisciplinary problems, requiring for their solution the co-operation of scientists and experts in different fields.

Like water, land is also essential for development as it provides the basic factor in agricultural production, in transportation infrastructures and in building. Water and land use are closely related owing to a large number of interactions of physical, chemical and biological nature, and this makes necessary the planning in a co-ordinated way of the use of water resources in each river basin or region. In a number of countries, and particularly in the more developed countries, the principles and standards for water resources and land use planning are established in a single law and administered by the same institution, or by different institutions duly co-ordinated.

The policy often adopted in the past, by which land-use planning decisions were made first and the corrective measures to face the impacts of these decisions on the water resources were only taken afterwards, must be abandoned. Instead, water resources planning and land use planning must be integrated processes pursued simultaneously at all levels, right from the beginning.

Renewability and mobility are two specific characteristics of water resources which distinguish them from other natural resources and are extremely important elements in water resources use and control.

In practical terms, water resources *renewability* implies that there is not a definite and irreversible consumption as is the case with other natural resources, but a consumption that on average can be recovered yearly. Thus, while the stock of most mineral resources is limited, water resources reserves can be more or less recovered yearly by the hydrological cycle.

Mobility of water resources is also a very important characteristic as regards water use. The separate water uses that take place at different points of the same river basin are strongly interdependent, since the upstream uses usually condition the downstream uses both in quantity and quality. The interdependence of the several water uses in the same river basin is extremely relevant both from an inter-regional and international point of view. The importance of this international interdependece is obvious if one considers that 40 % of the world's population is located in river basins shared by two or more countries.

Measures, Strategies and Policies

The demand for water of good enough quality and with an adequate distribution both in space and time has become increasingly important in many river basins. As water availability is limited, critical situations often arise, generating conflicts of different types possible between users, decision makers and social groups. This makes it necessary to face these situations and ensure an optimum water use through effective water resources management by putting into prctice several measures of technical, economic and institutional nature.

The implementation of a set of measures of the same type corresponds to the formulation of what is called a water resources management strategy. A coherent combination of these strategies corresponds to what can be called a water resources management policy.

The above-referred *technical measures* involve change in and expansion of the existing systems of water resources infrastructures, as well as the creation of new infrastructures. These can be of different types and have been extensively used in the past with different degrees of success. Among the more important technical measures, one can refer to the storage of surface water in reservoirs or of groundwater in aquifers, eventually resorting to aquifer recharge, the desalination of salt and brackish waters, the reuse and recirculation of sewage, the control of evaporation and evapotranspiration, the use of low water consumption industrial technologies, the reduction of water wastage, the control of losses in water distribution systems and, last but not least, the control of water pollution.

Among the *economic measures* associated with water resources management in river basins, one can mention the adoption of adequate pricing systems for water consumption, the adoption of pollution charges for waste river rejection and the subsidies, credits and tax exemptions given for the benefit of water users or institutions in charge of water management who would invest in technological procedures favouring a rational water use.

Finally, *institutional measures* can be related to the preparation and publication of laws and regulations which govern water resources use and with the definition of institutional frameworks responsible for water resources management.

Policy design consists of the generation of strategies and the preliminary screening of these strategies. This preliminary screening is based upon their engineering, social and economic feasibility. Those strategies considered unfeasible, on this a priori basis, can be screened out, the selected strategies being called promising strategies. The promising strategies, combined in a coherent manner, give rise to an alternative water resources policy.

The main aspects of policy design for a given river basin are as follows:
— identification and characterization of existing and potential water users in the basin and of their requirements;
— identification and evaluation of alternative water sources, both within and outside the basin, and of their characteristics;
— definition and specification of alternative water distribution systems, connecting the water users to the water sources;
— characterization of the physical, chemical and biological conditions in the water resources system in order to ensure its adequate ecological balance;

— control of water quality by means of treatment of effluent discharges originated by the water users;

— specification of the operating rules for each water resources system,

— setting prices for the uses of water, including the use of water bodies for receiving polluted effluents, in order to optimize the use of the bodies receiving water and to control the environmental impacts.

The selection of water resources policies is the task of the decision makers, based on alternative scenarios of the water resources system, and on the assumptions on the water resources system itself. The alternative scenarios correspond to the specification of certain consistent combinations of the more relevant factors, corresponding to future conditions of economic and social development, technological progress, standards of living and public policies. The system assumptions describe the characteristics of the water resources systems, and are aimed at including the behaviour of users and managers of the systems.

The Role of Models

Regional water resources management involves the need to establish policies concerning the development and use of the resources. These policies will have economic, social and environmental impacts that will vary over space and over time. The development of models that can predict these varying impacts remains a challenge, especially if such models are to be used in the policy making process. Such models should be easy to use and to understand, and must be adaptive to a wide variety of possible policy alternatives. In particular, they must be able to provide information that is credible in a minimum amount of time. Furthermore, they should be designed so as to provide an unbiased guide which helps policy makers in taking their decisions.

In using models to predict the behaviour of the water resources systems for management purposes it is essential to simplify the complex reality by means of a schematization of the real phenomena. A model should always be built with a set of specific objectives, and different objectives should imply different models. Once the schematization of the system to be modelled is ensured and the model is built, it is necessary to use it with the purpose of trying to solve the problem under consideration. However, these two steps will only be consequent if the transposition of the solution obtained in the model to the real system is possible.

The degree of success in the achievement of the three steps previously stated, i.e. schematization of the real system, resolution of the problem on the model and transposition of the solution to the real system, has a decisive influence on the quality of the final result.

Models are a powerful tool for water resources management. A management model is meant to define the set of decision variables $X(t)$ for $t = 1, 2 \ldots$ that optimizes certain objective functions $Y(t)$, being subject to a set of constraints.

The decision variables $X(t)$ correspond to the controllable inputs of the system (e.g. controllable discharges, controllable pollution loads, biological activity, pollution abatement measures) as opposed to the uncontrollable inputs (e.g. precipitation, evaporation, tide effects, salinity).

The objective functions or outputs $Y(t)$ for $t = 1, 2 \dots$ can be of different types, e.g. water levels at different points, concentrations of several pollutants, some economic criteria such as cost-benefit or cost-effectiveness ratios which may include, implicitly or explicitly, the consideration of risk.

Depending on the problem, a single objective or multiple objectives can be considered. Several techniques are available to formulate and use multi-objective models.

The constraints can also be of many types, usually a distinction being made between technological constraints — like constraints on flood levels, ground water levels or water quality — and institutional constraints, for instance of a legal, administrative, social and political nature. These constraints condition the state transition functions and the output functions.

The state transition function f_s expresses the state of the system at time $t+1$ as a function of the state at time t and of the decision variables $X(t)$:

$$S(t+1) = f_s[S(t), X(t)]$$

The output functions f_o make it possible to determine the output variables $Y(t)$ as a function of the state variables and the decision variables:

$$Y(t) = f_o[S(t), X(t)]$$

Several types of uncertainties may effect the already mentioned elements of the model, namely $Y(t)$, $Y(t)$, $S(t)$, f_s and f_o. Of these the following are stressed:

— uncertainty related to natural phenomena that condition the quantity and quality of water resources;

— uncertainty related to deficiencies of the available information very often met with as regards hydrologic, economic, social and environmental data;

— uncertainty associated with the difficulty of predicting the future as regards evolution of water use and of economic factors such as costs of construction, operation and maintenance of works, and costs of coping with floods, droughts and pollution;

— uncertainty related to the limitations of knowledge of the technological evolution concerning water resources use and control;

— uncertainty related to the validity of the models used for analysing water resources management problems.

Evaluation of Alternative Water Resources Policies

The development and application of policy oriented models to problems involving conflicts in water resources management in river basins have taken place for over twenty years. But judging from the current literature on policy modelling and from discussions at meetings on this subject, the impact of all this effort has been less than analysts might have wished or expected. While there have been a few notable successes, there have been many more disappointements.

Systems analysts have been generally successful in developing comprehensive models of complex systems, and in inventing efficient algorithms to solve these models. Future improvements in modelling system complexity and in algorithms for solving such models, while possible and of scientific value, will not, we believe, have much of an impact on policy makers. This is certainly true at least until the analysts learn more about how to implement models, intended for policy making, within the policy making process. It is in this area where the gap is the widest between what analysts have been doing and what is needed: the demand for information that can be derived from models exists, but the analysts have to learn more about how to meet that demand with the type of models that are generally available today. Most of these models are well suited to the needs of planners and consulting engineers, but very few are oriented toward policy makers.

To improve this situation, the aim should be to develop an interactive approach focused on reducing as much as possible this implementation gap. The proposed approach is not one that will increase the linkelihood that model solutions will be implemented. Rather it is one that should increase the linkelihood of the models themselves being implemented. Models, if used in the policy making process, are used because they provide information that is considered relevant to the problems being addressed.

Analysts and policy makers at several levels should work together in order to build interactive command programmes, linking several interdependent models, to permit a sequential model selection and solution analysis in order to assist policy makers in their task of selecting the most desirable policy. This process of interactive use of a series of linked strategy/impact-prediction models for evaluating alternative management policies in river basins is displayed below (Fig 1).

The first step is obviously the full definition of the system to be studied. Then follows the selection of the problems to be solved and of the adequate models related to these problems, specifying in both cases the appropriate spatial and temporal scales for the model. To face the existing problems, different strategies, or sets of measures, will be formulated, the selected models being used to predict the impacts of these strategies. These impacts can be of a technical, economic and environmental nature. The next step is the evaluation of these impacts which will determine the revision of the established strategies or lead to further analysis of the problem, which in turn, will determine possible directions in the process of selection of the best water resources policy.

Fig 1 A schematic presentation of sequential steps in the process of selection of the best water resources policy.

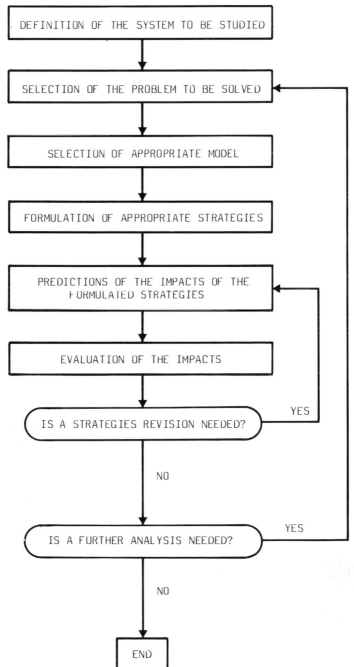

Lundqvist, J.; Lohm, U. and Falkenmark, M. (eds.):
Strategies for River Basin Management, pp. 337–346
© *1985 D. Reidel Publishing Company*

Problems Related to Growing Urban Systems from a River Basin Perspective

Lindh, Gunnar, Dept. of Water Resources Engineering, Lund Institute of Technology/ University of Lund, Box 725, S-22007 Lund, Sweden

Abstract: Because of the rapid growing of many urban areas it becomes more and more important to analyse the interactions caused by these areas with other parts of the river basin. In order to carry out such a study the author tries to apply a systems approach where especially interactions between the rural and the urban areas are considered. The analysis is complicated by the fact that not only water related factors have to be taken into consideration but also factors that indirectly exert influence on the development of the river basin. Human goals and aspirations as well as socio-political decisions may come to play an important role if we want a harmonic and sound development of urban eco-systems in conjunction with natural ecosystems. The paper mainly focuses on water flows processes between the different subsystems.

Introduction

The basic fact that justifies the subject of this paper as being an important problem is that the movement of people from rural to urban areas is still growing in all countries (Lindh 1983a). Half of the world's population will be living in urban areas at the turn of the century, thereby causing a continuous change in land occupancy and use, cf. OECD (1979). Thus, the concentration of urban populations occurs within a small space, very often in the order of less than 4%. However, as Kuprianov (1975) has pointed out, this area corresponds to about 15% of the whole territory intensively used by man. Consequently it may be expected that the human activities concentrated on these urban regions may cause considerable changes in the hydrological behaviour of a river basin as well as causing various, important environmental impacts. These effects are not only local in character but regional as well.

The transformation of a natural environment to a man-made one creates consequences for the qualitative and quantitative aspects of the hydrological regime. At first, one might indirectly consider these man-made changes to be the same as those occuring in urban areas, but, nevertheless, there are far-reaching repercussions of increased urban activities. For example, one important consequence is the transfer of water from remote areas in order to cover shortage of water in urban areas. Such long distance transfers often occur not only within one country but also take place between countries. Thus, one must not only consider both water problems particular to the urban areas within the river basin but also those originating from emerging conflicts due to competing interests.

Another important aspect to keep in mind is that, when focusing upon problems from a river basin perspective, it is easy to restrict one's interest to problems related to the hydrological aspects, since the concept of a river basin is naturally associated to hydrological phenomena. However, in order to make a comprehensive analysis of problems due to urban growth one must, as pointed out by Lindh (1983 b) among others, also be aware of other aspects that are related to hydrological factors but may not be hydrological by nature. Such aspects may be associated with socio-economic, cultural or other values that must be taken into account when one tries to comprehend the interrelated phenomena within the river basin. Such an observation will, of course, make it more difficult to determine which general approach might be applied to cover the most relevant aspects.

It is also useful to regard the remark by Maione (1982) that methodologies developed for industrial countries are seldom transferable to developing countries due to economic, social and political dissimilarities. The method applied in the present paper is the use of a systems approach. The difficulties are by no means surmounted by such a method because the problem remains to define this system and its subsystems in order to include all relevant aspects. The question about a systems approach will be dealt with in the next section.

A Systems Approach

Conceptually, as stated previously, a river basin is determined by its hydrological properties. However, as also was pointed out above, consequences due to growing urban settlements in a basin is not only caused by direct hydrological factors but also by secondary and higher order effects related indirectly, for instance, to social needs and demands. Such effects may often be different in character, termed interactive, diachronic or cumulative by Vlachos & Hendricks (1977). It is often a characteristic feature of urban settlements that economic planners have concentrated on achieving a rapid increase in national income through accelerated growth, see for instance Standke & Anandakrishnan (1980). This has sometimes resulted in a considerable deterioration of rural settlements as well as an unbalanced growth of the urban areas. One consequence thereof has been that cities have grown beyond their carrying capacities with a noticeable decline in the quality of urban life. Such an uncontrolled growth has also had a negative growth rate effect upon, as well as disruption of rural economies.

The conclusion to be drawn is that development of urban and rural areas calls for integrated spatial and economic planning (Friedmann & Wulff 1976). In a systems approach, economic and physical planning means that physico-ecological as well as socio-economic systems have to be considered, including demographic, economic, social and physical and environmental aspects of the society. Thus one could consider the urban situation from the river basin perspective as part of a socio-ecological system. Such an integrated approach has been tested previously (Hamilton et al. 1969; Richardson & Tauber 1979).

In analyzing an ecosystem it is helpful to identify and isolate component parts, a fact often repeated as one of the fundamentals of systems analysis (Haggett 1980). Since urban systems, as well as rural systems, consume goods, services and energy, responding to the demand of the human population, these systems must consist of a least three main segments: the man-made energy and material pathways; the natural environment; and the socio-economic-political systems (Sterns & Montag 1974). To construct a system out of this information is not an easy task. We know that the essential spatial parts are the urban areas and the rural areas in the river basin, creating the physical boundary of our system. Outside this boundary we have "the rest of the world", according to systems analysis terminology (Fig 1). This concept does not make a system unless we can define which relationships exist between the various interacting components. We are here faced with a great difficulty because the whole system is characterized by these interactions, which are part of the definition of a system (West Churchman 1968).

However, these interactions cannot be described from the behaviour of any of the components viewed in isolation. Another factor to consider when using the term ecosystem is that any natural ecosystem contains positive as well as negative feedback mechanisms. Negative feed-back thereby acts as a dominant damping mechanism. In the man-made urban/rural ecosystem there may exist some sort of control mechanism that is derived from man's unique property of a social and cultural organism. This necessitates social and institutional controls that are, however, not always satisfactory in avoiding deteriorating situations. Such severe problems as solid waste disposal in sanitary landfills or discharge of liquid wastes into water courses in the river basin are examples of such uncontrolled feedback effects within a socio-ecological system. The natural environment (within and outside of the river basin regions) has also been regarded as "the environment

Fig 1 The river basin as an ecosystem consisting of urban and rural subsystems as well as the natural environment (not influenced by man)

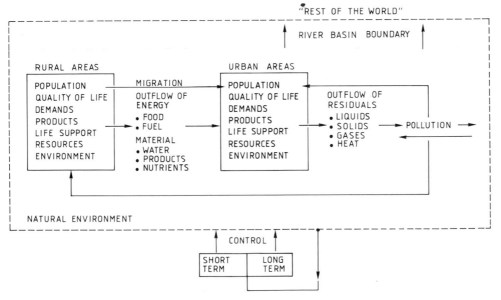

of the human social system" meaning that it is the source of resources and the sink of wastes. Stress in this areas has been aggravated mainly because of an imbalance between positive feedback dynamics of a human economy and negative feedback from an insufficient resilience of the limited properties of a natural environment. These factors, combined with a new social strategy, require us to extend the definition of ecosystem to "urban ecosystem", as discussed by Linville & Davis (1976).

Thus, as stressed above, the complete system to be analyzed is one where the urban system — regarded as a subsystem — must be characterized by both natural and social/institutional control mechanisms. These latter mechanisms will become increasingly predominant. This means particularly that human goals and aspirations as well as socio-political decisions must harmonize with the sound development of urban ecosystems in conjunction with natural ecosystems.

As mentioned above, a characteristic feature of an urban ecosystem, as a component of the river basin socio-ecological system, is the extraction of resources for manufacturing and consumption at one end of the energy and material flows, and the discharge of wastes into the natural environment at the other end. Such a flow system has been pictured in Fig 2. The underlaying assumption is that, when talking about energy and material flow-processes, there is an interchange between rural areas and urban/industrialized areas. Because we are mainly interested in flow component characteristics of a river basin, the flow of water is of paramount importance. Moreover, since the main task is to analyse the consequences of urban growth on existing water resources in the river basin, we may now proceed to a study of water processes in the urban area.

Water Processes in the Urban Area

In considering the water situation in an urban area, say a city, we have to deal with socio-economic as well as urban hydrological aspects. The urban hydrological aspects consist of two parts: the inner system, which is related to the sources of water supply and the outer system, which involves pure hydrological processes within the city (Fig 3 and Lindh 1979). There are two aspects of the urban hydrological system that are extremely important when considering the effects of a growing city from a river basin perspective. First, there is the extraction of water from surface water or ground water reservoirs (or both in conjunction) for water supply, and secondly, there is the discharge of residual water from water use as well as discharges due to natural or man-influenced effects in the outer subsystem. Today it is possible in principle to dertermine the amount of water demanded as well as the amount of wastes discharged from a city (for example, Hogland & Berndtsson 1983; Niemczynowicz & Hogland 1981). Moreover, it seems plausible to estimate the increase in water use due to a growing city population and thus, consequently, also the increased waste water discharges to rivers in the river basin.

At this stage of our analysis it seems logical to move on to the socio-economic aspects of the development of an urban system. Generally we are then confronted with such concepts as the "quality of life" or "social well-being", (Vlachos 1978; Lindh 1976). It may be too difficult to enter into details when it comes to such sociological concepts. It will

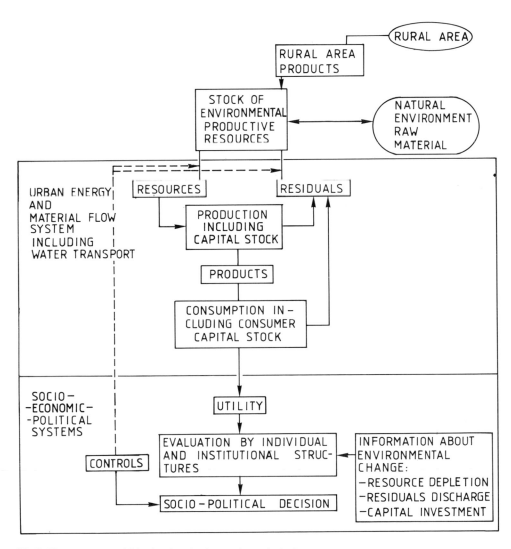

Fig 2 Flow patterns within the river basion socio-ecological system

suffice to refer to Fig 4 which aims to focus on some important factors that determine the development of water resources in a city. As shown in this Fig., a systems way of thinking is used to delineate the water problems of a city. Sociodemographic, economic, and physical factors as well as norms and values may be considered as inputs to a process which seeks to enhance the water supply as well as the quality of life in connection with an increased productivity as one form of developmented change. This development is dependent on social institutions, physical infrastructure and the rules of operation that are available.

342

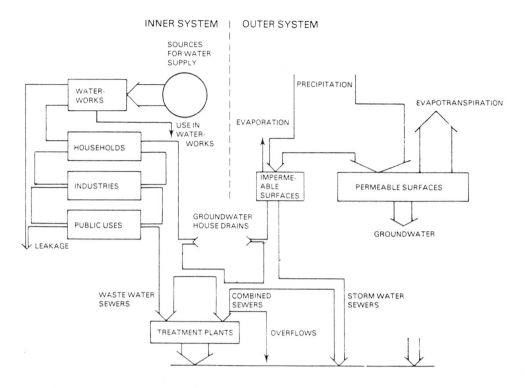

Fig 3 The inner and the outer hydrological system of a city

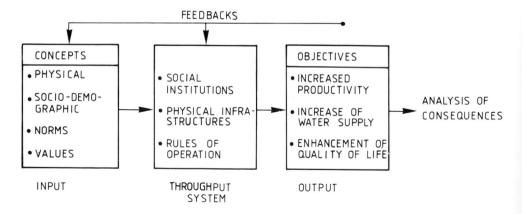

Fig 4 Socio-economic aspects of water resources problems as shown in a "black-box" model

We have here tacitly understood that this model should apply to the development of the city of the urban area. However, if a river basin is considered as the basic hydrological unit, river basin comprehensive planning must include a strategy where urban interests and demands cannot be satisfactorily isolated from other demands in the basin. This situation apparently leads to a series of intermingled conflicts between rural and urban areas which are considered in the next section.

Conflicts between Rural and Urban Areas

In this paper it has been suggested that the problems of growing urban areas within a river basin should be studied by a systems approach method. As one section of such a system, the urban water processes were considered as one part of the flow process between the urban area and its environment. Water processes were especially studied because water plays an important role in river basin flow dynamics. However, in order to gain more knowledge of these flow processes, we have to broden the concept of water flow processes, with due respect to intermingled effects, in a manner that is schematically shown in Fig 5. Despite the weakness of such a schematic illustration, a series of interactions between urban areas and their "rural hinterlands" that may give rise to serious conflicts can be shown. For instance, water use and demand in the urban areas may cause conflicts with the needs of water for irrigation in the rural area. It is a wellknown fact that as a global average the need of water for irrigation is about 80 % of the total water use and it is certainly very easy to see that a conflicting situation could arise. We many also imagine

Fig 5 Some important links between rural and urban areas (after an idea by White & Burton, 1983)

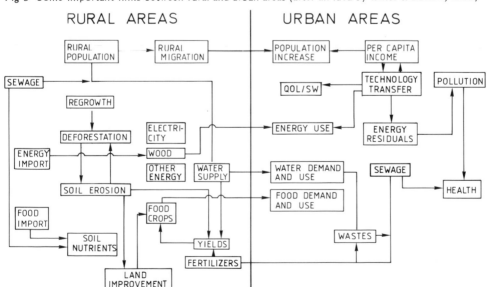

that the urban need of surface water and ground water may be so high that it necessitates an import of water from a more or less remote river basin. This in turn causes water quality problems which are well documented from, for example, California and Colorado in the USA as well as from many water transfers in Germany (Falkenmark & Lindh 1976; McPherson 1974).

The discharge of waste water and solid wastes have already been mentioned as activities that result in increasing deterioration of the environment. Many western countries are trying to tackle these problems with more or less success. Deterioration of water courses as well as polluting of the ground water due to sanitary landfills are other well-known examples of problems in developed countries. In the long run, such activities may cause health damage, but they can, of course, also be benficial if wastes are recycled into the urban and rural economy.

Moreover, Fig 5 shows that changes in the use of one resource may have influences on the other resources. As an example the destruction of forests for wood and a charcoal supply may be mentioned. In fact such an activity may cause effects on the rural as well as on the urban water supply. Obviously crop yield may be affected; siltation of power-generating dams in the river basin will increase and the carrying capacity of agricultural land will be affected. Migration, that is the movement of people from rural to urban areas, will cause a series of stresses on energy and material flows. Many rapidly growing cities in the world, perhaps mainly in the developing countries, show that this type of development leads to an excessive need of water and energy that cannot be maintained by the city authorities and that logically causes severe consequences in the rural hinterlands. "Squatters" settlements and shanty towns grow up on the fringes of the urban settlements with disorganised water and energy supplies. Thus, very often migration causes severe socio-economic effects, not to mention the ineffective use of material transport to the city as well as the increasing need for agricultural development and food production for the people.

Concluding Remarks

In the preceeding chapters we have tried to establish a logical scheme using a systems analysis approach in order to reveal interrelated activities in the river basin that may have consequences of hydrological character. Carrying through such an analysis is a delicate task. One way of checking the usefulness of such an analysis would, of course, be to compare it with observations made and reported in the literature. There are several observations published which clearly indicate the applicability of the systems approach (cf. Zuidema 1979; Colenbrander 1980; Saha & Barrow 1981; Zaman 1983; White & Burton 1983; Delleur & Torno 1983). The overall conclusion that can be drawn from such a literature review is that the general systems approach must be modified according to the special river basin study at hand. Particularly the relative importance between urban and rural areas may play a decisive role in the performance of an analysis. As a whole there is agreement in the literature that an integrated approach has to be adopted. This not only means that there has to be an integrated land and water approach, but also that all important factors that may indirectly affect the hydrological aspects have to be taken into account. Such an approach will also strengthen the contacts between planners and decision makers, thus reducing the credibility gap that otherwise could develop (Biswas 1978).

References

Biswas, A. K.: Systems approach in water management. McGraw-Hill 1976.

Colenbrander, H J.: Casebook of methods of computation of quantitative changes in the hydrological regime of river basins due to human activities. Studies and reports in hydrology 28, Unesco, Paris (1980)

Delleur, J. W.; Torno, H C.: Proc of the international symposium on Urban hydrology. Amer. Soc. Civ. Engnrs. NY, USA, 1983.

Falkenmark, M.; Lindh, G.: Water for a starving world. Westview Press, Colorado, USA 1976.

Friedmann, J.; Wulff, R.: The urban transition: comparative studies on newly industrializing societies. In: Board, Ch.; Chorley, R. I.; Haggett, P. & Stoddart, D. R. (eds). Progress in geography. Edward Arnold, London 1976.

Hamilton, H. R.; Goldstone, S. W.; Milliman, J. W.; Pugh, A. L.; Roberts, E. B.; Zellner, A.: Systems simulation for regional analysis: an application to river basin planning. The MIT Press, Cambridge, Massachusetts 1969.

Hogland, W.; Berndtsson, R.: Quantitative and qualitative characteristics of urban discharge to small river basins in the South West of Sweden. Nordic Hydrology 14, 155–166 (1983)

Haggett, P.: Systems analysis in geography. Clarendon Press, London 1980.

Kuprianov, V. V.: Hydrological aspects of urbanization. In: The hydrological characteristics of river basins and the effects of these characteristics of better water management, Proc. of the Tokyo Symposium, IAHS-AISH. Publication 117, 1975.

Lindh, G.: Socio-economic aspects of urban hydrology. Ambio 7, 16–22 (1976)

Lindh, G.: Socio-economic aspects of urban hydrology. Studies and reports in hydrology 27, Unesco, Paris (1979)

Lindh, G.: Water and the city. Unesco, Paris 1983a.

Lindh, G.: Planning aspects of integrated river basin development. In: Proc. IUGG Conference Hamburg 1983 (to be published).

Linville, I.; Davis, R.: The political environment. An ecosystems approach to urban management. American Instiute of Planners, Washington DC 1976.

Maione, U.: ,Planning and optimization of river basin development. Memorie e studie dell'Institute di Idraulica e Costruzioni Idrauliche, N. 298, Politecnico di Milano 1982.

Niemczynowicz, J.; Hogland, W.: Urban water budget and impact on the receiving waters, Lund case study. Nordic Seminar on "Transport, omsättning og effekter af regnafledning till recipienter". The University of Aalborg, Denmark 1981.

OECD: Interfutures. Facing the future. Organisation for economic co-operation and development, Paris 1979.

McPherson, M. B.: Hydrological effects of urbanization. – Studies and reports in hydrology 18, Unesco, Paris (1974)

Richardson, R. W.; Tauber, G.: The Hudson river basin. Academic Press, London 1979.

Saha, S. K.; Barrow, Cr. J.: River basin planning. Theory and practice. John Wiley and Sons, Chichester 1981.

Standke, K.-H.; Anandakrishnan, M.: Science, technology and society. Pergamon Press, New York, NY 1980.

Sterns, F.; Montag, T.: The urban ecosystem: a holistic approach. Dowden, Hutchinson and Ross, Stroudsburg, Pennsylvania 1974.

West Churchman, C.: The systems approach. Dell Publishing Co, New York, NY 1968.

White, R.; Burton, I.: Approaches to the study of the environmental implications of contemporary urbanization, MAB Technical notes 14, Unesco Press, Paris (1983)

Vlachos, E.: The future in the past: toward a utopian syntax. In: Maryama, M.; Harking, A. M. (eds): Cultures of the future. Montan Publishers, The Hague 1978.

Vlachos, E.; Hendricks, D. W.: Technology assessment for water supplies. Water Resources Publication, Fort Collins, Colorado, USA 1977.

Zaman, M. (ed.): River basin development. Tycooly International Publishing Limited, Dublin 1983.

Zuidema, F. C.: Impact of urbanization and industrialization on water resources management. Studies and reports in hydrology 26, Unesco, Paris (1974)